International Conference on Resource Sustainability (ICRS 2021)

IOP Conference Series: Materials Science and Engineering
Volume 1196

Dublin, Ireland
19 – 23 July 2021

ISBN: 978-1-7138-4655-0
ISSN: 1757-8981

Printed from e-media with permission by:

Curran Associates, Inc.
57 Morehouse Lane
Red Hook, NY 12571

Some format issues inherent in the e-media version may also appear in this print version.

Institute of Physics
Dirac House, Temple Back
Bristol BS1 6BE UK

Phone: 44 1 17 929 7481
Fax: 44 1 17 920 0979

techtracking@iop.org

Additional copies of this publication are available from:

Curran Associates, Inc.
57 Morehouse Lane
Red Hook, NY 12571 USA
Phone: 845-758-0400
Fax: 845-758-2633
Email: curran@proceedings.com
Web: www.proceedings.com

TABLE OF CONTENTS

IOP Conf. Series: Materials Science and Engineering 1196 (2021) 011001 doi:10.1088/1757-899X/1196/1/011001

Preface

Title:

Materials and their role in a circular economy: engineering applications

Sustainable development depends on significant improvements in the efficient use of resources. It requires a holistic global life cycle perspective from producers to end consumers, especially in a closed loop fashion. Consequently, communities, governmental and business organisations have sought methods for the reuse, recycling, remanufacturing and recovery of products and materials, thereby extending their useful life and reducing quantities of waste. The United Nations' Sustainable Development Goals (SDGs) target "doing more and better with less", which means that decreasing resource utilization, degradation, and pollution along the entire life cycle, together with an increase in lifestyle quality, can improve the social benefits of economic activities. The resulting paradigm shift has been popularly and broadly termed as the circular economy (CE).

The series of International Conferences on Resource Sustainability (icRS) allows researchers and practitioners to exchange ideas on latest development in resource sustainability. icRS 2021 virtually held from University College Dublin (Ireland), 6th in the series, aims to provide rich opportunities for collaboration between researchers and industrial practitioners in order to advance the critical role of materials in CE as it pertains to underdeveloped, developing and developed economies. At icRS 2021, we will particularly welcome interdisciplinary contributions that involve natural sciences, and engineering applications.
We define resources broadly to include physical resources, biological resources, and "misplaced" resources:

- physical resources: metals, non-metallic minerals, energy, water, etc.
- biological resources: food, forestry, land, ecological systems, etc.
- "misplaced" resources: air emissions, water pollutants, solid waste

Five keynote speakers provided their plenary speeches at icRS 2021. Cllr Lettie McCarthy, An Cathaoirleach of Dún Laoghaire-Rathdown County Council, Ireland, gave a talk with the title of "driving innovation to safeguard the best that the county has to offer and to create sustainable connected communities and a smart business environment". In his keynote address, Prof. Arpad Horvath, Professor of Civil and Environmental Engineering at the University of California, Berkeley, USA, highlighted the need to manage the life cycles of infrastructure components. Prof. Fu Guo, Professor of Materials Science and Engineering at Beijing University of Technology, Beijing, China, discussed the current status of development of waste printed circuit boards (WPCBs) recycling in China. Finally, the role of resources in a circular economy towards a carbon neutral and net zero future was presented by Prof. Lenny Koh, Professor in Operations Management at the University of Sheffield's Management School.

The list of committee members and editors are provided in the following:

Conference Chairs

- icRS2021 Conference Chair: Pezhman Ghadimi, University College Dublin Conference
- icRS2021 Conference Co-chair: Michael Gilchrist, University College Dublin
- icRS General Chair: Ming Xu, University of Michigan

Local Organizing Committee

- Pezhman Ghadimi, University College Dublin
- Michael Gilchrist, University College Dublin
- Eoin Casey, University College Dublin
- Nikolaos Papakostas, University College Dublin
- Vincent Hargaden, University College Dublin
- Cathal Heavey, University of Limerick

Program Committee

- Chair: Chao Wang, Beijing University of Technology, China
- Co-Chair: Nan Zhang, University College Dublin

Award Committee

- Chair: Amir Hossein Azadnia, Letterkenny Institute of Technology, Ireland
- Co-chair: Mohammad Rajaeifar, Newcastle University

Advisory Committee

- H. Eugene Stanley, Boston University
- John W. Sutherland, Purdue University
- Joseph Sarkis, Worcester Polytechnic Institute
- Alexandre Dolgui, IMT Atlantique
- Angappa Gunasekaran, California State University
- Sami Kara, University of New South Wales
- Arpad Horvath, University of California, Berkeley
- Stefan Seuring, University of Kassel
- Kannan Govindan, University of Southern Denmark
- Yasser Dessouky, San Jose State University
- Anthony S. F. Chiu,De La Salle University Manila
- Michael Bruen, University College Dublin
- H. Scott Matthews, Carnegie Mellon University
- Brad Allenby, Arizona State University
- Bhavik Bakshi, The Ohio State University
- Christoph Herrmann, TU Braunschweig
- Gordon Huang, University of Regina

- Rui Cunha Marques, University of Lisbon
- Mark Pagell, University College Dublin
- Donna Marshall, University College Dublin
- Fu Zhao, Purdue University
- Stefanie Hellweg, Swiss Federal Institute of Technology
- Ming Lang Tseng, Asia University
- Hwong-wen Ma, National Taiwan University
- Sangwon Suh, University of California, Santa Barbara
- Arnold Tukker, Leiden University
- John L. Zhou, University of Technology Sydne
- Hung-Suck Park, University of Ulsan
- Lenny Koh, University of Sheffield

Scientific Committee

- Fengzhou Fang, University College Dublin
- Magnus Fröhling, Technical University of Munich
- Dongyang Deng, North Carolina A&T State University
- Mohammad Rajaeifar, Newcastle University
- Shen Qu, University of Michigan
- Mert Celikin, University College Dublin
- Ming K. Lim, Chongqing University
- Donal Finn, University College Dublin
- Benny Tjahjono, Coventry University
- Piwowar Arkadiusz, Wroclaw University of Economics
- Jian Zuo, University of Adelaide
- Anne Scheinberg, Springloop Cooperatie NL
- Hamed Fazllolahtabar, Damghan University
- Akira Otsuki, Université de Lorraine
- Nan Zhang, University College Dublin
- Rajesh Singh, Management Development Institute Gurgaon
- Maria Holgado Granados, University of Sussex
- Amirhossein Azadnia, Letterkenny Institute of Technology
- Yaqian Zhao, Xi'an University of Technology
- Alessandra Neri, Politecnico di Milano
- George Onofrei, Letterkenny Institute of Technology
- B. Krishna Prapoorna, Indian Institute of Technology Tirupati
- Bruno de Andrade, University College Dublin
- Surya P. Singh, Indian Institute of Technology Delhi
- Ankit Gupta, Banaras Hindu University
- Maryam Johari, Iran University of Science and Technology
- Muhammad Asif Hanif, University of Agriculture, Pakistan
- Latifah Abd Manaf, Universiti Putra Malaysia
- Marlene Arens, Fraunhofer ISI
- Yudi Fernando, Universiti Malaysia Pahang

Due to Covie-19 travel restrictions, the conference was postponed once from June 2020 to July 2021 with the hope of restriction being eased. However, that never happened, and it was decided to be held in a fully virtual format. Further postponement of the conference would have damaged the novelty of the submitted works for 2020 and 2021 and would have created frustration among the delegates and the community. The conference was held virtual from University College Dublin, Ireland from 19th to 23rd of July 2021. icRS 2021 was held in five days from 14:00 to 17:30 Irish Standard Time to accommodate presenters from all over the world, in 36 parallel sessions, a poster session, and five keynote speech sessions. The keynote

speeches were done in around 45 minutes each day. The scientific presentations were done back-to-back in slots of 10 minutes followed by 5 minutes Q&As. Overall, 220 scholars participated icRS 2021. Zoom Webinar was used to broadcast the Keynote speeches and Zoom Meetings to perform the parallel and poster sessions. The conference was a huge success technically. Five full conference papers were chosen for the best paper awards and 10 presentations were selected for the best presentation awards. As the organizing committee, we would like to thank our sponsors, authors and co-authors, scientific committee, the keynote speaker, the session chairs, whose contributions have made possible the accomplishment of icRS 2021.
The published papers with IOP conference series: Materials science and engineering are all related to the role of materials and resources (as described above) in a circular economy with application related to civil, mechanical, material, electrical, energy, and industrial engineering.

29 September 2021

Pezhman Ghadimi
Michael Gilchrist
Ming Xu

IOP Conf. Series: Materials Science and Engineering 1196 (2021) 011002 doi:10.1088/1757-899X/1196/1/011002

Peer review declaration

All papers published in this volume of IOP Conference Series: Materials Science and Engineering have been peer-reviewed through processes administered by the Editors. Reviews were conducted by expert referees to the professional and scientific standards expected of a proceedings journal published by IOP Publishing.

- **Type of peer review: Single-blind**

Single-blind reviewing.

117 submitted full conference papers were reviewed independently by the 30 members of the scientific committee. The six members of the local organising committee also helped with the review process. The acceptance was granted to the papers with the positive reply of two independent reviewers. The reviewers were asked to rate each manuscript with regards to the following criteria: technical, presentation, quality, and English language. The submissions with low ratings were rejected with some comments to improve for their future publication endeavours in other avenues. 10 papers were desk rejected by the conference science committee chair and co-chair due to being out of scope. In 69 instances, the submitted papers were asked to be revised and submitted for another round of reviews. The second round of the review process was conducted by the chairs of various icRS2021 committees to make sure that the comments of the original reviewers were considered by the authors.

- **Conference submission management system:**

  Aventri

- **Number of submissions received:**

  117

- **Number of submissions sent for review:**

  107

- **Number of submissions accepted:**

  73

- **Acceptance Rate (Number of Submissions Accepted / Number of Submissions Received X 100):**

  62%

- **Average number of reviews per paper:**
  2

- **Total number of reviewers involved:**

  36

- **Any additional info on review process:**

  -

- **Contact person for queries:**

- Name : Pezhman Ghadimi
  Affiliation: University College Dublin
  Email : icrs2021dublin@gmail.com

Evaluating two interventions for scaling up upcycling: Community event and upcycling plaza

K Sung[1*]

[1] School of Art, Design and Architecture, De Montfort University, Leicester, LE1 9BH, UK

*Corresponding author Email: kyungeun.sung@dmu.ac.uk

Abstract. Upcycling – creation or modification of a product from used or waste materials, components, and products for higher quality or value – incorporates multiple strategies for increasing material efficiency and provides other environmental, economic and social benefits. With a growing concern for the environment, the past few years have seen a growth of upcycling practitioners and businesses, yet despite such a growth, upcycling largely remains a niche and requires scaling-up to have a significant impact on the environment and society. For scaling up upcycling, several promising interventions have been suggested. This paper evaluates two of these: community event and upcycling plaza. Two case studies (on "Art with Upcycling" co-creation family event in Leicester, UK, and Seoul Upcycling Plaza in Seoul, South Korea) have been conducted using questionnaire, observation and interview studies. The results show that the community event made positive psychological impact on participants, and that many programmes run by the upcycling plaza were perceived as successful by stakeholders. Based on the aforementioned positive impact, these two interventions are recommended for prototyping, piloting and monitoring in other cities and countries for scaling up global upcycling.

Keywords: Upcycling, Sustainable development goals, waste materials, material efficiency

1. Introduction

The United Nations' Sustainable Development Goal (UN SDG) No. 12, Responsible Consumption and Production (i.e. Ensure sustainable consumption and production patterns), aims at "doing more and better with less" [1]. Material efficiency – providing material services with less material production and processing [2] – is well aligned with this UN SDG 12. There are four major strategies for increasing material efficiency: longer-lasting products, modular design, component reuse for remanufacturing, and products designed with less materials [2]. Upcycling – creation or modification of a product from used or waste materials, components and products which is of equal or higher quality or value than the compositional elements [3-4] – is an umbrella concept which incorporates multiple strategies for increasing material efficiency. Upcycling extends lifetimes of existing products (e.g. refurbished and redecorated furniture), reuses components (e.g. remanufactured electronics from reused components), or utilises waste materials (using less new virgin ones) for production of new products (e.g. clothing and fashion items from waste clothes and textiles). It is the amalgamation of a variety of 'creative' forms of repair, reuse, repurpose, refurbishment, upgrade, remanufacture and recycling, all of which are important part of the circular economy – the alternative to the linear economy of take-make-use-and-dispose which is restorative and regenerative by intention and design [5-6].

The benefits of upcycling are not limited to reducing environmental impact – extending lifetimes of materials, components and products [4,7], thereby increasing material efficiency and reducing waste and energy consumption [4,8]. Businesses based on upcycling can be financially sustainable and create new jobs [9-11]. Individuals who engage in upcycling in households could gain socio-cultural and psychological benefits such as learning new skills, relaxing with mindful craft activities and being empowered by gaining control over products [12]. With a growing concern for the environment, resource scarcity, and increasing volumes of waste, the past few years have seen a growth of upcycling practitioners and businesses [13-15]. Despite such a growth, upcycling largely remains a niche activity and requires scaling-up – the dynamic process of transitioning from niche to mainstream (or regime) [16-17] – to have a significant impact on the environment and society. For scaling up upcycling, several promising interventions have been suggested such as community events, TV and inspirational media, government procurement change, upcycling centre/plaza, curriculum enrichment, and financial incentives for upcycling businesses and initiatives [4,8]. This paper evaluates two of these: community event and upcycling plaza.

2. Methods

Two case studies [18] were conducted between July 2018 and January 2019. The case studies were based on questionnaire, observation, and interview studies.

2.1. Case study one: A community event in Leicester, UK

A one-day family event, "Art with Upcycling," was organised in November 2018 at New Walk Museum and Art Gallery in Leicester, UK [19]. The event was an art and craft co-creation workshop using upcycling. The event invited families with young children to contribute to making art pieces and create small upcycled artefacts. The practice of upcycling art and craft was demonstrated by the sustainable artist and sculptor, Michelle Reader [20], and assisted by volunteering students from De Montfort University (Figure 1). The event was part of Being Human Festival, a festival of the humanities led by the School of Advanced Study at the University of London in partnership with Arts and Humanities Research Council and The British Academy [21].

Figure 1. "Art with Upcycling" co-creation family event in Leicester, UK.

IOP Conf. Series: Materials Science and Engineering 1196 (2021) 012001 doi:10.1088/1757-899X/1196/1/012001

(photo of New Walk Museum and Art Gallery: https://d3d00swyhr67nd.cloudfront.net/_source/LAMS_location_image_1.jpg; and photo of Michell Reader: https://www.michelle-reader.co.uk/artist-profile.html)

The researcher asked all the participants to complete a paper-based, one-page questionnaire when they were about to leave the event venue. There was no selection rule for respondents: some participants filled in the questionnaire whereas some did not when everyone was asked. Parents completed questionnaires for their children. The questions were about: a) whether the event increased participants' awareness of upcycling (0 = not at all; 5 = to a very great extent); b) whether the event made participants think that upcycling is un/pleasant, bad/good, worthless/worthwhile, harmful/beneficial, or un/enjoyable (0 = unpleasant, bad, worthless, harmful, unenjoyable; 5 = pleasant, good, worthwhile, beneficial, enjoyable); c) whether the event inspired participants to upcycle (0 = not at all; 5 = to a very great extent); d) whether the event gave participants confidence to upcycle (0 = not at all; 5 = to a very great extent); and e) if participants will try upcycling at home after the event (0 = very probably not; 5 = definitely). Forty-seven participants completed the questionnaire. The responses were analysed by employing descriptive statistics and bivariate correlational analysis (Spearman's Rank Order Correlation), using IBM SPSS (Statistical Package for the Social Sciences) Version 25.

2.2. Case study two: An upcycling plaza in Seoul, South Korea

An upcycling centre/plaza envisioned by a number of UK experts in upcycling was a halfway house between a recycling centre and a charity shop which is aligned with the existing waste collection and management system, and provides gatekeepers to check the state of the materials that are entered into the centre as well as fixers and upcyclers to offer quick, affordable services [8]. The existing centre/plaza closest to this vision was identified as Seoul Upcycling Plaza (SUP), the world's largest upcycling cultural complex space in Seoul, South Korea [22] (Figure 2). The plaza was visited by the researcher in July 2018 for observation, and subsequent email interviews were conducted between August 2018 and January 2019. The interview questions were about: a) whether the programmes (facilities and services) provided by the plaza were perceived as un/successful and why; and b) what efforts/actions by the plaza and other actors are required for improving the plaza therefore contributing ultimately to scaling up upcycling. Twenty-nine upcycling artists, designers, makers and entrepreneurs (identified from the brochure found at SUP reception), sixteen organisers and managers (identified from the SUP website), and six randomly selected visitors (asked in person) were contacted via email. Five of them responded to the email interviews. The results were synthesised and interpreted.

Figure 2. Seoul Upcycling Plaza in Seoul, South Korea.

(photos: https://korean.visitseoul.net/comm/getImage?srvcId=MEDIA&parentSn=17440&fileTy=MEDIA&fileNo=1&thumbTy=L; https://t1.daumcdn.net/thumb/R720x0/?fname=http://t1.daumcdn.net/brunch/service/user/2RB/image/x-N5v-_Psnncipv93ni7NazQdAI.jpg; https://m.young.hyundai.com:444/upload/CMS_NEWS_IMAGE/2018/06/11/CMS_NEWS_IMAGE_FkFKbbadr5YQslTbKgsr.jpg; http://newsroom.etomato.com/userfiles/180506_193.jpg)

3. Results

3.1. Case study one: A community event in Leicester, UK

The respondents (part of the participants in the upcycling co-creation family event) were mostly residents in East Midlands, UK (n=41, 93%). The majority of them were females (n=33, 70%). The biggest age groups were 0-15 years (n=14, 30%) followed by 35-44 years (n=13, 28%), 45-54 years (n=7, 15%) and 25-34 years (n=6, 13%). Simply put, the majority consisted of mothers with their children in East Midlands. The descriptive statistics showed that the event increased the participants' awareness of upcycling to a moderate to great extent (M=3.41; SD=1.15). Throughout the event, the participants thought that upcycling was pleasant (M=4.40; SD=0.90), good (M=4.50; SD=0.83), worthwhile (M=4.45; SD=0.88), beneficial (M=4.44; SD=0.87), and enjoyable (M=4.56; SD=0.63). The event inspired the participants to upcycle in the future to a moderate to great extent (M=3.64; SD=0.87). The participants believed that the event gave them confidence to upcycle in the future to a moderate to great extent (M=3.64; SD=0.87). It is probable or very probable that the participants will try upcycling at home in the future (M=3.66; SD=0.96). See Table 1 for all results.

Table 1. Descriptive statistics.

	Max	Min	Mean	Std. Deviation	Skewness	Kurtosis
Awareness	5	1	3.41	1.15	-0.33	-0.50
Pleasant	5	1	4.40	0.90	-1.79	3.87
Good	5	1	4.50	0.83	-2.39	7.61
Worthwhile	5	1	4.45	0.88	-2.01	5.00
Beneficial	5	1	4.44	0.87	-1.98	4.97
Enjoyable	5	3	4.56	0.63	-1.13	0.28
Inspire	5	2	3.64	0.87	-0.03	-0.64
Confidence	5	2	3.64	0.87	-0.24	-0.50
Intention	5	1	3.66	0.96	-0.32	-0.07

The correlation analysis results showed that the perception of increased awareness is statistically significantly related to the perceived gain of inspiration and confidence. All attitude items (pleasant, good, worthwhile, beneficial, enjoyable) turned out to be correlated with each other. The perceived pleasantness, goodness, and enjoyability of upcycling were correlated to the perceived gain of inspiration. The perceived goodness, worth and enjoyability of upcycling were correlated to the perceived gain of confidence. The perceived gain of inspiration was correlated with the perceived gain of confidence. Perceived gain of inspiration and confidence was correlated with intention to upcycle in the future (Table 2). Simply put, increased awareness of and positive attitudes towards upcycling had statistically significant correlations with perceived gain of inspiration and confidence to upcycle. The perceived gain of inspiration and confidence to upcycle were then also correlated with intention to upcycle in the future.

Table 2. Correlation Coefficient (Spearman's Rank Order).

	Awareness	Pleasant	Good	Worthwhile	Beneficial	Enjoyable	Inspire	Confidence	Intention
Awareness	1.00	0.22	0.27	0.17	0.25	0.25	0.62**	0.35*	0.22
Pleasant	-	1.00	0.80**	0.77**	0.83**	0.63**	0.32*	0.25	0.11
Good	-	-	1.00	0.90**	0.94**	0.66**	0.37*	0.34*	0.10
Worthwhile	-	-	-	1.00	0.95**	0.67**	0.24	0.36*	0.17
Beneficial	-	-	-	-	1.00	0.62**	0.29	0.30	0.16
Enjoyable	-	-	-	-	-	1.00	0.37*	0.35*	0.26
Inspire	-	-	-	-	-	-	1.00	0.63**	0.59**
Confidence	-	-	-	-	-	-	-	1.00	0.49**
Intention	-	-	-	-	-	-	-	-	1.00

**. Correlation is significant at the 0.01 level (2-tailed)
*. Correlation is significant at the 0.05 level (2-tailed)

3.2. Case study two: An upcycling plaza in Seoul, South Korea

The five stakeholders who responded to the email interview were one manager in Seoul Upcycling Plaza (SUP), two upcycling entrepreneurs, one upcycling artist, and one visitor. The stakeholders were asked to evaluate seven programmes provided by SUP: a) dream factory – a fabrication laboratory for citizens providing design-and-make equipment including CNC (Computer Numerical Control) routers, laser cutters, 3D (3-Dimensional) printers, and carpentry equipment; b) material library – a permanent exhibition on waste and sustainable materials; c) upcycling academy – upcycling educational programmes for all age groups; d) upcycling flea market – a cultural space to experience upcycling and share upcycled goods; e) upcycling gallery – a space for special exhibitions; f) upcycling stores – stores to promote and sell upcycled products; and g) upcycling studios – space for upcycling entrepreneurs and artists.

According to the stakeholders, material library and upcycling gallery are the most successful programmes (regarded successful by four stakeholders). The material library was considered as an "informative, educational space with sufficient interesting resources" including "daily recycling wastes and new environmentally-friendly materials" (one said don't know). Upcycling gallery was viewed as a "good exhibition space where upcycling-based artists and small companies could communicate and showcase their art pieces and products to citizens". One stakeholder however believed that it was not very successful as "the space was often empty without any exhibitions".

Upcycling stores, academy and studios were considered successful by three stakeholders. Regarding upcycling stores, three stakeholders believed that "the stores showcase and sell upcycled products well", whereas two others had "doubts about actual sales volume, profits and how lucrative each store and all the upcycling businesses involved have been". Upcycling academy was considered successful as "the room, environment and facilities are great", and "families with children and groups of varied people have been participating" (two said they don't know). Upcycling studios were seen as successful since they "promote the concept and practice of upcycling and raise awareness" and "the upcycling artists and small businesses in the studios are passionate about what they do". One however pointed out, "proper management is not in place" and another stated, "most studios were shut during weekends".

Dream factory and upcycling flea market were considered least successful. Two stakeholders thought dream factory was successful as citizens could use all the equipment for varied activities yet one stated "It has high-tech facilities and equipment which I don't believe are suitable for scaling up or expanding upcycling businesses" (two answered that they don't know). Two stakeholders considered upcycling flea market successful as "it is a good place where citizens could easily engage in upcycling" whereas other two stakeholders thought it was unsuccessful since "there is a lack of things to do, see or play with regarding upcycling." One thought it was very unsuccessful as "there seemed to be no expert in planning,

IOP Conf. Series: Materials Science and Engineering 1196 (2021) 012001 doi:10.1088/1757-899X/1196/1/012001

developing and delivering upcycling-related activities. They [Seoul Design Foundation] seemed to care more about the number of participants rather than the quality or activities. I believe they need proactive communication with and engagement of citizens in order to plan and deliver better activities".

When the stakeholders were asked about what efforts/actions by SUP and other actors are required for improving SUP and therefore contributing ultimately to scaling up upcycling, the most important action turned out to be making meaningful partnerships with relevant businesses (four out of five answered). One particularly mentioned, "I would like the plaza to communicate closely with the upcycling artists and entrepreneurs in the studios more often in order to recreate the plaza as the true citizen's cultural complex without having to use a large budget. The city council invested 50 billion Korean won [approximately £34 million] to build and launch the plaza but due to the lack of abilities of planners, organisers and contractors, the budget has been wasted to some extent and the majority of the operations seemed to be just for show." Two stakeholders suggested improvement of the current programmes provided by SUP: one stated "they [SUP] need to put more effort into creating a variety of programmes for different experiences and increasing overall citizen's engagement and participation." and another said "they need to focus more on the programmes that could make actual changes in the society. The changes to influence people's behaviours and impact businesses' performances". One suggested "more support from the city council".

4. Discussion and conclusions

The art and craft co-creation family event using upcycling in Leicester, UK was successful in the sense that it increased the participants' awareness, made the participants think that upcycling is pleasant, good, worthwhile, beneficial and enjoyable, inspired them and gave them confidence to upcycle, and that the participants expressed their high intention to upcycle at home in the future. Community events have been used and proven to promote sustainable behaviours/practices such as cycling [23], recycling [24] and sustainable food consumption [25]. The first case study showed that upcycling is not an exception. It confirms the suspected potential positive impact of community events for scaling up upcycling [8,26]. The correlation between intention and perceived gain of inspiration and confidence implies that the emphasis of upcycling community events should be on providing inspiration and increasing confidence. Throughout indirect correlations (e.g. increased awareness correlated with perceived gain of inspiration which is correlated with intention), all tested variables except for one attitude item, harmful-beneficial, appear to have potential impact on increasing intention. It suggests that future research could use the same questions and items (maybe removing harmful-beneficial item if one wishes) to measure the impact of upcycling events.

The upcycling plaza in Seoul, South Korea could be considered successful in that it has been providing some useful and meaningful programmes such as material library, upcycling gallery, stores, academy and studios which have contributed to awareness-raising and skills training for citizens as well as promotion and commercialisation of upcycled art, craft, and products. Such programmes could be replicated in other cities and countries as interventions to scale up upcycling. If the plaza takes appropriate actions on building partnerships with relevant businesses, improving their current programmes focusing on increasing citizen's participation and making changes in the society (e.g. change consumer behaviour, help small upcycling businesses to grow), and getting more support from the city council, their success and impact could become more significant. Especially regarding partnerships, meaningful partnerships between governments, businesses, NGOs (Non-Governmental Organisations), experts and general public have been regarded important for sustainable development [27] and transition to low-carbon future [28]. As the second case study revealed only how successful each intervention programme has been perceived and why, and how to improve it, future studies could measure the impact of these interventions on consumer behaviour and/or upcycling-based businesses.

According to the recent study on developing interventions for scaling up upcycling [8], community events are one of the prioritised interventions for scaling up upcycling in consumption domain where buying new products is the current regime (i.e. mainstream practice). The Leicester upcycling event case study clearly showed that the event affected consumers' awareness, attitude, confidence, and intention for upcycling. Such effect on consumers, when scaled up in multiple locations, could impact on creating a niche-cluster (e.g. local networks of passionate hobbyists and activists for upcycling) which could develop into a niche-regime (e.g. regional or national network and social movement). Upcycling centre or plaza is one of the prioritised interventions for scaling up upcycling in production domain where mass production based on virgin materials is the current regime [8]. The Seoul Upcycling Plaza case study showed that the plaza has potential for impacting on scaling up upcycling in both consumption (e.g. by consumer education) and production (e.g. by upcycling gallery and stores) domains. This plaza could become a niche-regime – a regional, national or international network of companies, academic institutions, NGOs, and citizens for industrial symbiosis, knowledge transfer and marketing. These two interventions with proven positive impact are therefore recommended for prototyping, piloting and monitoring in other cities and countries for scaling up global upcycling.

The results from the first case study may not be generalisable to the overall UK population, and those from the second case study to other cities and countries due to the limited sample size sought in the restricted geographical areas. Since Seoul Upcycling Plaza opened in September 2017 and its development is still an ongoing process, the findings in this paper may turn out to be different in the future, which calls for follow-up studies. Despite the limitations, this paper presented the first two case studies on evaluating community event and upcycling plaza as interventions for scaling up upcycling. It extends our understanding of how a community event affects participants' awareness, attitudes, inspiration, confidence, and intention, and to what extent each intervention programme as part of an upcycling plaza has been perceived un/successful for what reason and how the plaza as a whole could be improved. The paper therefore contributes to the development of upcycling theory, which could lead to practical development on upcycling (i.e. scaling-up). It is the author's hope that the findings from these two case studies will inform academic researchers for further studies, and motivate and enable relevant stakeholders to contribute to scaling up upcycling in the UK, South Korea, and many more countries.

5. References

[1] United Nations 2019 Goal 12: Ensure sustainable consumption and production patterns (https://www.un.org/sustainabledevelopment/sustainable-consumption-production/)

[2] Allwood JM, Ashby MF, Gutowski TG and Worrell E 2011 Material efficiency: A white paper *Resources, Conservation and Recycling* **55**(3) pp 362-381

[3] McDonough W and Braungart M 2013 *The upcycle: Beyond sustainability--designing for abundance* (New York: North Point Press)

[4] Sung K 2017 Sustainable production and consumption by upcycling: Understanding and scaling up niche environmentally significant behaviour PhD thesis (Nottingham: Nottingham Trent University)

[5] Ellen MacArthur Foundation 2016 *Circular economy* http://www.ellenmacarthurfoundation.org

[6] Stahel WR 2016 The circular economy *Nature* **531**(7595) pp 435-438

[7] Cooper T 2010 *Longer lasting products* (Abingdon: Routledge)

[8] Sung K, Cooper T and Kettley S 2019 Developing interventions for scaling up UK upcycling *Energies* **12**(14) 2778

[9] Cumming D 2017 A case study engaging design for textile upcycling *Journal of Textile Design Research and Practice* **4** pp 113-128

[10] Han SL, Chan PY, Venkatraman P, Apeagyei P, Cassidy T. and Tyler DJ 2016 Standard vs. upcycled fashion design and production *Fashion Practice* **9** pp 69-94

[11] Sung K and Cooper T 2015 Sarah Turner–Eco-artist and designer through craft-based upcycling. *Craft Research* **6**(1) pp 113-122

[12] Sung K. Cooper T. and Kettley S. 2014 Individual upcycling practice: Exploring the possible determinants of upcycling based on a literature review. *Sustainable Innovation 2014 Conference*, Copenhagen pp 237-244

[13] Merriman C 2019 *Google pledges to do more upcycling and to ship products carbon-neutrally* (https://www.theinquirer.net/inquirer/news/3080111/google-pledges-to-do-more-upcycling-and-ship-it-carbon-neutrally)
[14] Small K 2019 Sustainable art and design: How Calgary artists, entrepreneurs are repurposing and upcycling (https://globalnews.ca/news/5304295/sustainable-art-calgary-repurpose-recycle-upcycle/)
[15] Tesler R 2019 The thrifting revolution: How retailers are upcycling for big gains (https://www.foxbusiness.com/retail/thrifting-retailers-upcycling-depop-app)
[16] de Haan JH and Rotmans J 2011 Patterns in transitions: Understanding complex chains of change *Technological Forecasting and Social Change* **78**(1) pp 90-102
[17] Smith A 2007 Translating sustainabilities between green niches and socio-technical regimes. *Technology Analysis & Strategic Management* **19**(4) pp 427-450
[18] Yin RK 2013 *Case study research: Design and methods* (London: Sage publications)
[19] Sung K 2018 Art with upcycling (https://beinghumanfestival.org/past-festivals/) (https://www.slideshare.net/skesung/art-with-upcycling-origins-and-endings)
[20] Reader M 2019 Michelle reader: Art from waste (http://www.michelle-reader.co.uk/)
[21] Being Human 2019 Being human, A festival of the humanities (https://beinghumanfestival.org/)
[22] Seoul Design Foundation 2019 Seoul upcycling plaza (http://www.seoulup.or.kr/eng/index.do)
[23] Rissel CE, New C, Wen LM, Merom D, Bauman AE and Garrard J 2010 The effectiveness of community-based cycling promotion: Findings from the cycling connecting communities project in Sydney, Australia *International Journal of Behavioral Nutrition and Physical Activity* **7** 8(2010)
[24] Mozo-Reyes E, Jambeck JR, Reeves P and Johnsen K 2016 Will they recycle? design and implementation of eco-feedback technology to promote on-the-go recycling in a university environment *Resources, Conservation and Recycling* **114** pp 72-79
[25] Alkon AH 2008 From value to values: Sustainable consumption at farmers markets *Agriculture and Human Values* **25**(4) pp 487-498
[26] Sung K, Cooper T and Kettley S 2017 Individual upcycling in the UK: Insights for scaling up towards sustainable development *Sustainable development research at universities in the United Kingdom* ed WL Filho (Cham: Springer) pp 193-227
[27] Presas T 2001 Interdependence and partnership: Building blocks to sustainable development. *Corporate Environmental Strategy* **8**(3) pp 203-208
[28] Parthan B, Osterkorn M, Kennedy M, Bazilian M and Monga P 2010 Lessons for low-carbon energy transition: Experience from the renewable energy and energy efficiency partnership (REEEP) *Energy for Sustainable Development* **14**(2) pp 83-93

Acknowledgments
This research was funded by De Montfort University (DMU) – VC2020 Fund and funding from DMU Square Mile. Many thanks to Guy Bingham, Kelley Wilder, Siobhan Keenan, Kate Cheyne and Deborah Cartmell for supporting this research at DMU. I would also like to acknowledge Michelle Reader for organising the community event, and Annabelle Herbert, Hashini Dissanayake, Hetal Mistry, Monika Basara, Renata Budzynska and Ross Kelly for helping to organise the event.

Upcycling of cementitious wastes in one-part alkaline cement binders

M H Samarakoon[1] and P G Ranjith[1,*]
[1] Department of Civil Engineering, Monash University, Building 60, Melbourne, Victoria, 3800, Australia
[*] Corresponding author: E-mail: ranjith.pg@monash.edu, Phone/Fax: 61-3-9905 4982

Abstract. Cementitious wastes generate at an increasing rate and are generally disposed of in landfills, creating harmful environmental impacts. Therefore, enhancing potential pathways to recycle such wastes is critical. In this study, two ways of introducing cementitious wastes in one-part alkaline cement binders (ACBs) are presented: as solid precursors and as activators. A mixture of fly ash and blast furnace slag as the solid precursors and a blend of thermally treated sodium hydroxide and waste glass as the alkali activator were used to cast ACBs by just adding water similar to ordinary Portland cement. Glass powder-based activator showed 70% of efficiency in creating crystalline sodium silicate which dissolves in the binder phase to enhance the production of hydration products. Reasonable compressive strengths of the hardened paste were reported resulting in a dense microstructure with the curing age. Therefore, in this study, the increasing involvement of cementitious wastes in one-part ACBs is confirmed.

Keywords: cementitious wastes, one-part alkaline cement binders, partial activator, strength and microstructure.

1. Introduction

Alkaline cement has become a vast technology over time and can replace ordinary Portland cement overcoming its inherent disadvantages. Recent studies have pointed out its capability to consume cementitious wastes and hence favours much attention in waste recycling points of view. ACBs mainly consisted of two components: solid precursors and activators. Materials with significant silicon, aluminium, or calcium minerals serve as solid precursors with several combinations. Coal ashes, metal slags, municipal solid waste incinerator ashes, mine tailings, construction debris, waste glass are the most commonly reported solid precursors in ACBs [1]. Commercial alkali hydroxides and alkali silicates, both in liquid and solid forms, are widely used as activators in ACBs. In the meantime, few researchers studied the potential use of waste materials such as rice husk ash and waste glass as alkali activators of ACBs [2, 3]. Depend on the physical form of the activator: solid or liquid, ACBs are mainly divided into two categories: one-part and two-part ACBs, respectively. However, two-part ACBs, consisting of solid precursors and liquid activators, have several drawbacks over one-part ACBs. One-part ACBs use both precursors and activators in solid form and cast similar to ordinary Portland cement (OPC) by simply adding water. Solid activators are preferred over liquid ones because activators in liquid form require serious attention on handling, transporting and storing aspects [4]. The activator solutions such as a blend of commercial sodium hydroxide (NaOH) and sodium silicate (Na_2SiO_3) with

Na_2O/SiO_2 = 1 are considered as corrosive materials [5]. Therefore, liquid activators restrict the cementing applications in in-situ conditions, especially in bulk production.

In most of the studies, commercially available alkali hydroxides and silicates have been used as the solid activator in one-part systems, although they have used cementitious wastes as solid precursors [6]. These commercial activators provide high alkaline conditions, pH > 12, to accelerate the dissolution of solid precursors. However, it is challenging to create high alkaline conditions using the activators with waste materials alone. On the other hand, the initial dissolution of solid precursors is relatively low in one-part binders compared to two-part binders since the liquid alkaline solutions provide better alkaline media in two-part systems [7]. In one-part systems, dissolution of activator in water is the most critical step to create alkaline conditions to subsequent leach out of minerals of solid precursors. Hence, it is advisable to partially replace the solid activator with potential waste blended with a commercial alkali supplier to enhance the initial dissolution of aluminosilicate precursors. However, those commercial activators, especially alkali silicates, have significant adverse effects on the cost, energy and emission points of view [8]. Moreover, a limited number of studies are available regarding the one-part ACBs consisting of waste materials that can substitute for the solid activator. Therefore, it is necessary to incorporate potential waste materials into the activator of one-part ACBs, enhancing the recycling rate of such types of wastes. This study mainly focuses on the possible inclusion of waste glass powder to replace commercial alkali silicates in the solid activator. Finally, it is expected to increase the involvement of potential waste materials in one-part ACBs for both activator and precursors by minimizing the inclusion of commercial materials.

2. Experimental work

2.1. Materials and mix design

Class F fly ash, and ground granulated blast furnace slag (GGBFS) were sourced from Independent Cement & Lime Pty Ltd, Australia. The crushed glass was purchased from Let's Clean Pty Ltd, Australia and further pulverized into a powder using a laboratory-grade ring mill. NaOH micro-pearls were obtained from Redox Pty Ltd, Australia. The chemical compositions with the mean particle diameters of the raw materials are shown in Table 1.

Table 1. Oxide compositions of raw materials with median diameters.

Wt.%	SiO_2	Al_2O_3	Fe_2O_3	CaO	MgO	K_2O	Na_2O	SO_3	P_2O_5	TiO_2	LOI[a]	D_{50}[b]
FA	62.7	27.1	2.6	2.1	0.4	1.26	0.59	<	0.3	1.0	2.5	17.6
GGBFS	32.7	13.0	0.3	43.0	5.8	-	0.4	1.4	-	-	0.7	15.3
CG	72.43	1.24	0.43	13.43	0.69	0.45	10.81	0.11	-	-	0.03	11.5

[a] Loss on ignition.
[b] Median particle diameters (μm).

The activator was prepared by mixing 52% crushed glass and 48% NaOH solids by mass and adding a small amount of water. These mass ratios were selected to adjust the alkali-modulus of the activator by considering SiO_2 and Na_2O compositions of crushed glass and NaOH as 70% and 77.5% by mass, respectively. Next, the paste mixture was heated in a laboratory-grade oven at 150 °C for 2 hours based on the optimum conditions suggested by Vinai and Soutsos [9] and then the heated residue was ground down to a powder using mortar and pestle before use as the solid activator. The binder was prepared using 60% fly ash and 40% GGBFS by dry mixing with the solid activator first and then adding water. Alkali content of the binder (Na_2O%) was kept constant at 7.5% by considering the efficiency (EF%) of formed crystalline Na_2SiO_3 in solid activator as 70% [10]. These results may be further described in detail in section 3.2. The effective water to binder ratio (W/B) was calculated by accounting for the crystalline water inside NaOH. Mix design parameters of both activator and binder are given in Table 2.

Table 2. Mix design details.

Parameter	Value
Na_2O%	7.5
AM	1.0
EF%	70%
Dry activator/Binder	0.21
Effective W/B ratio	0.5
FA	60%
GGBFS	40%

The fresh ACB slurry was poured into cylindrical PVC moulds (38 mm× 76 mm cylinders with aspect ratio = 2.0) by three layers with 30 seconds of compacting by vibration on each layer. Casted samples were covered by a polythene wrap to avoid evaporating moisture in low humid environments. Then the covered samples were placed in a curing chamber with 20 °C and 95% relative humidity. After 1 day, samples were de-moulded, again sealed by a polythene wrap and placed in the same conditions for further curing until the required testing age of 1, 7 and 28 days.

2.2. Testing methods
Uniaxial compressive strength (UCS) tests were carried out using the Instron 50 kN compression frame with a 0.2 mm/min displacement rate. Three replicates were used for a particular test age, and all the samples were face-ground to make the compressive loading purely axial. X-ray diffraction (XRD) analysis was conducted for powder samples using Bruker D8 Cobalt diffractometer (Co Kα1-1.7889 Å radiation and 10–90° 2θ range) to identify the mineral phases of raw glass, solid activator, and hardened ACB paste after particular curing age. Semi-quantification analysis for the solid activator was conducted using DIFFRAC.EVA V4.3 software. FEI Quanta 3D scanning electron microscope (SEM) was used to visualize the morphology of the raw materials and reaction phases in the hardened ACB paste. Hardened paste samples were collected from the fracture surface after compressive testing, treated with acetone/ethanol to avoid further hydration and kept in the oven for 1h at 60°C before XRD or SEM analysis [11].

3. Results and discussion

3.1. Compressive strength
The strength properties of ACBs at different curing ages were investigated to find the effectiveness of the waste glass-based solid activator. Figure 1 shows the compressive strength development and corresponding Young's modulus of the ACB paste up to 28 days of curing. At the early age of 1 day, ACBs showed lower values in strength and Young's modulus. Furthermore, the early age cured samples showed more ductile behaviour than the brittle nature of 28 days cured samples. The 7 and 28 days strengths were around 17 and 30 MPa, which is acceptable to use as a binder in construction applications [12, 13]. The strength development with the curing age is also related to the improved microstructure described in section 3.3. According to the compressive strength results, glass-based solid activator ACBs showed comparable strengths over ACBs with commercial activators, such as NaOH and Na_2SiO_3.

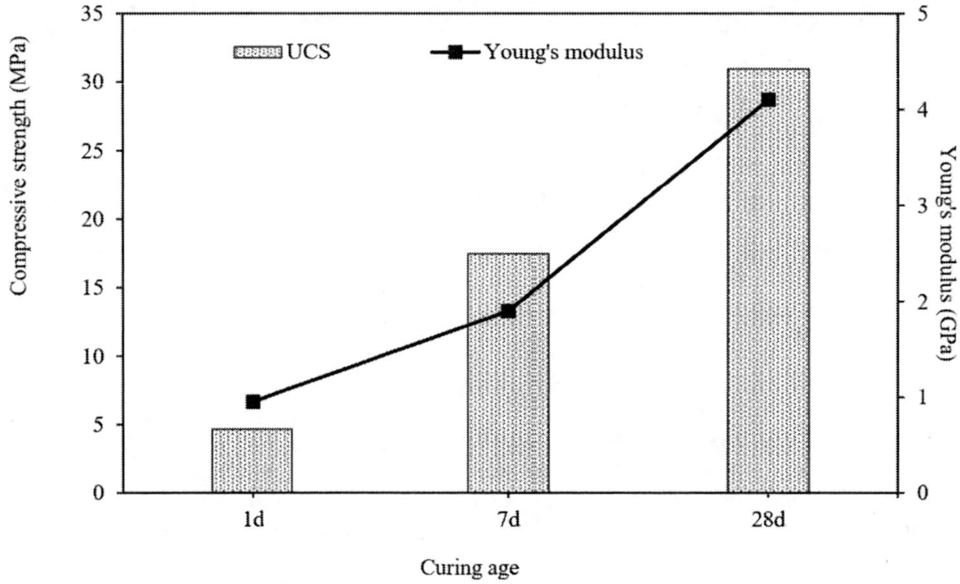

Figure 1. Variation of UCS and Young's modulus of ACBs.

3.2. Mineralogy

XRD tests were conducted on both solid activators to quantify the efficiency of forming Na_2SiO_3 and hardened ACBs to identify the mineral phases formed with the curing time. Figure 2(a) compares the nature of raw glass powder with the solid activator and shows a total amorphous nature of raw glass powder. Solid activator shows more crystalline nature with moreover flatted baseline compared to raw glass powder. According to semi-quantitative analysis, more than 70% of crystalline Na_2SiO_3 was formed by the total weight of phases. In addition, it shows minor crystalline phases of sodium carbonate (Na_2CO_3), different forms of sodium silicates (Na_4SiO_4, $Na_6Si_2O_7$) and calcium aluminium oxide ($CaAl_2O_4$). Hence, the efficiency of forming Na_2SiO_3 was assumed as 70% for mix design calculations of the ACB paste. Figure 2(b) shows the effect of curing age on the formation of mineral phases of hardened ACB paste. A noticeable increment in the crystalline peak can be observed for 28 days compared to 1-day ACB paste between the 2θ range of 32 and 37° which indicates the higher formation of calcium silicate hydrate (C-S-H) products in the binder phase [14]. This formation of C-S-H peaks also attributes to the strength increment with the curing age of ACBs described in section 3.1.

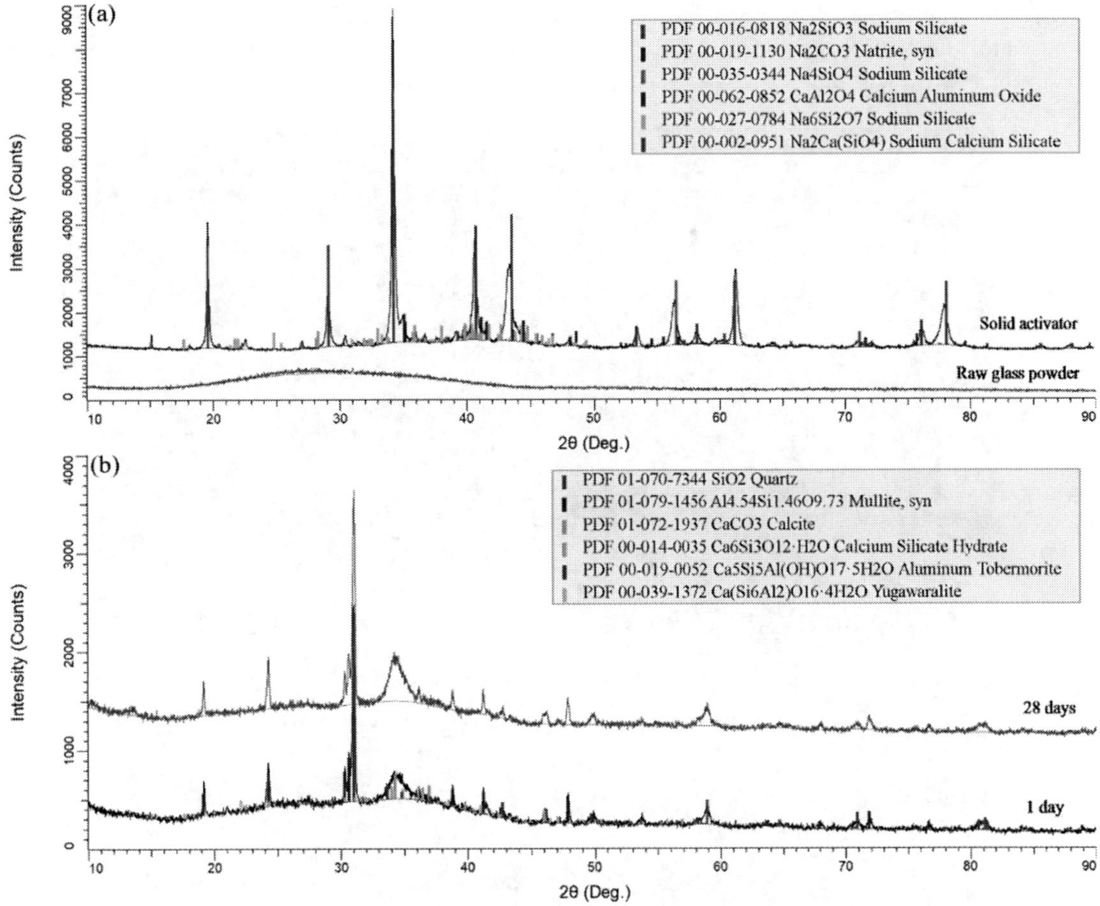

Figure 2. XRD patterns of (a) raw glass and solid activator and (b) hardened ACB paste (For better visualization of the colour legend, please refer to the online version of the article).

3.3. Microstructure

Surface morphology characteristics of raw glass powder and solid activator are shown in secondary electron images of Figures 3(a) and (b), respectively. The irregular particle morphology of the solid activator was affected by the odd shape of the raw glass particles. However, particles in solid activator show agglomerated nature which indicates a better reactive phase formed by micro-scale glass particles and NaOH micro-pearls due to the heat-treatment process. These micro-particles in the solid activator may dissolve in water and create basic media to leach out minerals of the solid precursors. Figures 3(c) and (d) show the backscattered electron images of hardened ACB paste at 1 day and 28 days curing age, respectively. After 28 days, the paste shows a dense structure with fewer unreacted phases than 1 day, which reduces the porosity and increases the strength characteristics.

4. Conclusions

This study elaborates the potential pathways to incorporate cementitious wastes in alkaline cement binders (ACBs) by involving fly ash, blast furnace slag, and crushed glass. Fly ash and blast furnace slag act as the solid precursors, while crushed glass behaves as the partial activator for the binder. Crushed glass powder can be successfully blended with commercial NaOH to produce a solid activator for one-part ACBs after thermal treatments. Glass powder-based activator showed 70% efficiency in creating crystalline sodium silicate, which is the major alkali-silica source for the activation of solid precursors. Moreover, the experimental results on strength, mineralogy, and microstructure of alkaline cement paste show crushed glass substituted solid activator plays a similar role in one-part ACB systems compared to commercial activators. Finally, this study opens up the effective ways of recycling potential cementitious wastes, which produce a cement binder with better mechanical and microstructural properties.

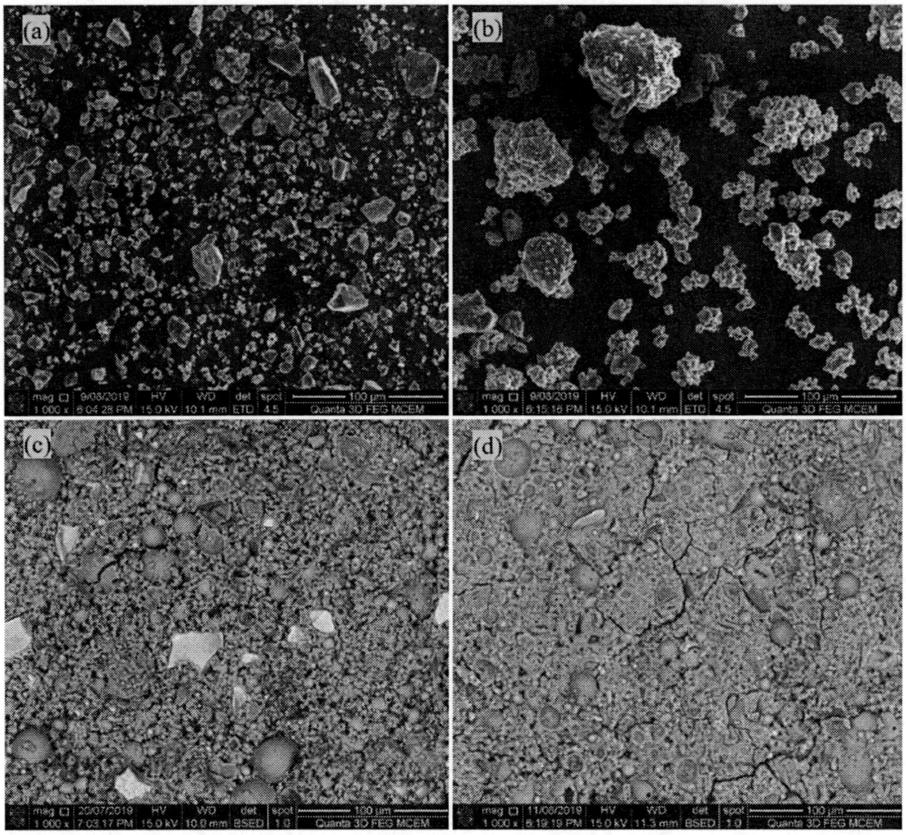

Figure 3. SEM images of (a) raw glass powder, (b) solid activator, (c) ACB paste – 1 day curing and (d) ACB paste – 28 days curing.

5. References

[1] T Zhang, P Gao, P Gao, J Wei, Q Yu, Effectiveness of novel and traditional methods to incorporate industrial wastes in cementitious materials—An overview, *Resources, Conservation and Recycling* **74** (2013) 134-143.

[2] H K Tchakouté, C H Rüscher, S Kong, E Kamseu, C Leonelli, Geopolymer binders from metakaolin using sodium waterglass from waste glass and rice husk ash as alternative activators: A comparative study, *Construction and Building Materials* **114** (2016) 276-289.

[3] F Puertas, M Torres-Carrasco, Use of glass waste as an activator in the preparation of alkali-activated slag. Mechanical strength and paste characterisation, *Cement and Concrete Research* **57** (2014) 95-104.

[4] C Ma, B Zhao, S Guo, G Long, Y Xie, Properties and characterization of green one-part geopolymer activated by composite activators, *Journal of Cleaner Production* **220** (2019) 188-199.

[5] H A Gawwad, S A El-Aleem, A Ouda, Preparation and characterization of one-part non-Portland cement, *Ceramics International* **42**(1) (2016) 220-228.

[6] T Luukkonen, Z Abdollahnejad, J Yliniemi, P Kinnunen, M Illikainen, One-part alkali-activated materials: A review, *Cement and Concrete Research* **103** (2018) 21-34.

[7] X Ke, S A Bernal, N Ye, J L Provis, J Yang, One-Part Geopolymers Based on Thermally Treated Red Mud/NaOH Blends, *Journal of the American Ceramic Society* **98**(1) (2015) 5-11.

[8] C Ma, G Long, Y Shi, Y Xie, Preparation of cleaner one-part geopolymer by investigating different types of commercial sodium metasilicate in China, *Journal of Cleaner Production* **201** (2018) 636-647.

[9] R Vinai, M Soutsos, Production of sodium silicate powder from waste glass cullet for alkali activation of alternative binders, *Cement and Concrete Research* **116** (2019) 45-56.

[10] M Samarakoon, P Ranjith, W H Duan, V De Silva, Properties of one-part fly ash/slag-based binders activated by thermally-treated waste glass/NaOH blends: A comparative study, *Cement and Concrete Composites* **112** (2020) 103679.

[11] M Torres-Carrasco, F Puertas, Waste glass in the geopolymer preparation. Mechanical and microstructural characterisation, *Journal of Cleaner Production* **90** (2015) 397-408.

[12] G F Huseien, J Mirza, M Ismail, S Ghoshal, A A Hussein, Geopolymer mortars as sustainable repair material: A comprehensive review, *Renewable and Sustainable Energy Reviews* **80** (2017) 54-74.

[13] M H Al-Majidi, A Lampropoulos, A Cundy, S Meikle, Development of geopolymer mortar under ambient temperature for in situ applications, *Construction and Building Materials* **120** (2016) 198-211.

[14] S Zhang, A Keulen, K Arbi, G Ye, Waste glass as partial mineral precursor in alkali-activated slag/fly ash system, *Cement and Concrete Research* **102** (2017) 29-40.

Acknowledgments

This study was supported by the ARC Linkage grant (Grant ID: LP150100103). The authors are grateful to Independent Cement & Lime Pty Ltd, Australia, for providing the materials used in this study. The authors would also like to acknowledge the use of the facilities of the Monash X-ray Platform and the Monash Centre for Electron Microscopy – a Node of Microscopy Australia.

Rainwater Harvesting in the Rainforest: A Technical and Socioeconomic Review of Community Approaches in Brazil

Amy Farrell[1*], Dr Andrew Swan[1], Dr Ronaldo Lopes R. Mendes[2]

[1]Leeds Beckett University, UK. [2]Universidade Federal do Pará, Brazil.

* **Email:** amy.farrell@arup.com, A.D.Swan@leedsbeckett.ac.uk, rmendes@ufpa.br

Abstract. This paper reviews potable water issues affecting river dwelling communities and assesses the use of Rainwater Harvesting techniques to help meet potable water requirements. The study specifically explores river communities on the periphery of the city of Belem, where there is no potable mains water supply. This interdisciplinary study reviews the socioeconomic context of the different organisational stakeholders as well as analysing technical data to ascertain the effectiveness of the systems in practice. Surveys indicate nearly half of residents consume water from sources potentially contaminated by raw sewage and mining effluent with detrimental impacts to public health. This paper reviews previous local studies and describes on-going attempts to derive locally appropriate rainwater collection and treatment processes. This includes attempts to construct and trial organic material-based rainwater filters that comprise of different mixes of sand, gravel and activated carbon derived from by-products of local fruit processing industries, namely acai harvesting.

Keywords: Rainwater Harvesting, Peripheral communities, Water filtration, Governance.

1. Introduction

This study explores the water issues affecting river island communities in the Amazon region of Brazil, and on-going attempts to assess the viability of employing Rainwater Harvesting techniques and locally sustainable filtration methods using organic materials to help meet local water requirements. The study specifically considers river communities on the periphery of the city of Belem (Figure 1) in the State of Para. Belem's municipal area includes a number of inhabited river islands – the total population of the city is estimated to be 1.45 million [12], whilst the population of the surrounding river-dwelling communities, known in Portuguese as 'Riberenhos', was reported at 3049 in the 2010 census.

Belem municipality experiences a range of significant water and sanitation challenges. For example, only 59.9% of the city's population are connected to the mains clean water supply [3]. In terms of sanitation, Bland [3] reports that only 6.3% of the city's population are served by a sewer network and that the vast majority (95%) of Belem's raw sewage is pumped into surrounding rivers, Guamá and Tocantins. Some of this river water is abstracted for drinking supplies by surrounding communities [9][21]. Belem experiences Brazil's highest rates of hospitalisation due to diarrhoeal diseases for children under the age of five (Fundacao Abrinq, 2018). In addition to the direct public health impacts, these issues lead to wider economic pressures: the lack of sanitation in one of Belem's neighbouring districts (i.e. the municipality of Ananindeua) resulted in R$19,5 million (approximately US$5.1 million) of medical care costs between 2007 and 2015 (Globo, 2017).

Figure 1: Inhabited river islands around the City of Belem in Brazil's Para State.

2. Background context

2.1. Sustainable Development Goals

The United Nation's Sustainable Development Goal 6 (UN, 2010) aims to ensure access to water and sanitation for all, stating that lack of access is partially attributable to economics or poor infrastructure. These factors cause millions of people to die every year, many of them children, from water related diseases. The SDGs highlight the importance of integrated action to address economic, social, and environmental development. In this context, it is worth considering the range of organisations that play a role in relation to the supply of clean water within Belem.

2.2. Key stakeholders

Complexities in Brazil's political structures impact the governance of water. Brazil is a Federal Republic, where the states, such as Pará, are autonomous sub-national entities with their own governments. The states are sub-divided into municipalities. Each municipality has an autonomous local government [4]. Different parts of this federal system dictate the legal aspects of water supply to communities in Brazil, organisations in water management exist at the federal and municipal level, for example, the ANA (National Water Agency) and SINGREH (National System of Water Resources Management – incorporates several institutions including ANA) [1]. The urban water supply in Belem is managed by Cosanpa (i.e. the Pará water and sanitation company), but this agency does not serve the river islands in the municipality of Belem [2]. Other organisational roles in the provision of clean water for Belem's island communities are highlighted in Table 2. These organisational relationships were previously reviewed by Cardoso-Castro *et al.* [5].

Table 1. Organisational roles linked to clean water provision on Belem's River islands.

Policy	Federal Government (e.g. FUNASA – National Public Health Authority)
Regulation	Municipal Government e.g. AMAE – Water and Wastewater Regulator
Research	Federal University of Para; Federal Institute of Science Education and Technology
Finance	NGOs e.g. CARITAS; Federal Government e.g. MDS; State Government (SEASTER)
Implementation	Island Community Groups; NGOs e.g. CARITAS; State Government (SEASTER)

2.3 Local water issues

The vast majority of Belem's island communities are not served by mains water supplies or sewer networks [2]. It has been reported that this is partially due to their remote locations but may also have resulted from a lack of enforcement and enactment of Federal Government guidelines on water quality for domestic supplies [13]. At the municipal level, Belem residents are not receiving the Federal standards of water [7][17]. The water company for Para state 'Cosanpa' is suffering financial problems (Fundacao Abrinq, 2017) stating that they are not currently able to make further investments to improve water infrastructure (Globo, 2017).

Communities are forced to rely on a range of alternate water sources including river abstraction. Previous studies [21][9] explored the different water sources used on two case study islands (Ilha Murutucu and Ilha Grande), through community survey exercises. These studies reported that 11.4% of residents on '*Ilha Murutucu*' and 5.7% on '*Ilha Grande*' were consuming water directly drawn from the river. It was highlighted in the previous section that this tidal river receives untreated sewage from the city of Belem [3]. This is a common occurrence, it has been reported that on average 86% of the sewage discharged from the cities in the Amazon basin is not treated (Fundação Abrinq, 2018). The river may also contain other pollutants, which have emanated from further upstream in the Amazon basin. For example, there are china clay mining activities in the adjacent municipality of Barcarena (Imerys, 2018) and waste spills emanating from bauxite processing at alumina refineries [11](Nathanson, 2018).

Figure 2: River-based vendors on the Amazon [9][2].

Veloso [21] and Gonçalves [9] highlighted that a significant percentage of the surveyed communities (i.e. 42.7% of residents on Ilha Murutucu and 42.9% of Ilha Grande) purchased jerry cans of water from river-based vendors (Figure 2). The containers of water sold by these river traders at a price of R$2.00 (US$0.52) are not considered to emanate from a clean source and often exhibit high levels of turbidity [10].

2.4 Rainwater harvesting by river dwelling communities

Almost half of the island populations surveyed by Goncalves [9] and Veloso [21] were reported to be consuming potentially contaminated water (e.g. sourced from the river vendors or abstracted from the river). In response to these challenges a range of previous studies [9][21][8][19] have sought to explore the merits of alternative water sources that are appropriate to this context, and less likely to be contaminated. Given that Belem has a wet season that lasts for 6 months per year [22], with rain continuing to fall during the dry season, Rainwater Harvesting has been considered as a viable alternative water supply.

Veloso [21] undertook community survey work on a number of river islands which indicated that between 50-61% of residents were in favour of using a RWH system. However, some were reportedly opposed as they did not like the taste of rainwater compared to the water they usually bought from local vendors. The study concluded that although RWH systems appeared to be both environmentally and socially viable, local residents could not afford to install the systems due to their incomes being lower than the minimum wage [21].

A separate exercise conducted by Dias [8] undertook a sustainability appraisal of RWH systems in the Amazon context using five indicators (Environmental, Social, Economic, Politico-institutional, Technical). This study suggested that the politico-institutional aspects of implementation were unsustainable despite the technical scalability of RWH [8]. More recently, Silva [19] sought to explore the quantity and quality of rainwater collected from these systems. This study attempted to define an efficient yet easily repeatable procedure for treating this rainwater [19]. Further studies have explored the merits of employing filter beds that utilise local materials to treat collected rainwater.

These RWH field-trials have compared water quality indicators from sampled raw (untreated) roof water, with water treated using a sand/activated carbon bed. The five common water quality indicators used in these tests are: pH; Turbidity; Colour; Coliform Bacteria; E. Coli. The Federal Government of Brazil standards set out by the Ministry of Health (2011) in Decree 2.914/11 are consistently referred to as the criteria required for potable water in Brazil throughout the studies this report has reviewed (See Table 4).

Table 2: Standards for collective systems supplying < 20,000 inhabitants, [13].

Parameter	Recommended Maximum Value
pH	6-9 for guidelines to be followed
Turbidity	1 NTU across 95% of samples, to a maximum 5 NTU
Colour	15 Hazen units
Coliform Bacteria	No more than 1 positive test per month permitted in structured testing routine
E. Coli	0/100ml

2.5 Application of locally sourced Acai as a water filter medium

The acai palm is a tree species that is farmed across the Amazon region for its fruit, palm, leaves and wood. This activity generates a substantial amount of waste material. The acai fruit has a relatively thin mesocarp (~1mm thick) which is used for the fruit pulp. The majority of the fruit consists of the seed, of which large volumes are discarded as waste during processing. For example, Belem and Para State produced 706.488 tonnes of acai pulp in 2010, representing 86.9% of Brazil's annual production [14](SAGRI, 2010). Acai harvesting and production is said to employ more than 25,000 persons directly in the north of Brazil, and constitute 70% of river island communities' incomes [14](Lopez and Santana, 2005). Research into the use of activated carbon from fruit processing by-products, namely acai for filtration has been applied in various contexts in the removal of water contaminants such as dyes, sewerage and heavy metals by Brazilian researchers in recent years [18][17][16](Cunha *et al.*, 2012).

2.6 Review of previous water filter strategies

A number of previous studies have investigated alternative filtering/treatment strategies that might be used by river island communities to improve the quality of water collected from RWH systems. These studies have explored the merits of various filters that comprise different mixes of sand, gravel and in three cases activated carbon from acai seeds. Figure 5 lists previous research carried out by local researchers in the state of Para (PA) and the state of Espirito Santo (ES) (i.e. with the use of a larger sand and gravel filter to supply potable water to isolated communities there).

Case	Filter Type 1	Filter Type 2	Filter Type 3	Filter Type 4	Filter Type 5
Author	Goncalves (2012a)	Goncalves (2012a)	Oliveira et al (2017)	Cruz, Teixeira and Menezes (2017)	Goncalves (2012b)
Location	Ilha Murutucu, PA	Ilha Grande, PA	UFPA Belem, PA	Curio - Utinga, PA	UFES, Vitoria, ES

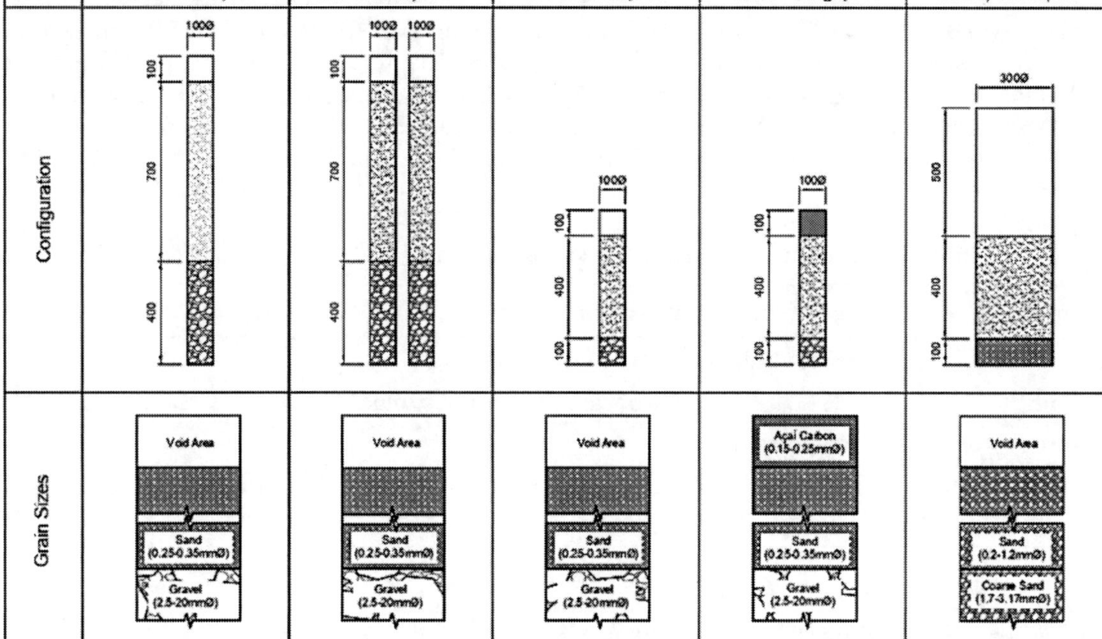

Sample Results										
	Raw	Filtered	Raw	Filtered	Raw	Filtered	Raw	Filtered	Raw	Filtered
Turbidity (NTU)	1.04 Median	0.5 Median	1.28 Median	0.47 Median	0-5.6 Range	0-6.8 Range	2.3 Mean	0.5 Mean	1.13 Median	0.06 Median
Colour (HU)	27 Median	11 Median	20.5 Median	9 Median	1-67 Range	3-72 Range	7 Mean	0 Mean	15.23 Median	3.12 Median
pH	5.5 Median	5.4 Median	6 Median	5.6 Median	6,4 Mean	6 Mean	7.5 Mean	7 Mean	N/A	N/A
Coliform Analysis	E-coli present	E-coli present	E-coli present	E-coli present	49.7% average coliform reduction in filtered case		41.6 coliforms NMP/100ml Mean	4.8 coliforms NMP/100ml Mean	262 E-coli NMP/100ml Median	7.2 E-coli NMP/100ml Median

Figure 3: Summary of publications investigating rainwater filter configurations in remote/peri-urban contexts in Brazil.

2.6.1 Summary of findings from case studies 1-5

Case studies 1 and 2 at the Federal University of Para [9] trialled two filter configurations on the islands of Murutucu and Grande. These filters used sand and gravel to treat water collected from tiled roofs. Units were installed in August and were in use from September. Fifteen water samples were collected during consecutive site visits conducted between January and April. Tests indicated that the filtered water was generally not of a potable standard without additional treatment using sodium hypochlorite, although reductions were achieved in turbidity, colour and pH. Case Study 3 at the Federal University of Para [15] explored the effectiveness of acai filters using a small sand and gravel filter tested on campus, which was new at first use with the materials disinfected with 2% chlorine bleach for 2 hours prior to assembly. 60 Sample sets were taken during March to August 2014 and 17 from October to January 2015. Results found minimal reductions in the tested parameters compared to the raw water case, and in some cases worse results post-filtration. It concluded sand and gravel filters alone were not effective for the filtration of water. Case Study 4 [6] constructed two similar rigs, to those in case study 1 and 2, in riverside communities in Belem using filters the same size as case study 3 but containing a mixture of sand and activated acai carbon. The units were constructed between March and April, and were operational for 2 weeks prior to the first sample set. Five sample sets were taken insitu from household units during April and May 2014. A secondary UV filter was added in order to eliminate remaining coliform bacteria from the sand and acai filter to make the water potable. Case Study 5 was undertaken at the Federal University of Espiritu Santo, Vitoria, which is located in a region of Brazil with differing local sand composition [10]. This exercise utilised a sand and gravel filter with a larger volume and smaller grain sizes than those in Para State. Two identical filters were operated daily, typically filtering $8m^3/m^2$ per day, on the University campus over the course of 110 - 130 days. This filter performed well with a reported 80% reduction in turbidity and colour. Coliforms were not completely eliminated and UV light was used for post-filtration disinfection.

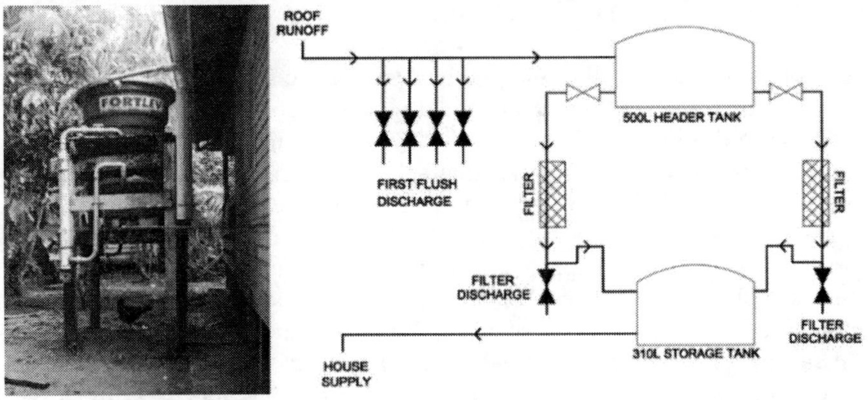

Figure 4: Rainwater Harvesting (RWH) system on Ilha Grande.

2.7 Other water filter studies employing activated acai carbon

Another study at the Federal University of Para [17], which did not directly relate to rainwater harvesting, utilised activated acai carbon for the filtration of heavy metals from river water. This study highlighted that these fruit by-products require heating to temperatures of around 400°C to be viable for filtration. Samples were taken at 19 different locations at 3 depths per location. This study found that the use of activated carbon reduced heavy metals from river water samples by 67.46% for aluminium, 41.67% for chromium and 65.15% in the case of zinc. Pereira [16] also highlighted the potential of activated acai carbon as secondary filtration for domestic tap water, in this case in the town of Moju, Para where the acai filter was found to reduce iron content to acceptable levels and eliminated detectable presence of e-coli.

3. Results and discussion

The key results associated with the removal of turbidity, colour, pH and coliforms for the five studies highlighted in Figure 5 are presented in Figure 7 in terms of their percentage reduction (i.e. from raw to filtered water samples). The associated publications either detailed a reduction in coliform numbers or describe whether e-coli presence was found on a yes or no basis. The inconsistencies in these previous bacteriological studies, has prevented the author from plotting a comparative graph/dataset. All of the aforementioned publications reported either the presence of e-coli or coliform in the filtered samples. Given that Brazil's Health Ministry requires the removal of e-coli and coliforms, all of these publications recommended either post-filter chlorination or UV treatment to remove bacteria prior to consumption.

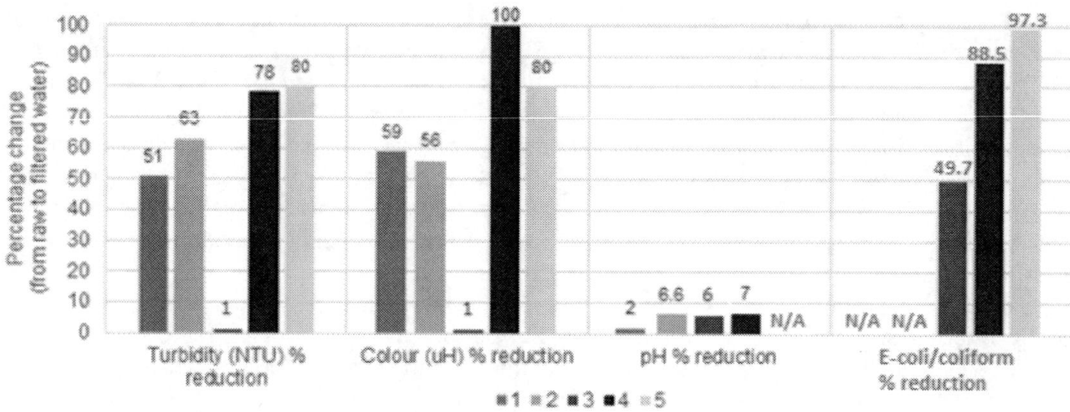

Figure 5: Average % reduction for RWH filter cases where comparable results available.

With the exception of Filter 3, all of the previous trials gave a high reduction in terms of dissolved solids, with Filters 4 and 5 performing exceptionally well in colour and turbidity reduction. A small reduction in pH occurred across the filters, this data was not available for Filter 5 but a similar change might be expected as the other sand and gravel filters. In terms of the reported testing routines, the length of time that the filters were in use for prior to samples being taken varied. Filters 1,2 and 5 were in use for ~90 days prior to the sample sets being taken, whereas 3 was new and 4 was insitu for 2 weeks prior to testing. Older filters would be expected to exhibit lower flow rates and worsening contaminant removal. The poor performance of Filter 3 might relate to filter materials passing through which had not properly been flushed before first use. All of the reported raw water samples failed to meet the Brazilian Ministry of Health standards for turbidity (i.e. of 1 NTU). This failure pattern, with the exception of Filter 4, is once again repeated for the Ministry's standards for colour. In terms of the filtered water samples, all filters with the exception of Filter 3, achieved the Ministry's standards for both turbidity and colour. Filter 3 gave satisfactory results for colour only within its first three months of use, however so did the raw water in this sample set. Filter 4 maintained the same physical size as Filter 3, however, 100mm of Acai carbon was added to the filter medium. It appears that this produced comparable percentage changes from the raw water case to Filter 5 with much less material volume, however, filter 5 was 75 days older at this point and would therefore not have been in optimum condition. Further testing as filter 4 ages could indicate that the powdered nature of the carbon particles enhance the filtration process in a small volume filter. Filters 2 and 3 [9] concurrent field trial yielded similar results, Filter 2 removed 12% more solids, but used double the volume of filter materials.

4. Conclusions

This paper has reviewed potable water issues experienced by a number of river island communities to assess whether Rainwater Harvesting techniques might be used to meet local water requirements where communities are not served by the mains water supply. Previous community surveys have indicated that nearly half of local residents may be consuming water from potentially contaminated sources.

On-going attempts to derive locally appropriate rainwater collection and treatment processes constructed and trialled water filters comprising of different mixes of sand, gravel and activated carbon derived from by-products of local fruit processing industries. The studies also highlighted the need for post-filtration disinfection. The length of time that each of these filters had been operated prior to the reported samples being taken varied. Older filters would be expected to exhibit lower flow rates and worsening contaminant removal. It appears that filter type 4 with sand, gravel and activated carbon was efficient at reducing colour and dissolved solids from the raw water state versus its physical volume, however this was a relatively new filter. To ascertain if filters with finer materials such as the powdered nature of the acai carbon or finer Espírito Santo sand improve performance, these filter types should be

IOP Conf. Series: Materials Science and Engineering 1196 (2021) 012003 doi:10.1088/1757-899X/1196/1/012003

tested concurrently to measure performance. Using a layer of finer material could improve efficiency where coarser sands and gravels are only available locally. Further controlled testing is required to measure the effectiveness of the acai to a similar volume of medium with a finer sand. None of the rainwater harvesting filter tests examined nutrient or metals content.

In terms of sustainability, the carbonisation process utilises a considerable amount of energy, and it should be ascertained whether it is the carbon that optimises the filter, or if other available fine material in a smaller diameter filter would achieve the same effect by slowing the filtration process. It would be useful to compare cost and performance against a commercially produced filter in order to examine the overall sustainability of producing carbon from fruit processing waste for rainwater filtration, moreover where access to financial support is limited. Lack of resources and political barriers impact peripheral communities' access to services, therefore support from NGOs, in this case principally Caritas, local community and local university groups has been required to successfully execute pilot projects. Cuts to budgets for education and community organisations would have an effect in terms of scaling up these projects for greater impact.

References

[1] A N A, 2017. Water Resources Management in Brazil. [online]. Available at: <http://hidroweb.ana.gov.br/doc/WRMB/part2.htm> [Accessed 3 Sep. 2017].

[2] Belém Municipality. 2014. Plano municipal de saneamento basico de abastacimento de agua e esgotamento sanitario de Belém – Pará, PMSB Volume I. Prefeitura Municipal de Belém. Belém, Brazil.

[3] Bland D, 2013. Brazil's Belém in critical need of water, sewage system improvements. [online] BNamericas. Available at: <http://www.bnamericas.com/en/news/waterandwaste/brazils-belem-in-critical-need-of-water-sewage-system-improvements> [Accessed 22 June. 2017].

[4] Brasil Governo Federal, 2017. [online] How the structure of the Brazilian State works. Available at: <http://www.brazilgovnews.gov.br/federal-government/state-structure> [Accessed 6 Sep. 2017].

[5] Cardoso-Castro P, Swan A, Mendes R (2017) Stakeholders' Structural Factors Affecting the Implementation of Rainwater Systems in the Amazon: the case of Belém. *SEEDS conference*, 13-14 September 2017, Leeds, UK.

[6] Cruz J, Teixeira L and Menezes H (2018). Verificação da potabilidade de água de chuva através de filtro lento e desinfecção ultravioleta. [online] Gvaa.com.br. Available at: http://www.gvaa.com.br/revista/index.php/RBGA/article/view/4952 [Accessed 12 Jun. 2018].

[7] Diario do Para, 2012. [online] Belém sofre com deficiência de água potável. Available at: <http://www.diarioonline.com.br/noticia-194004-belem-sofre-com-deficiencia-de-agua-potavel.html> [Accessed 6 Sep. 2017].

[8] Dias A D, 2013. A Sustentabilidade De Tecnologias Sociais De Abastecimento De Agua Da Chuva: O Caso de Comunidades Insulares de Belem-PA. MSc, NUMA, Universidade Federal do Para, Belem.

Fundação Abrinq. (2018). Acesso à água potável e esgotamento sanitário ainda é crítico no Brasil. [online] Fundação Abrinq. Available at: https://www.fadc.org.br/noticias/181:acessoa-a-agua-potavel-e-esgotamento-sanitario-ainda-critico-no-brasil [Accessed 14 Mar. 2018].

Globo. (2017). G1 - Doenças causadas por saneamento ruim custam R$ 19,5 mi a Ananindeua - notícias em Pará. [ONLINE] Available at: http://g1.globo.com/pa/Pará/noticia/2017/02/doencas-causadas-por-saneamento-ruim-custam-r-195-mi-ananindeua.html. [Accessed 01 December 2017].

[9] Goncalves C 2012a. Aproveitamento De Agua Da Chuva Para Abastecimento Em Area Rural

Da Amazonia. Estudo De Caso: Ilhas Grande E Murutucu, Belem-Para. NUMA, Universidade Federal do Para, Belem.

[10] Goncalves R F, 2012b. Potabilização de água de chuva através de filtração lenta e desinfecção ultravioleta para abastecimento descentralizado de comunidades. Universidade Federal do Espirito Santo, Vitoria.

[11] Hydro (2018). Hydro in Barcarena, Brazil. [online] Hydro.com. Available at: https://www.hydro.com/en/About-Hydro/Hydro-worldwide/Brazil/Barcarena/ [Accessed 12 Jun. 2018].

[12] IBGE, 2017. Instituto Brasileiro de Geografia e Estatística. [online] Cidades Pará Belém. Available at: <http://ibge.gov.br/cidadesat/xtras/perfil.php?lang=_EN&codmun=150140&search=para%7Cbelem> [Accessed 3 Sep. 2017].

Imerys (2018). Kaolin Operations - Europe and Middle East. [online] Europe and Middle East. Available at: http://imerys-kaolin.com/europe-middle-east/en/about-imerys/kaolin-operations/#Barcarena [Accessed 12 Apr. 2018].

[13] Ministry of Health Brazil. (2011). Portaria Nº 2.914, de 12 de Dezembro de 2011 Dispõe sobre os procedimentos de controle e de vigilância da qualidade da água para consumo humano e seu padrão de potabilidade. [online] Available at: http://bvsms.saude.gov.br/bvs/saudelegis/gm/2011/prt2914_12_12_2011.html [Accessed 21 Feb. 2018].

[14] Nogueira A, Santana A and Garcia W (2013). A dinâmica do mercado de açaí fruto no Estado do Pará: de 1994 a 2009. [online] **60**(3), **pp**.324-331. Available at: http://www.scielo.br/pdf/rceres/v60n3/04.pdf [Accessed 10 Jul. 2018].

[15] Oliveira D C, Correa M S, Sousa M S de, Tavares L C, Teixera L C, (2016) Avaliação do filtro de areia em um sistema de captação e tratabilidade de água de chuva. 10o Simpósio Brasileiro de captação e manejo de água de chuva. Available at: http://10sbcmac.ufpa.br/anais/ARQUIVOS/GT1-1-64-20161120160401.pdf [Accessed 10 Jul. 2018].

[16] Pereira E (2013). Carvão Do Caroço De Açaí (Euterpe Oleracea) Ativado Quimicamente Com Hidróxido De Sódio (NaOH) E Sua Eficiência No Tratamento De Água Para O Consumo. Xxvii Prêmio Jovem Cientista. Ministério Da Ciência, Tecnologia, Inovações E Comunicações, Brasil. **pp**.204-222.

[17] Pereira S, Maciel A, Santos D, Mares E, Oliveira G, Bittencourt J and Oliveira J (2018). Remoção de metais de águas superficiais usando carvão vegetal de açai (euterpe oleracea mart). [online] XIII International Conference on Engineering and Technology Education. Available at: http://copec.eu/congresses/intertech2014/proc/works/89.pdf [Accessed 12 Jun. 2018].

[18] Schodermayr D, Costa A, Araujo F, Melo L, Sobrinho M and Filho R (2018). Adsorção Do Corante Azul De Metileno (Ci 52015) No Carvão Do Endocarpo Do Açaí Ativado Termicamente. [Online] 12 Encontro Brasileiro Sobre Adsorção. UFSM. Available at: http://scheneventos.com.br/eba/envio/files/269_arq1.pdf [Accessed 12 Jun. 2018].

[19] Silva R Da 2016. Estudo Do Potencial De Aproveitamento De Água Da Chuva Para Abastecimento Domiciliar Na Área Urbana De 05 Municípios Da Região Metropolitana De Belém Do Pará. MSc. Universidade Federal do Para.

[20] UN (2010). Goal 6 [online] Available at: https://sustainabledevelopment.un.org/sdg6 [Accessed 5 Dec. 2017].

[21] Veloso N (2012). Agua Da Chuva E Desenvolvimento Local: O Caso Do Abastecimento Das Ilhas De Belem. MSc. NUMA, Universidade Federal do Para. Belem

[22] WP (2018), Long-term historical weather –Washington Post [online], Available at: http://www.washingtonpost.com/wp-srv/weather/longterm/historical/data/belem_brazil.htm [Accessed 20 March. 2018].

Integrated Management of Municipal Solid Waste in Brazil: A Case Study in São Paulo City

Michel Xocaira Paes[1,*], **Jose A Puppim de Oliveira**[1,2,3]

[1]Fundação Getulio Vargas (FGV), São Paulo School of Management (FGV/EAESP), São Paulo, Brazil
[2]Fundação Getulio Vargas (FGV), Brazilian School of Public and Business Administration (FGV/EBAPE), Rio de Janeiro, Brazil
[3] Institute for Global Public Policy, Fudan University, China

* **Email:** michelxocaira@gmail.com

Abstract. Municipal Solid Waste (MSW) Management Systems are constantly changing and innovating around the world. Thus, this work aimed to develop a methodology that allows identifying and understanding the main drivers, barriers and possible pathways for innovative waste management that accelerate the urban sustainability transition and how the political, social and institutional factors can impact operational, economic and environmental indicators in a municipality. The work was developed through a case study in the city of São Paulo (Brazil). It made it possible to identify the main actors involved in the processes of elaboration and conduct of public waste management policies, as well as operational impacts (97.79% of waste sent landfill), economic (US$126,51/ton.msw) and environmental (370.06 kg CO2eq/ ton.msw), contributing to the definition of criteria for the construction of Integrated Solid Waste Management methodology. The results also made it possible to highlight guidelines and practices for the improvement of MSW public policies.

Keywords: municipal solid waste, public policies, transition to sustainability, integrated waste management.

1. Introduction

Significant technological and management innovations in waste management have emerged in the last decade in order to face the growing demand for materials and to confront the increasing ecological and social impact of the disposable consumer economy. Although some guidelines aim at reforming and improving traditional waste management frameworks, others are fundamentally designed to reconceptualise and reformulate them completely [1][9].

For the year 2016, official estimates [24] indicate a daily average generation of 161,400 t Municipal Solid Waste (MSW) in Brazil, where 59% were disposed in landfills, 19.9% were dumped in soil without any treatment, 3.4% were treated and recovered in sorting, composting and recycling units and 17.7% were without information.

In some European Union (EU) countries, such as Germany, Austria, Belgium, Denmark, the Netherlands and Sweden, public policies have raised the rates of reusing, recycling and incineration with energy recovery and/or composting to 95% [4][28]. Even having other destination technologies, sending MSW to landfills is still a major practice in the USA. In 2015, this option was used for 52.5% of the solid waste, followed by incineration with energy recovery (12.8%), recycling (25.8%) and composting (8.9%) [3].

The best indicators can be attributing to the technological advances of the last decades, as well to the management models used, where the effectiveness of public policies is directly linking to the processes of planning, social participation and local control of waste management. In addition how to institutions act in intergovernmental relations, such as sectoral policies, policies at other levels of government, civil society, academic and universities, private initiative and within itself [13][17][18][26].

Authors also highlight that social, political and institutional maturity are crucial for the success of transition and innovation public policies for sustainability, which generate economic and environmental benefits for society [16][27].

Even with some theoretical advances that allow to evaluate the performance of social actors or stakeholders in certain public policies, some methods have not yet been developed for these diagnostics and analysis in waste management policies [7][6][17][25]. Such methods also don't allow to evaluate how such measures of institutional innovation are related to the other dimensions and functional principles of integrated waste management, such as economic and environmental aspects [5][8][10][15][23].

Therefore, this paper aims to develop a method for assessing levels of innovation in waste management, which allows understanding how and why innovations in waste management happen and how these innovations may affect the other aspects (environmental and economic) that make up the concepts of integrated waste management.

2. Method

The works were developed through field work and mostly present primary data, obtained through interviews and technical visits to the sorting, composting and landfill facilities - conducted between May and October 2019 -, as well as public secondary data, such as waste generation and composition data and GHG reduction targets (i.e. PMSP, 2014, 2017). Bibliographic review works were also performed to define the criteria for analysis of Integrated Management of MSW.

For the analysis of the political, social and institutional dimensions, the method known as "*multi-level perspectives (MLP)*" was used [6][7] which propose that the analyses be performed at three different levels, being: 1) niches, which can be understood as those that arise directly from society or even from individuals, as ideas and initiatives from the scientific / academic environment, non-governmental organizations and private initiative, which are independent of local politics. 2) socio-technical regimes, those directly linked to public policies and government management. 3) socio-technical landscapes, which can be understood as the broader context, capable of influencing the other two levels, as it would be something that sustains a society, such as customs, ideologies, values and beliefs, which would be beyond individual control.

For the environmental analysis of the impacts of climate change, was used the tool "CO2ZW (R)", developed by the Institute of Environmental Sciences and Technologies (ICTA) of the Autonomous University of Barcelona (UAB) and by the research company Inèdit [22], already adapted to the Brazilian reality [14]. This tool was developed based on Life Cycle Assessment (LCA) concepts and thus considers direct emissions from these systems - from waste treatment and disposal facilities - as well as indirect emissions - from electricity and diesel consumption - and emissions avoided by the recovery and recycling of waste. [22].

In the economic dimension, also based on complex and broad studies on the themes [11][19][14], the operating and investment costs of the São Paulo city were considered.

After social, political and social analysis and environmental and economic calculations, these results were relating and analysing together.

3. Results and Discussions

3.1. Management System Characterization

The MSW Management System (MSWMS) of the São Paulo city is basically divided into two contractual and management models with the private sector. First, through a Utilities Concession Agreement, two companies provide the services of common collection and transportation, transhipment and final disposal in landfill, in addition to selective collection, transportation and sorting of Household Waste. Through a Service Agreement, six more companies are responsible

for Public Cleaning Waste, which consists of the selective collection of organic waste from the free fair and sent to the five composting facilities, besides the collection and disposal in landfill of rubble and inert, sweeping, pruning waste, removal of waste disposed of in inappropriate places, dead animals and expired food.
Figure 1 presents the main activities of the MSW Management System - composed of the generation, collection, transportation, treatment, destination and generation of co-products - in addition to the quantities generated and managed by the São Paulo City Hall.

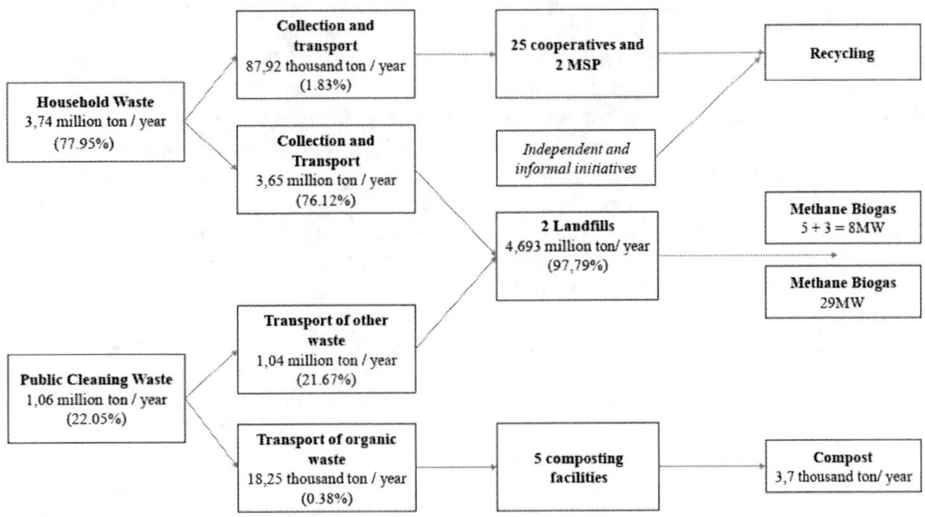

Figure 1: MSWMS of the São Paulo city and the quantities of MSW Generated and Managed in 2017.*MSP: Mechanized Sorting Plants.*

It is important to highlight, through Figure 1, that the city of São Paulo has been seeking information on independent and informal initiatives, which could increase the amount of waste generation and recycled materials in the city.

3.2. Institutional, Political and Social Dimensions

Through the *Multi-level Perspective (MLP),* applied to MSWMS, it was possible to identify some of the main drivers of public waste management actions and policies, as well as some possible barriers and pathways.
It was possible to understand that the initiatives of some *niches* (which could be quantified) are still timid and have little impact on the efficiency of MSW reuse. However, it is noteworthy that some important actors for the management of MSW in developing countries cannot be identified and quantified in this study, such as informal waste pickers.
The *landscape*, which can be understood here as cultural aspects, values, beliefs, ideologies and customs, has also had little impact on the results of public waste management policies, which can be noticed by the low participation of the population in formal selective collection processes, sorting and recycling. The relationship with indices of human development, education and region of the country can also make to infer that there is a need for greater investment in mobilization and education of the population for such actions to be effective, as highlighted by Mazzola [12].
Following are some of the main drivers, barriers and possible pathways that could be identified within the *sociotechnical regime*.
Drivers: São Paulo City Hall itself, its municipal secretariats and municipalities, such as the Municipal Authority for Urban Cleaning (*Amlurb*), as well as the Councils. The City Council; the State Control Institutions, such as the Public Prosecution Service and the Courts of Auditors. The State through the institution of environmental licensing. The Union through financing agencies,

such as the Ministry of Labor and Employment and the National Bank for Economic and Social Development (BNDES). Civil Society and NGOs through participation in councils, public hearings and specific actions. Universities and Research Centers in Projects and Partnerships. Companies and Service Providers. Recycling and Remanufacturing Industries. The Informal Sector and Independent Initiatives. As well as International Organizations (such as UN and World Bank) through projects and funding.
Barriers: Change of public offices (technical and strategic) during the changes of Municipal Governments. Poor integration between the wishes of the population and service providers. Poor public and population engagement in waste separation. Institutions that allow low change - such as the environmental licensing processes of the São Paulo State Environmental Company (*Cetesb*) and long contracts with service providers that do not encourage the best environmental performance. Low integration of MSWMS with Cooperatives. Little knowledge of Independent and Informal Waste Picker Initiatives and scrap dealers. Mechanized Sorting Plants operating below capacity. There is no specific fund for waste management in the city.
Possible Pathways: More effective staff, beyond qualification, for Increasing Learning Ability. Greater harmonization between Selective Collection, Composting and Landfill, through contracts that stimulate the environmental improvement of the systems. Strengthening Participatory Councils and Environmental Education Actions (formal and informal). Decentralization in MSW Operation and Management through actions and planning via the sub-municipalities. Increased knowledge and integration of Independent Initiatives and Informal Waste Pickers with MSWMS. Strengthening of Follow-up Working Groups on the Municipal Waste Management Plan and the GHG Reduction Targets Program (which aims to reduce 131,000 t.CO2 by Dec. 2020). Use of economic instruments, such as the "waste tax", for greater participation of the population. Need for political prioritization for mobilization and overcoming the city's waste challenges.

3.3. Integration of Analyses for Integrated Management

From the analysis of social, political and institutional dimensions and their operational results, a relationship was then established with the other dimensions (environmental and economic) that make up the concepts of Integrated Waste Management, as can be seen in Figure 2.

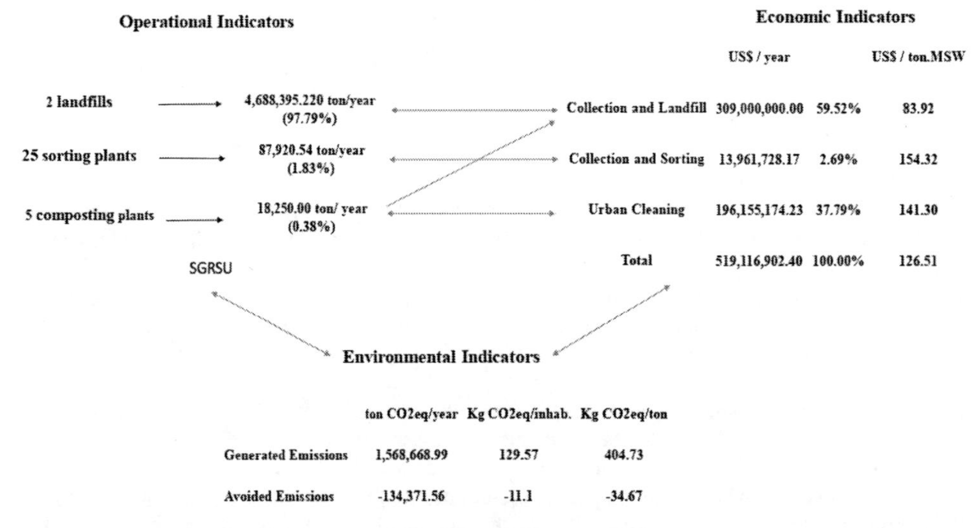

Figure 2: Integration between MSWMS operational, economic and environmental results.

It can be noted, therefore, a lower cost per ton of waste from common collection and landfill actions, while urban cleaning and selective collection sorting actions have higher costs. When analyzing GHG emissions, it is also noted a low avoided emission (only 8.6%) due to the low rates of waste recovery, but the energy cogeneration (through the reuse of methane) of the two landfills contributes to reduce the emissions generated and the system balance when compared to other locations [14].

When results are related, it can be inferred that institutional, social and policy improvements - identified through the drivers, barriers and possible pathways - could contribute to more efficient indicators, lower costs and emissions. Thus, the need for scale-up in the selective collection and sorting systems can be highlighted, which could occur through the expansion of public campaigns and greater participation of the population in the separation of waste, in addition to the use of economic instruments that encourage separation at source. The installation of Voluntary Delivery Points could also contribute to this increase in scale, while more decentralized composting systems and urban agriculture in neighbourhoods and schools could also avoid costs and emissions with traditional collection and disposal systems.

Therefore, the analysis of social, institutional and operational results, as well as economic and environmental aspects, allowed identifying some actions and guidelines that could be proposed for the system under study, thus filling gaps highlighted by researchers in the area [2][5][8][14][15][23]. Thus, these results may allow new advances and debates in the areas of institutional innovation in local public policies, transition to sustainability and integrated waste management.

4. Conclusions

Through the objectives of this work - to develop a method for assessing levels of innovation in waste management, which allows understanding how and why innovations in waste management happen and how these innovations may affect the other aspects (environmental and economic) that make up the concepts of integrated waste management - it can be concluded that the method has identified the main drivers, barriers and possible pathways and how these aspects impact the operating results, operating and investment costs and GHG (CO2eq) emissions of MSW management systems. Through the integration of analysis and results, this paper aims to contribute to the advance of knowledge for the definition of criteria for a sustainability analysis method and the dimensions of Integrated Management of MSW, also enabling the proposition of actions and guidelines for municipal public policies.

5. References

[1] DjuricIlic D, Eriksson O, Odlund L and Åberg M (2018), No zero burden assumption in a circular economy, *Journal of Cleaner Production*, Vol. **182**, **pp**. 352-362.

[2] El Bilali H, 2019. The Multi-Level Perspective in Research on Sustainability Transitions in Agriculture and Food Systems: A Systematic Review. *Agriculture*, **9**, 74, https://doi.org/10.3390/agriculture9040074

[3] EPA, 2016. *Advancing sustainable materials management: assessing trends in material generation, recycling and disposal in the United States, United States Environmental Protection Agency.* Available at: https://www.epa.gov/sites/production/files/2016-11/documents/2014smmfactsheet_508.pdf (Accessed December 12, 2017).

[4] Eurostat, 2014. *Waste Generated and Treated in Europe, Office for Official Publications of the European Communities – Environment Statistics, Luxembourg.*

[5] Fernández-Braña Á, Sousa V, Dias-Ferreira C, 2019. *Are municipal waste utilities becoming sustainable? A framework to assess and communicate progress.* Environmental Science and Pollution Research. https://doi.org/10.1007/s11356-019-05102-4 pp 1–12.

[6] Geels F W, 2012. A socio-technical analysis of low-carbon transitions: introducing the multi-level perspective into transport studies. *Journal of Transport Geography*, **24**, 471–482.

[7] Geels F W, Schot J, 2007. Typology of sociotechnical transition pathways. *Research Policy*. **36**, 399–417.

[8] Guerrero L A, Maas G, Hogland W, 2013. Solid waste management challenges for cities in developing countries. *Waste Management*. **33**, 220-232.

[9] Lauridsen E H and Jørgensen U (2010), Sustainable transition of electronic products through waste policy, *Research Policy*, Vol. **39** No. 4, **pp**. 486-494.

[10] Khandelwal H, Dhar H, Thalla A K, Kumar S, 2019. Application of life cycle assessment in municipal solid waste management: A worldwide critical review. *Journal of Cleaner Production*, **209**, 630-654.

[11] Massarutto A, Carli A, Graffi M, 2011. Material and energy recovery in integrated waste management systems: A life-cycle costing approach. *Waste Management*, **31**, 2102 – 2111.

[12] Mazzola L. Niches of grassroots innovation in municipal solid waste management: the case of sustainable initiatives in the favelas of Rio de Janeiro. 70 f. Thesis (Master Degree in Administration) - Brazilian School of Public Administration And Business, Fundação Getulio Vargas (FGV/ EBAPE), Rio de Janeiro, 2019.

[13] Paes M X, Mancini S D, Medeiros G A, Bortoleto A P, Kulay L A, 2018. Life cycle assessment as a diagnostic and planning tool for waste management – a case study in a Brazilian municipality. *The Journal of Solid Waste Technology and Management.* **44** (3), 259-269.

[14] Paes M X, Medeiros G A, Mancini S D, Ribeiro F M, Puppim De Oliveira J A, 2019. Transition to circular economy in Brazil: A look at the municipal solid waste management in the state of São Paulo. *Management Decision*, https://doi.org/10.1108/MD-09-2018-1053

[15] Paes M X, Medeiros G A, Mancini S D, Bortoleto A P, Puppim de Oliveira J A, Kulay L A, 2020. Municipal solid waste management: Integrated analysis of environmental and economic indicators based on life cycle assessment. *Journal of Cleaner Production*, v. **254**, 2020 https://doi.org/10.1016/j.jclepro.2019.119848

[16] Puppim de Oliveira J A, 2013. Learning how to align climate, environmental and development objectives in cities: lessons from the implementation of climate co-benefits initiatives in urban Asia. *Journal of Cleaner Production.* **58**, 7-14.

[17] Puppim De Oliveira J A, 2017. Breaking resilience in the urban system for improving resource efficiency: the case of the waste sector in Penang, Malaysia. *International Journal of Urban Sustainable Development,* Vol. **9**, No. 2, 170–183.

[18] Roppongi H, Suwa A, Puppim De Oliveira J A, 2016. Innovating in sub-national climate policy: the mandatory emissions reduction scheme in Tokyo. *Climate Policy*. DOI: 10.1080/14693062.2015.1124749

[19] Petit-Boix A, Llorach-Massan P, Sanjuan-Delmas D, Sierra-Perez J, Vinyes E, Gabarrell X, Rieradevall J, Sanye-Mengual E, 2017. Application of life cycle thinking towards sustainable cities: A review. *Journal of Cleaner Production.* **166,** 939-951.

[20] PMSP – Prefeitura Municipal de São Paulo, 2017. Programa de Metas 2017-2020: Objetivos Estratégicos, Metas e Iniciativas para o biênio 2019-2020. Available at: http://programademetas.prefeitura.sp.gov.br/assets/up/Programa%20Metas%202019-2020_texto.pdf (Accessed July 24, 2019).

[21] PMSP – Prefeitura Municipal de São Paulo, 2014. Composta São Paulo. Available at: https://compostasaopaulo.eco.br/resultados2014/ Acesso (accessed July 24, 2019).

[22] Sevigne Itoiz E, Sevigna Itoiz E, Gasol C M, Farreny R, Pons J R and Durany X G. 2013. CO2ZW: Carbon foot print tool for municipal solid waste management for policy options in Europe. Inventory of Mediterranean countries. *Energy Policy*, **56**, 623 – 632.

[23] Silva A, Rosano M, Stocker L, Gorissen L, 2017. From waste to sustainable materials management: Three case studies of the transition journey. *Waste Management*, **61**, 547–557.

[24] SNIS, 2018. Diagnostico do Manejo de Resíduos Sólidos (National system of information on sanitation: solid waste management inventory), Sistema Nacional de Informações em Saneamento Básico Brasília. Available at: www.snis.gov.br/diagnostico-residuos-solidos (accessed December 12, 2018).

[25] To L S, Seebaluck V, Leach M, 2018. Future energy transitions for bagasse cogeneration: Lessons from multi-level and policy innovations in Mauritius. *Energy Research & Social Science*. 2018, **35**, 68–77

[26] United Nations, 2017. New Urban Agenda.
Available at: http://habitat3.org/wp-content/uploads/NUA-English.pdf. (accessed September 17, 2019).

[27] Van Den Bergh J C J M, Van Leeuwen E S, Oosterhuis F H, Rietveld P, Verhoef E T, 2007. Social learning by doing in sustainable transport innovations: Ex-post analysis of common factors behind successes and failures. *Research Policy*. **36**, 247–259.

[28] World Bank, 2013. *What a waste: a global review of solid waste management*, available at: http://go.worldbank.org/BCQEP0TMO0 (accessed October 9, 2018).

IOP Conf. Series: Materials Science and Engineering 1196 (2021) 012005 doi:10.1088/1757-899X/1196/1/012005

Green and Lean? – Understanding ecological and environmental implications in the light of Industry 4.0

Johannes W. Veile[1,*], Hendrik S. Birkel[2], Marie-Christin Schmidt[1], Marc Rücker[1], Julian M. Müller[3,], Evi Hartmann[2], Kai-Ingo Voigt[1]**

1 Friedrich-Alexander-University Erlangen-Nürnberg (Germany), Chair of Industrial Management
2 Friedrich-Alexander-University Erlangen-Nürnberg (Germany), Chair of Supply Chain Management
3 Kufstein University of Applied Sciences (Austria)

* **Email:** johannes.veile@fau.de
** **Email:** julian.mueller@fh-kufstein.ac.at

Abstract. Research on Industry 4.0 has only begun to analyse its environmental and social impacts, while technological and economic analyses dominate the current academic debate. However, merging the current debate for a more sustainable industrial value creation with the developments in the context of Industry 4.0 is vital given its benefits beyond merely economic considerations. In response to the scarcity of extant research, this paper analyses research publications and gives insights using an empirical approach. Qualitative-empirical data from 33 expert interviews within German manufacturing enterprises are used to extend the current state of literature in this field. In particular, company-internal and company-external benefits from an environmental and ecological point of view are uncovered, along with drivers and requirements to approach the potentials. Further, the results are discussed against the background of extant publications in the field. The paper closes with recommendations for managerial practice and suggestions for future research.

1. Introduction

The concept of Industry 4.0 promises numerous opportunities for companies to succeed in value creation and global competition. Based on digital interconnection in horizontal (across the supply chain) and vertical (across functional departments) directions and end-to-end transparency of data (from sourcing to recycling), the initial concept behind Industry 4.0 already shows potential paths towards environmental sustainability, or a circular economy [1]. However, in extant literature on Industry 4.0, many authors solely focus on technical and technological aspects of Industry 4.0 [2]. In recent years, however, economic, environmental and social opportunities have become significantly more important against the backdrop of globalization, changing environmental circumstances, and social pressure. Yet, in the context of Industry 4.0 aspects of environmental and ecological importance have been little regarded, such as recycling aspects and requirements towards a circular economy [1,3,5]. Further, as sustainability dimensions of economic, ecological, and social dimensions, referring to the Triple Bottom Line of sustainability, require to balance the three dimensions against each other and especially consider

IOP Conf. Series: Materials Science and Engineering 1196 (2021) 012005 doi:10.1088/1757-899X/1196/1/012005

interdependencies [2,3]. In response to this research gap, this paper is devoted to the following two research questions:

1) *What potentials does Industry 4.0 pose for the environmental or ecological dimension?*
2) *What interdependencies with economic or social aspects do those potentials have?*

2. Background

This paper is based on the literature review that was conducted for the study and that is explained in the method. Due to limited space in this short paper, only a brief summary is given in the following, focusing on a short overview of potentials associated with Industry 4.0 from an environmental and ecological perspective.

Given a scarcity of resources, climate change, and sustainability awareness in society, the importance of ecological aspects expands to the industrial context. Industry 4.0 can help to improve the ecological sustainability of industrial value creation through intelligent processes and thus increase the resource efficiency of companies [7,8]. Reasons for that are manifold but they are mainly caused by the intelligent interconnection of value chains and networks alongside the process transparency. Both processes and products can be optimized in an ecologically sustainable way. For instance, processes and the entire ecological footprint of companies can be monitored, controlled, and managed, using data analyses and simulations based on data exchange across the entire supply chain [7,8]. In the same manner, energy consumption can be optimized through intelligent energy systems [9]. By taking a holistic view and managing the product life cycle, resources can be used optimally [2,10]. This also includes using renewable energies, the reusing and recycling of resources or products as well as retrofitting [11]. The latter refers to the retrofitting, modernization or expansion of existing plants, machines and equipment. Doing so requires significantly lower investments compared to new purchases. Furthermore, Industry 4.0 enables to reduce resource consumption both for resources used during operations (e.g., materials, energy, and water) just like for by-products generated during operations, e.g., waste, emissions such as greenhouse gases, and chemical residues [11,12].

In addition, Industry 4.0 opens up ecological opportunities in transportation and logistics. Superfluous material flows and transportation processes can be avoided with the help of end-to-end data transparency throughout the value chain. In addition, transportation routes and capacity usage can be made more efficient, reducing waiting and storage times, and decreasing incorrect deliveries [3]. Further, reshoring, i.e. the relocation of production sites back to where a product is consumed, is a further opportunity to simplify transportation and logistics, and also in turn, reduce environmental pollution [13]. New technologies, such as additive manufacturing, help to reduce resource consumption, avoid waste, and improve logistics processes, e.g., in spare parts delivery [14].

3. Material and Methods

This paper applies a mixed method approach combining the results of a qualitative empirical study with those of a literature review. The methods are explained in the following.

First, a qualitative, empirical research design was chosen in order to extend the existing research on Industry 4.0 regarding environmental and ecological aspects. Semi-structured interviews with experts from business practice were conducted forming the empirical basis. The shape and form of the interviews ensure a scientific structure, whereas at the same time, the procedure does leave space for deviating, individual questions. Given this openness and flexibility, the semi-structured interviews allow gathering new and unexpected information and grasping the entire context.

Between January and May 2019, 33 semi-structured interviews were conducted with experts from various German industrial companies. The interviews lasted between 25 and 60 minutes and were conducted in German, the mother tongue of the interviewer and the interviewee alike. All experts are selected for their knowledge, experience, and awareness of Industry 4.0 opportunities. For confidential reasons, the names of the companies and the interviewees were anonymized in the present work. The sample companies stem from the following manufacturing industry sectors: Automotive industry, mechanical and plant engineering, electrical and electronics engineering, and information and communication technologies.

The interviews were recorded, transcribed, and then examined applying a qualitative content analysis. We followed an inductive approach, this means the categories are not created before viewing data, but they are derived from the data material, i.e. directly from the transcripts. The aim of this procedure is to bundle and summarise the collected data without distorting the interviewees' statements and intentions [15]. In a first step, the interviewees' statements were precisely recorded in the category formation and their statements are paraphrased. In a second step, these categories were then synthesized into superordinate categories. Doing so generalizes the statements and raises them to a more abstract level. Hereafter, the categories are coded consolidating homogeneous categories into subcodes. In a last step, related subcodes are summarized and transferred into topcodes.

Second, a literature review is conducted in order to grasp the current state of research and to uncover similarities and differences compared to our qualitative empirical results. The analysis bases on three steps (1) preparation, (2) selection and classification, and (3) analysis, as is common in research practice [16]. First, we defined search terms (e.g., Industry 4.0, IIoT, sustainability, ecology), determined the time horizon (2011-present), selected databases (ABI/Inform, EBSCO, Science Direct, Scopus), and selected publication types (peer-reviewed, published high-quality academic journals). Second, the selected articles are classified and categorized. In a last step, we compared, contrasted, critically evaluated, and discussed the articles against the backdrop of our empirical results. The entire process is summarized in Figure 1. The following section presents the findings and discusses them respectively.

1. Preparation	2. Selection & Classification	3. Analysis
(1) Defining search terms (2) Determining time horizon (3) Selecting databases (4) Selecting publication types	(5) Selecting articles (6) Classifying articles	(7) Analyzing selected articles

Figure 1. Approach for literature review.

4. Results

Table 1 presents the results of the coding process and indicates the topcodes and subcodes developed from the empirical data. It also includes references that confirm the results. In general, most potentials can be attributed to higher data transparency and data sharing across the supply chain and across product lifecycles, optimized products and production, predictive capabilities, optimized logistics processes, and eased management of processes. Hereby, multiple occurrences per interviewee are possible, the number in brackets refers to the number of mentions out of 33 interviews.

Table 1. Data structure.

Topcode	Subcode	Description and enablers	References
Reduction of resource consumption and waste generation (22)	Reduction of production input resources (11)	Load balancing and data sharing across the supply chain, early recognition of errors.	[4,14,11,17]
	Reduction of waste and emissions (11)	More efficient processes can be established through self-optimizing and interconnected production.	
	Reduction of travel, paper and documentation (8)	Communication and documentation can be transferred to digital means or become automated.	

Resource efficiency (17)	Energy and resource-efficient production and process (13)	Better load and energy balancing across the supply chain, more individual and customer-specific products which last longer due to predictive capabilities, and better product design capabilities.	[3,9,10,14]
	Reuse, renew, and recycle (5)	Recycling information and chances for a "second life" can be identified more easily.	
Management, controlling and supervision (7)	Environmental and energy management (3)	Smart energy distribution, demand forecast sharing and control and usage of renewable energy sources.	[2,6,11,17]
	Certification (2)	Environmental certificates can be shared easily across the supply chain.	
	Life-Cycle Assessment (1)	Data transparency across the lifecycle.	
	Spare part management (1)	New technologies, such as additive manufacturing, allow to reduce the impact of spare part production.	
	Carbon footprint documentation (1)	Documentation and visualization across the supply chain.	
Reduction of transport and logistics processes (7)	Reduction of logistics processes (6)	Better data transparency and forecast data sharing across the supply chain. Reshoring of value generation closer to the customers.	[3,18,19]
	More efficient transport routes (1)	Optimization of transport routes based on real-time data availability.	

5. Discussion

In terms of the environmental benefits of Industry 4.0, our empirical analysis has shown that increasing resource efficiency is highly relevant, based on the efficient design of processes and products throughout the product lifecycle, and recycling of products. In the existing literature, these aspects are also listed and supplemented by the chance of retrofitting [11,18]. Furthermore, the reduction of resource consumption, e.g., the reduction of materials, CO2 emissions, and energy was identified as a significant ecological opportunity both in empirical analysis and in existing literature. The experts emphasized the reduction of paper consumption due to digital processes in their companies.

There is research in the literature on the extent to which additive manufacturing contributes to reducing resource consumption [13]. However, this aspect is not confirmed in the empirical results as it was not mentioned by any of the interviewed experts. Furthermore, the empirical analysis found the reduction of transportation and logistics processes to be an opportunity, which is also theorized by some authors in the literature [2,10]. The interviews and literature emphasize the reduction of material flows and transportation processes, which reduces waiting and storage times just like incorrect deliveries [3]. The experts consider that an orientation towards ecological goals as relevant, for example through the development and expansion of an environmental and energy management system. Another opportunity is the need for a holistic management across the entire product lifecycle [2].

As far as interlinkages between the dimensions are concerned, one of the most notable interdependencies results from the effects of energy efficiency on cost efficiency potentials. Since electrical energy accounts for a large part of the costs in industrial value creation, Industry 4.0 offers several potentials. With smart energy distribution, improved process efficiency, and improved design efficiency, the eco-effective production of the future can also make a significant contribution to the economic dimension [2]. For example, intelligent energy systems are capable of predicting a company's energy consumption adjusting it to the energy production. Through intelligently planning and scheduling energy peaks, companies can make use of times of cheap energy costs and in doing so, strategically adapt production speed to the energy supply [3]. In contrast, some authors argue that Industry 4.0 leads to higher energy consumption, for instance, as digital infrastructure and data centres consume a great amount of energy [10]. Furthermore, creating new technologies and setting up machines related to Industry 4.0 requires a great deal of additional resources [3]. Comparable interdependencies can be found between reducing logistics processes and waste reduction, which can be combined to generate both, economic and ecological benefits. More environmental-friendly processes can also lead to social benefits at the same time, such as hazardous or physically challenging tasks in which humans can be assisted by technical solutions [2,3,10].

For the development towards a circular economy, especially two barriers can be identified. First, the integration of developing countries, which are responsible for sourcing and recycling activities in many industrial processes. However, companies from these countries often do not have the means and technology to allow data transparency across the supply chain, and across the entire lifecycle of products [20]. Comparably, small and medium-sized enterprises are vital contributors to global value chains, but do not have the level of standardization or resources to provide data transparency required for Industry 4.0 [18,19].

6. Conclusion

By integrating the empirical insights gained from the cases of 33 German industrial companies with extant literature, this paper provides an overview of potentials related to ecological or environmental aspects within the Triple Bottom Line of sustainability. Further, the paper shows interdependencies with other dimensions, such as the economic and the social dimensions. This allows to shape future research in the field of Industry 4.0, and to design managerial strategies aligning the dimensions of the Triple Bottom Line. In particular, the understanding of how data transparency and data sharing across the supply chain and across the lifecycle of products does generate economic and ecological benefits is vital to handle the struggles of technical interfaces and standardization.

The research results of the present work include some limitations given their qualitative character. The empirical analysis was limited to 33 German industrial companies of different industry sectors and company sizes. Although the research results of the empirical analysis were stem from semi-structured interviews to guarantee a high degree of reliability, robustness, comparability and to allow replicability, they should not be generalized to other situations or circumstances without further consideration. In addition, experts from non-governmental organizations, research institutes, industry associations and politics could also be interviewed in the course of further research. Other recommendations include focusing on supply chains and business ecosystems, and approaches that statistically quantify sustainability aspects of Industry 4.0. This could help to balance the three dimensions against each other, also relating to their perceived importance, or interdependences.

7. References

[1] De Sousa Jabbour A, Jabbour C, Godinho Filho M and Roubaud D 2018, Industry 4.0 and the Circular Economy: A proposed Research Agenda and Original Roadmap for sustainable Operations, *Annals of Operations Research*, **270** 273-286

[2] Kiel D, Müller J M, Arnold C and Voigt K I 2017, Sustainable industrial Value Creation: Benefits and Challenges of Industry 4.0, *International Journal of Innovation Management*, **21**

[3] Birkel H S and Müller J M 2021, Potentials of Industry 4.0 for Supply Chain Management within the Triple Bottom Line of Sustainability–A Systematic Literature Review, *Journal of Cleaner Production,* 125612

[4] Rahman S M, Perry N, Müller J M, Kim J and Laratte B 2019, End-of-Life in Industry 4.0: Ignored as before? *Resources, Conservation, and Recycling* in press

[5] Tseng M L, Tan R R, Chiu A S, Chien C F and Kuo T C 2018, Circular Economy meets Industry 4.0: Can Big Data drive industrial Symbiosis? *Resources, Conservation and Recycling* **131** 146-147

[6] Sarkis J and Zhu Q 2017, Environmental Sustainability and Production: Taking the Road less travelled *International Journal of Production Research,* **56** 743-759

[7] Müller J M, Veile J W, and Voigt K I 2020, Prerequisites and incentives for digital information sharing in Industry 4.0–An international comparison across data types, *Computers & Industrial Engineering,* **148** 106733

[8] Veile J W, Schmidt M C, Müller J M and Voigt K I 2020, Relationship follows technology! How Industry 4.0 reshapes future buyer-supplier relationships, *Journal of Manufacturing Technology Management* in press

[9] Shrouf F and Miragliotta G 2015, Energy Management based on Internet of Things: Practices and Framework for Adoption in Production Management. *Journal of Cleaner Production*, **100**, 235-246

[10] Stock T, Obenaus M, Kunz S and Kohl H 2018. Industry 4.0 as Enabler for a sustainable Development: A qualitative Assessment of its ecological and social Potential. *Process Safety and Environmental Protection*, **118**, 254-267

[11] Beier G, Niehoff S, Ziems T and Xue B 2017 Sustaina*bility Aspects of a digitalized Industry – A comparative Study from China and Germany International Journal of Precision Engineering and Manufacturing-Green Technology*, **4**, 227-234

[12] Waibel M W, Steenkamp L P, Moloko N and Oosthuizen G A 2017. Investigating the Effects of Smart Production Systems on Sustainability Elements. *Procedia Manufacturing,* **8** 731-737

[13] Oettmeier K and Hofmann E 2017. Additive Manufacturing Technology Adoption: An empirical Analysis of general and Supply Chain-related Determinants. *Journal of Business Economics,* **87** 97-124

[14] Müller J M and Voigt K I 2018. Sustainable industrial value creation in SMEs: A comparison between industry 4.0 and made in China 2025. *International Journal of Precision Engineering and Manufacturing-Green Technology*, **5** 659-670

[15] Gioia D A, Corley K G and Hamilton A L 2013, Seeking qualitative Rigor in inductive Research: Notes on the Gioia Methodology. *Organizational Research Methods,* **16** 15-31

[16] Tranfield D, Denyer D, and Smart P 2003, Towards a methodology for developing evidence informed management knowledge by means of systematic review. *British Journal of Management,* **14** 207-222

[17] Müller J M 2021, Data-based sustainable business models in the context of Industry 4.0. *Proceedings of the International Conference on Information Systems (ICIS)* (13.-16.-12.2020, Hyderabad (India), Online) AIS eLibrary

[18] Bressanelli G, Perona M and Saccani N 2018. Challenges in supply chain redesign for the Circular Economy: a literature review and a multiple case study. *International Journal of Production Research,* **23** 7395-7411

[19] Schmidt M C, Veile J W, Müller J M and Voigt K I 2020, Ecosystems 4.0: redesigning global value chains. *International Journal of Logistics Management* in press

[20] Kamble S, Gunasekaran A and Sharma R 2018, Analysis of the driving and dependence power of barriers to adopt Industry 4.0 in Indian manufacturing industry. *Computers in Industry*, **101** 107-119

Sewage sludge management options and civil society stakeholders' influence: first evidence from Italian cases

E Cagno, P Garrone, A Neri *

Department of Management, Economics, and Industrial Engineering, Politecnico di Milano, Milano, Italy

* **Email:** alessandra.neri@polimi.it

Abstract. Sewage sludge management is a pivotal process in the path toward sustainability, broadening the scope of wastewater treatment from removing pollutants to recycle and recover resources. Water utilities can choose among several management options and treatment technologies. The paper focuses on the possible influence of civil society stakeholders on the decision-making process of water utilities, and to investigated it, inductive case studies were conducted in Italian water utilities adopting different management options. The analysis allowed us to capture the richness of observations and make interesting variables possible shaping the relationship emerged, as the role of regulatory uncertainty. The first results show how the investigated water utilities look at other Countries for robust solutions and select the less risky in terms of civil society's acceptance. Communication and involvement of civil society appear to be an important asset to mitigate possible negative relationships.

Keywords. Sewage Sludge Management; Circular Economy; Wastewater Utility; Civil Society Stakeholders; Case study.

1. Introduction

Sewage sludge management is gaining momentum in scientific literature, in terms of treatment techniques, and associated industrial and institutional changes [1]. The production of sludge is increasing as water treatment technologies improve - also due to stringent regulation on wastewater releases [2], and as an always higher number of households is connected to the wastewater treatment plants (WWTPs) in charge of the treatment of the sewage [3]. Further, traditional sludge disposal options – landfill and open-water disposal – are losing ground owing to stringent environmental quality targets and bans [4]. Sludge must be disposed of in an environmentally sound way, being largely composed of offensive substances [1]. At the same time, the Waste Hierarchy steered lawmakers to maximize resource efficiency, moving from a perception of sewage sludge as a waste to a source of valuable resources. The imperative of WWTPs thus shifts from the removal of pollutants to the recycle and recovery of the resources embedded in sewage sludge [5], in line with the Sustainable Development Goals [6] and Zero Waste Europe [7].

The paradigm-shift in the wastewater treatment [8] is leading to major changes in the ways wastewater and sludge are disposed, to the point that the selection of the wastewater management system should be strongly driven by choices related to the sewage sludge management option [9]. WWTPs thus become resources and energy recovery facilities, leading to potentially marketable sludge-based added value products [10,11]. The options for transforming sludge into a resources are different [12]: 1) Reuse through spread on fields; 2) Recovery of energy or material, in terms of i) thermal energy; ii) nutrients (phosphorus, nitrogen); iii) construction material (cement, ceramic bricks). Other recoveries are possible during the sludge treatment process: i) biogas, from the anaerobic digestion of sewage sludge; ii) struvite, from dewatering side-streams or incinerated ashes. Selecting the type of recovery, the sustainable options for the long-term management of sewage sludge must be environmentally friendly, economically viable, and socially acceptable [13] – thus at the intersection of the three sustainability pillars [14].

Water utilities (WUs) are central players for the integrated management of water – water distribution, wastewater collection and treatment, sewage sludge treatment [15]. Making a decision on wastewater treatment technology, WUs might not limit themselves to comply with regulated standards [16] - traditionally quite innovation averse [17]. WUs managers, indeed, increasingly perceive society's representatives as an impactful player [18] and WUs may seek to be congruent with the concerns of the Civil Society Stakeholders (CSSs), or secondary stakeholders [19]. CSSs are stakeholders influenced by and/or that can influence a WU, without being counterparts of economic transactions (suppliers, customers) nor exerting the State's authority (regulators). Residents, politicians, media, or non-governmental organizations pressure and incite the WU to pursue environmental performances even beyond the regulated targets or toward unregulated aims - investing-up or over-performance. CSSs might change the WU's priorities and influence water infrastructure decisions [20], as the WU wants to obtain legitimacy from CSSs to justify its right to operate [21]. The topic started to be addressed for wastewater treatment technologies [16,18], but, aiming at integrating wastewater and sludge treatment technologies [9], the CSSs' influence on the selection of sewage sludge management options by WUs should be addressed as well. Anyhow, also following the legislative trend that is separating wastewater treatment from sludge treatment, it seems important to conduct an initially separate analysis for the sludge treatment processes. The social acceptance of the different recovery options began to be tackled, but mainly within the context of technical reports or focusing on a specific CSS or sludge destination at a time. In detail, sludge in agriculture is susceptible to public perception for the increasing awareness of health and environmental impacts of contaminants that might not be sufficiently tested before applying sludge on land [22]. As the contaminants might affect crop production [13], farmers may abandon this option [23]. As for biogas, CSSs' issues against the construction of new facilities were highlighted [24], while incineration mainly creates worries about pollution and toxic ashes [25], and possible negative impacts of the incineration plants on landscape and residents' lives [26], so that authorities seem reluctant in developing such facilities [27]. In this context, the landfill appears alas especially easy [28].

The legislation related to the sludge treatment is still evolving in many European countries, tailoring European Commission Directives into specific contexts. Parallelly, a relevant role in the selection of the sewage sludge management option seems to be played by the relationships between the WU and CSSs, who, with different mechanisms than policymakers, might encourage WUs to enhance specific options from an environmental and social perspective, so to ensure the highest sustainable value. Whether and how the CSSs' concerns may diverge from regulated standards, and how CSSs may stimulate the over-performance in WUs are untapped areas of research. The present study wants thus to investigate how and to what extent the selection of a sewage sludge management options by a WU is influenced by the CSSs. Particularly considered the critical role of sludge treatment from a sustainability perspective, the aim is to highlight the relationships between the WUs and CCSs, focusing on the impacts of the specific management options, and understanding who are the most relevant CCSs.

2. Materials and Methods

The aim of the empirical investigation was thus to understand the CSSs' influence on the adoption of different sewage sludge management options. We chose to perform inductive case studies [29], with semi-structured interviews and secondary materials, so that we could capture the richness of observations without being limited by theory and/or specific relationships between variables. As a WU can adopt different management options, the unit of analysis is the single configuration in the context of the WU. We relied on multiple case studies, treated as single ones with conclusions examined on their own [30]. Case studies were conducted in two Italian WUs (WU1 and WU2), located in different Regions but with similar characteristics – municipal ownership; serving >0.5 million equivalent inhabitants (e.i.); located in urbanized areas; owning about 50 WWTPs; high investment in research and development (R&D) – allowing a replication logic. Italy is deemed relevant for the study: i) it is the third European producer of sludge [31] - 41% of which still landfilled and only the 9% incinerated [32]; ii) the Italian legislation transposed the European Commission Directives into national laws, but the Regions can adapt the national laws to local situations; iii) sewage sludge is defined as a special, non-hazardous waste, thus being subjected to waste legislation; (iv) national laws set specific regulations for sludge landfilling and its reuse in agriculture, while sludge incineration falls under general incineration regulations [33].

We interviewed people involved in the decision-making process and acknowledgeable of the impacts of the WU's activities and feedbacks from stakeholders: the Technical Area Manager (WU1-I1), the R&D Manager (WU1-I2) and the Planning and Control Manager (WU1-I3) in WU1; the Sewage and Wastewater treatment Manager (WU2-I) in WU2. The WUs were contacted by e-mail or telephone, and secondary data (utility websites, reports, etc.) was collected, particularly regarding projects, initiatives, and similar activities toward sludge treatment. We supplemented the case study protocol - aimed at minimizing the impact of contextual effects [34] - with questions and free comments emerging during the interview [35]. Interviews lasted about 2 hours and were coded and corroborated with secondary data and other findings emerged during the interviews, like field notes [36]. We conducted a content analysis, applying emergent coding [37], and using an open code [38]. As for methodological rigor [39], triangulation of multiple sources of evidence and the creation of an electronic folder containing all the data collected for each case led to construct validity [40]. Multiple case studies and the case study protocol increased the reliability, while also contrasting possible researcher bias [41]. The specification of the population and the use of multiple case studies strengthen the generalization [40].

3. Results

The inductive case study allowed to identify important variables and the relationships and patterns among them [29]. The results from the two cases are here presented and will be discussed in Section 4.

3.1. Water Utility 1

WU1 owns around 50 wastewater treatment plants, several of which were acquired. Thus, the technologies used in the plants often depend on the choices made by the former owners. As the plants have different sizes, for more than half of them (>20,000 e.i.) is economically reasonable to produce and treat sludge, while the sludge from the smaller plants is moved to bigger ones. The treated sludge is produced as stabilized dewatered sludge, while in three specific plants further treatment processes are operated: WU1.1, equipped with a dryer owned by WU1 but operated by a third party, produces dried sludge then managed by an external cement factory; WU1.2 and WU1.3, where the sludge is transformed in calcium carbonate - the process is managed by a third party who then reuses the calcium carbonate as fertilizer. As for now, almost 80-90 tons of dehydrated sludge is produced. From the highest quality sludge, fertilizers are recovered with a limit determined by the production capacity of WU1.2 and WU1.3. The other destinations are agriculture - currently suffering from legislation's targets and uncertainty; cement factories - after a drying process; co-incineration; landfill - only 2% of the sludge.

WU1 is currently developing a new project for the integrated platform WU1.4. It will have a line for the sludge treatment aimed at the energy recovery and the production of fertilizers, and a line for the anaerobic digestion of organic fraction of municipal solid wastes and the production of biomethane. The project embroiled since the beginning the involvement of public opinion and the citizenry of all the neighboring municipalities affected by the project, asking CCSs to express their opinions and concerns about the project. Behind the project there is an anticipative strategy, particularly regarding the recovery of phosphorus, as better detailed by WU1-I1: "*the reason why we are investing* [...] *is because phosphorus is becoming a critical resource, and some European countries obliged utilities to mono-*

incinerate the sludge, producing ashes rich in phosphorus [...] *there is not such a legislation in Italy, but I'm sure we will arrive there soon*". They later added that another strong push came from market risks, as WU1-I3 said "*the uncertainty of the market pushes us to be independent*", adding that, although WU1's investment ranks high among Italy WUs, they are "*very far from European utilities, but* [...] *also in terms of tariffs*". Developing the project, a relevant issue was represented by the residents' ex-ante propensity to express opposition, especially if incited by local politics. As WU1-I2 highlighted, generally "*they start from considerations that are not technical and arrive at conclusions that do not have any technical value*". According to WU1-I1, this is an approach of the overall Country toward sludge: "*we export sludge to France, where the legislation on sludge in agriculture is looser than ours, and France is not the new guy in terms of environmental issues* [...] *we are the only Country exporting sludge*". The interaction with the CCSs highlighted the concern of municipalities and citizens about the environmental impact of the project, as WU1 was requested to i) apply a strict control and total transparency on emissions in the atmosphere, odor, preservation of soil, rivers, and biodiversity; ii) assure the project's compliance with legislation, while also providing in-kind compensation through development of affected areas; iii) perform epidemiologic analysis in the area on the project. Although some requests go beyond legal requirements, WU1 promised to conform to them.

Asked to deepen the relationship with CCSs, the interviewees recognized the citizenry and the municipalities as the most critical CCSs. To interact with the citizenry, WU1 uses social media and organizes transversal meetings to build a constructive dialogue. As for municipalities "*they own our utility, so every request coming from Majors is considered, even if it goes beyond legal requirements*", said WU1-I2. They also specify that "*the stakeholders' impact on our activities is extremely important*" and that WU1 "*deals with each CCS independently, with specific communication channels for each of them*", also given that "*there is a relationship between citizenry, municipalities and other local authorities, but they don't leverage each other*". WU1-I2 also arose a further interesting point as WU1 has to: "*try to convince citizenry that it is not in our interest to forgo without quality, but they* [CSSs] *don't trust us because they don't have knowledge* [...] *the first thing that they don't understand is that it is not the water service that decides what it is dangerous* [...] *they also don't understand that we are controlled in turn*". WU1-I2 thus believes that the over-performance implemented by WU1 is related to its investments in R&D and innovative solutions.

Regarding future scenarios and innovations, WU1-I1 stated: "*our fortune is that we have a European perspective, basically we are copying what they do in Europe*", more in the specific "*we are investing in the thermal treatment because they do it in Europe*". As for WU1-I3, considering the ban on landfill and the perception that the Ministry of the Environment will restrict spread in agriculture, "*they are pushing toward thermal treatment without making it explicit*".

3.2. Water Utility 2

WU2 owns about 60 wastewater treatment plants. Among them, WU2.1, the only plant equipped with anaerobic digestion and vertically integrated into the stabilization phase, manages alone the 50% of the wastewater treatment capacity. In the other plants – that were all purchased, so that WU2 is restricted to the already installed technologies - the last processes of the sludge pre-treatment are managed by a third party. The sludge is brought to two waste-to-energy plants (WEP) for the generation of thermal energy (80% of the sludge) or to a bioreactor for the recovery in agriculture (20% of the sludge), all operated by a utility belonging to the same Holding as WU2. So far, no sludge is landfilled or destined for internal energy recovery.

WU2 is currently developing improvements in WU2.1, as it would like to process all the produced sludge in WU2.1, where a co-generator is about to be started. Particularly, WU2 would likely implement a lysis process so to increase the production of biogas - aiming at the energy self-sufficiency of the plant and produce better-sterilized sludge appropriate for the recovery in agriculture. After the lysis, WU2 may implement a further oxidation or a mono-incineration. Among the reasons supporting the mono-incineration, there are the general positive attitude of other European Countries toward it and the possibility to have a higher integration of processes and thus better control over them. However, WU2-

I stated: "*I don't know whether the community would accept it*": although they are "*sure that if a plant is properly developed and built, it can operate safely everywhere* [...] *we have to consider also other logics*". WU2 thus would prefer the oxidation over the mono-incineration, sending then the dehydrated sludge to WEPs or brick industries. The improvements in WU2.1 started as WU2 is expecting a progressive toughening of regulations of sludge spread on fields "*I cannot say there is a direct causal link between WEPs and lack of recovery in agriculture, but undoubtedly there is*". As this uncertainty led other WUs converging to the same WEPs as WU2, the WU2's costs increased. The new configuration of WU2.1 should help in reducing OPEX, while also mitigating the risk of the WEPs not accepting WU2's sludge. From WU2-I's perspective, WUs "*should tackle this situation reducing the production of sludge, or at least, finding destinations other than WEPs, also from a circular economy perspective*".

According to the interviewee, WUs need help for legitimizing the sludge treatment in the eyes of the Society: "*there is the need of a socio-political effort to make* [CSSs] *accepting the presence of sludge disposal facilities other than landfill*". Parallelly, "*effective and safe investments are needed* [...] *I would prefer a safe and organized system at the cost of increasing the tariff, rather than be in the current situation, sending wastes abroad*". As the Holding spent € 80 million in plants' renovation, the "*the population's doubts about the chimney stacks are still difficult to manage*". Luckily, the Holding has "*great attention toward the environment and the relationship with the civil society*" and "*advertises everything that has a positive impact on the environment*", also using social media and fliers.

As for WU2-I, the recovery of energy and the reuse in agriculture are the best destinations for sludge; however, strict legislative thresholds are limiting the application in agriculture, while although WEPs are extremely important for saving energy, "*then we come back to the impact on citizenry*". As the Holding's vision is "*characterized by great attention and innovation*", there is a huge effort to make CSSs accept the WEPs, and "*understand that WEPs are an action to use something that otherwise would be only a waste* [...] *a strong communication campaign toward citizenry is necessary*".

Regarding the impacts of the WEPs, WU2-I said that the most perceived one by the CSSs is related to pollution. In WU2-I's opinion, the pollution is due to the surrounding industrial system but often associated with the WEPs as they are "*a political topic often rode*", underlining the need for an awareness campaign.

As for future scenarios and innovations, WU2's benchmark is Europe, above all for legislative targets such as phosphorus recovery. While being confident in the legislator and that targets will be set, so far, they are looking abroad: "*if they act in a certain way in Europe, I don't think they take a loss*".

4. Discussion and Conclusions

The results obtained are of great relevance to understanding the relationship between CSSs and WUs, providing new insights into the dynamics of this relationship. Notably, it emerged that CSSs influence the WUs choices related to the adoption of sludge management options, above all in a context of regulatory uncertainty, as the investigated one. Although the importance of a technological integration between water and sludge treatment technologies has been deemed as particularly important [9], both the WUs mainly treat sewage sludge in purchased plants- maintain the already existent technologies particularly for economic constraints, and have the sludge treatment process managed by a third party.

The innovative projects the WUs are currently implementing aim at increasing the vertical integration of the sludge treatment processes and make the WUs safer from possible market and regulation related risks. The projects embroil the adoption of technologies and processes widely accepted in Europe but are carried out in a period of legislative transient, with uncertainty about the timing of reforms for current regulations. In the development and implementation of these projects, the two WUs are facing problems with CSSs, that they are tackling in slightly different ways. WU1 is implementing a strong communication strategy involving municipalities, citizenry, and public opinion, focusing on their concerns on the project, that is improved and modified according to the requests and the suggestions received by the CSSs: the transparency is very high as is the WU1's willingness to make the CSSs perceive the importance of the project. WU2 is more prone to choose the option basing on CSSs influences, selecting the less chancy one in terms of CSSs' relationships. The WUs strongly highlighted the need for a higher national socio-political effort to make CSSs accept certain technologies and processes, also considering their widespread acceptance in Europe. Indeed, even if both the WUs are investing in technologies already adopted and applied in Europe, the WUs are facing some issues with CSSs' acceptance. The WUs, on the one hand, are implementing innovative investments that may not be covered by the water tariffs and without a clear regulatory framework; on the other hand, they have to legitimize the choice to the CSSs, although the investments are made also to provide a safer and more effective service to the community.

The most relevant and critical CSSs appeared to be the citizens (or association of them), followed by municipalities. The result may be influenced by the urban context in which the two WUs are located, as suggested by [16]. According to the WUs, the CSSs perceive as main impacts deriving from the sludge treatment process the emissions in the atmosphere, odor emissions, and health and safety repercussions on the population, confirming [25]. CSSs also ask WUs to overperform about specific legislation targets, endorsing [16]'s findings on the wastewater treatment process in urban areas. From the WUs' perspective, CSSs often seem to lack a specific knowledge to rightly influence technological decisions and to not understand that it is in the WU's interest to be compliant with the legislation.

The results underline the relevant role played by the CSSs in a context of a legislative transitory. Indeed, the investigated WUs take inspiration from other European Countries - some of which implemented the European Commission Directives in strict ways so that the management options implemented abroad seem to be less risky from a legislative viewpoint. The decision, however, has also to consider the CSSs' perspective and influence. Communication and involvement of CSSs appear important assets for the decision-making process. The WUs are somehow overlapping their role with part of the legislator's one, by informing and forming CSSs, and over-performing in the identification of robust solutions, tailoring them on internal strategies and focusing on the less risky ones from a legislative, technological, economic, environmental, and social perspective. Uncertain legislation, limited investment availability and somehow biased CSSs seem to make it tough for the investigated WUs to operate at the same level of as other European Countries.

To close the loop and provide theoretical replication, we deem as necessary to extend the present research to other European Countries so to better appreciate possible differences related to policies and social contexts. Similarly, it would be interesting to investigate WUs with different characteristics, as not located in urbanized areas or with a less innovative attitude than the two cases investigated.

Acknowledgments

This paper has been developed based on work performed for the "PerFORM WATER 2030" (RL project 240750) and "Project Ô" (EU project 776816) projects. We acknowledge the support from Regione Lombardia under the POR FESR 2014-2020 Asse I - Azione I.1.B.1.3 programme and the European Commission under the H2020 programme.

References

[1] Tchobanoglous G, Burton F, Stensel H and R. T 2014 *Wastewater engineering: Treatment and resource recovery* ed Metcalf and Eddy (New York: McGraw-Hill Education)
[2] Twardowska I, Schramm K-W and Berg K 2004 Sewage sludge Waste Management Series vol 4 pp 239–95
[3] Kelessidis A and Stasinakis A S 2012 Comparative study of the methods used for treatment and final disposal of sewage sludge in European countries Waste Manag. **32** 1186–95
[4] Rorat A, Courtois P, Vandenbulcke F and Lemiere S 2019 No Sanitary and environmental aspects of sewage sludge management *Industrial and Municipal Sludge* (Amsterdam: Elsevier) chapter 8 pp 155–80
[5] Tyagi V K and Lo S L 2013 Sludge: A waste or renewable source for energy and resources recovery? *Renew. Sustain. Energy Rev.* **25** 708–28
[6] United Nations 2015 *Transforming our World: The 2030 Agenda for Sustainable Development.*

Access mode: https://sustainabledevelopment.un.org/post2015/transformingourworld)

[7] European Commission 2019 Circular Economy. Implementation of the Circular Economy Action Plan. Access mode: https://ec.europa.eu/environment/circular-economy/index_en.html

[8] van Loosdrecht M C M and Brdjanovic D 2014 Anticipating the next century of wastewater treatment *Science.* **344** 1452–3

[9] Spinosa L, Ayol A, Baudez J-C, Canziani R, Jenicek P, Leonard A, Rulkens W, Xu G and Van Dijk L 2011 Sustainable and Innovative Solutions for Sewage Sludge Management *Water* **3** 702–17

[10] Peccia J and Westerhoff P 2015 We Should Expect More out of Our Sewage Sludge *Environ. Sci. Technol.* **49** 8271–6

[11] Papa M, Foladori P, Guglielmi L and Bertanza G 2017 How far are we from closing the loop of sewage resource recovery? A real picture of municipal wastewater treatment plants in Italy *J. Environ. Manage.* **198** 9–15

[12] Healy M G, Clarke R, Peyton D, Cummins E, Moynihan E L, Martins A, Béraud P and Fenton O 2015 Resource recovery from sewage sludge *Sewage treatment plants: economic evaluation of innovative technologies for energy efficiency* ed K Konstantinos and K P Tsagarakis (London: IWA) chapter 8 pp 139–62

[13] Wang H, Brown S L, Magesan G N, Slade A H, Quintern M, Clinton P W and Payn T W 2008 Technological options for the management of biosolids *Environ. Sci. Pollut. Res.* **15** 308–17

[14] Trianni A, Cagno E and Neri A 2017 Modelling barriers to the adoption of industrial sustainability measures *J. Clean. Prod.* **168** 1482–504

[15] Carter C R and Rogers D S 2008 A framework of sustainable supply chain management: Moving toward new theory *Int. J. Phys. Distrib. Logist. Manag.* **38** 360–87

[16] Romano G, Salvati N and Guerrini A 2016 An empirical analysis of the determinants of water demand in Italy *J. Clean. Prod.* **130** 74–81

[17] Cagno E, Garrone P M, Neri A and Bozzi M S 2018 Analysing the Relationship between Wastewater Utility and Civil Society Stakeholders : First Evidence from Two Cases *13th SDEWES* (September 30 - October 4 Palermo)

[18] Wehn U and Montalvo C 2018 Exploring the dynamics of water innovation: Foundations for water innovation studies *J. Clean. Prod.* **171** S1–19

[19] Garrone P, Grilli L, Groppi A and Marzano R 2018 Barriers and drivers in the adoption of advanced wastewater treatment technologies: a comparative analysis of Italian utilities *J. Clean. Prod.* **171** S69–78

[20] Mitchell R K, Wood D J and Agle B 1997 Toward a Theory of Stakeholder Identification and Salience : Defining the Principle of Who and What Really Counts *Acad. Manag. Rev.* **22** 853–86

[21] Lienert J, Schnetzer F and Ingold K 2013 Stakeholder analysis combined with social network analysis provides fine-grained insights into water infrastructure planning processes *J. Environ. Manage.* **125** 134–48

[22] Maurer J G 1971 *Readings in Organization Theory: Open-System Approaches* (New York: Random House)

[23] Gawlik B M and Bidoglio G 2006 *Background values in European soils and sewage sludges. Results of a JRC-coordinated study on background values. Part III, Conclusions, comments and recommendations.* Access mode: https://op.europa.eu/en/publication-detail/-/publication/f7081979-a8dc-4c2d-a7eb-63b4dfdd9b83/language-en

[24] Christodoulou A and Stamatelatou K 2015 Overview of legislation on sewage sludge management in developed countries worldwide *Water Sci. Technol.* **73** 453–62

[25] Capodaglio A G, Callegari A and Lopez M V 2016 European framework for the diffusion of biogas uses: Emerging technologies, acceptance, incentive strategies, and institutional-regulatory support *Sustain.* **8** 1–18

[26] Samolada M C and Zabaniotou A A 2014 Comparative assessment of municipal sewage sludge incineration, gasification and pyrolysis for a sustainable sludge-to-energy management in Greece *Waste Manag.* **34** 411–20

[27] Hou G, Chen T, Ma K, Liao Z, Xia H and Yao T 2019 Improving social acceptance ofwaste-to-energy incinerators in China: Role of place attachment, trust, and fairness *Sustain.* **11**

[28] European Commission 2001 *Disposal and recycling routes for sewage sludge. Part 3– Scientific and technical report*. Access mode: http://ec.europa.eu/environment/archives/waste/sludge/pdf/sludge_disposal3.pdf.

[29] UN HABITAT 2008 *Global atlas of excreta, wastewater sludge, and biosolids management: moving forward the sustainable and welcome uses of a global resource.* Access mode: https://www.pseau.org/outils/ouvrages/un_habitat_atlas_excreta_wastewater_sludge.pdf

[30] Eisenhardt K M 1989 Building theories from case study research *Acad. Manag. Rev.* **14** 532–50

[31] Handfield R B and Melnyk S A 1998 The scientific theory-building process: a primer using the case of TQM *J. Oper. Manag.* **16** 321–39

[32] Wiechmann B, Dienemann C, Kabbe C, Andrea S B, Vogel I and Roskosch A 2013 *Sewage sludge management in Germany*. Access mode: https://www.umweltbundesamt.de/sites/default/files/medien/378/publikationen/sewage_sludge_management_in_germany.pdf

[33] Utilitalia 2017 *Gestione dei fanghi di depurazione*. Access mode: https://www.utilitalia.it/atti_e_pubblicazioni/pubblicazioni?280caad3-30f1-4b10-855c-947acc0e8dab

[34] Rizzardini C B and Goi D 2014 Considerations About European Directives and Italian Regulation on Sludge from Municipal Wastewater Treatment Plants: Current Status and Future Prospective *Open Waste Manag. J.* **2** 17–26

[35] Patton M Q 1990 *Qualitative evaluation and research methods* (Newbury Park, California: SAGE)

[36] Remler D K and Van Ryzin G G 2014 *Research methods in practice: strategies for description and causation* (Thousand Oaks: SAGE)

[37] Voss C, Tsikriktsis N and Frohlich M 2002 Case research in operations management *Int. J. Oper. Prod. Manag.* **22** 195–219

[38] Stemler S 2001 An overview of content analysis *Pract. Assessment, Res. Eval.* **7**

[39] Saldaña J 2009 *The coding manual for qualitative researchers* (Thousand Oaks: SAGE)

[40] Yin R K 2009 *Case Study Research Design and Methods* (Thousand Oaks: SAGE)

[41] Beverland M and Lindgreen A 2010 What makes a good case study? A positivist review of qualitative case research published in Industrial Marketing Management, 1971-2006 *Ind. Mark. Manag.* **39** 56–63

[42] Barratt M, Choi T Y and Li M 2011 Qualitative case studies in operations management: trends, research outcomes, and future research implications *J. Oper. Manag.* **29** 329–42

Major Discussions on the Life Cycle Cost of Electricity Generation Systems

Mariane Bigarelli Ferreira[1], Guilherme Mateus Kremer[2], Bruno Silva Ribeiro[2], Murillo Vetroni Barros[1], Rodrigo Salvador[1], Antonio Carlos de Francisco[1], Cassiano Moro Piekarski[1]

[1] Universidade Tecnológica Federal do Paraná (UTFPR), Ponta Grossa Campus, Graduate Program in Industrial Engineering, Sustainable Production Systems Laboratory (LESP)
[2] Universidade Tecnológica Federal do Paraná (UTFPR), Ponta Grossa Campus, Sustainable Production Systems Laboratory (LESP)
* Corresponding Author's E-mail: marianebigarellif@gmail.com

Abstract. The aim of this paper was to discuss the use of Life Cycle Cost (LCC) in the assessment of electricity generation systems. To that end, it was conducted a systematic literature review. EndNote, Mendeley and Excel spreadsheets were used to manage references, while the VOSviewer and Microsoft Power Business Intelligence software tools were used to create visual maps. Europe seemed to stand out regarding the number of studies, while there was a lack of LCC studies in America, Asia and Oceania. Concerns seem to have switched from being only about the source for producing electricity to how to diversify the energy system including renewable energy in the energy mix. Overall, there seems to be a shortage of recent studies regarding the LCC of electricity generation systems, which requires further contributions to enable better decision-making.

1. Introduction

In order to analyze the financial feasibility, it is possible to use the Life Cycle Cost (LCC) method, which enables measuring the acquisition, operation, maintenance and final disposal costs of a given system. In LCC, different cost categories can be defined according to the study's scope. Several studies involving LCC, as in the interest of this article, can be found in the existing literature, including comparisons of different energy production strategies, among other approaches.

Studies found in the literature, such as [1, 2, 3] used methods and tools, such as figure of merit (FM), net present cost (NPC), and levelized cost of electricity (COE) to characterize the performance of electrical systems and establish a relationship between the current net present cost of electricity and future energy sales. Therefore, it is possible to obtain more accurate estimates of the payback time and compare costs and sales prices if different renewable sources were applied.

The focus of this review is to analyze qualitative and quantitative characteristics of studies on LCC of electricity systems. To the best of the authors' knowledge, this is one of the first reports on the existing literature on LCC of electricity systems. Thus, the present study assists in organizing, summarizing and pointing out future paths on this aspect. Therefore, the aim of this paper was to discuss the use of Life Cycle Cost in the assessment of electricity generation systems.

This study is structured as follows. This first section presented the initial considerations on the topic and stated the objective of this article. Section 2 presents the methods used to conduct this research. Section 3 unveils the main considerations on the LCC of electricity generation systems, as well as the bibliometrics on the field. Section 4 draws on this study's final considerations, as well as limitations and suggestions for future research.

IOP Conf. Series: Materials Science and Engineering 1196 (2021) 012007 doi:10.1088/1757-899X/1196/1/012007

2. Methods

The investigation in this study was based on a systematic literature review, which was conducted in steps.

Step 1: Definition of keywords

The set of keywords, and their combinations, was defined using BOOLEAN operators and wildcards. The reference query for the search was: ("Life Cycle Cost*" OR LCC) AND ("electric* grid" OR "electric* mix" OR "electric* system" OR "electric* product*" OR "electric* generat*" OR "electric* energy").

Step 2: Search and document collection

The base query was used to conduct searches in the ScienceDirect, Scopus, and Web of Science databases, seeking to capture all available literature fitting the query up to January 13, 2020. The reference query was then adapted to the particularities of each database. The EndNote software tool, Mendeley, and Excel spreadsheets were used to manage the references. Therefrom, a series of filters were applied to select the relevant studies.

Step 3: Portfolio Selection

The searches returned a considerable number of documents. The assessment of all studies would be impractical time- and resource-wise. Thus, the following filters were used to select the most relevant studies:

•Excluding duplicates: all duplicate entries were excluded;

•Excluding grey literature: only articles from peer reviewed journals were kept, thus documents from all other sources were ruled out;

•Title, keywords and abstract: all titles, keywords and abstracts of all articles were read, and the documents not fitting this study's scope were excluded;

•Full-reading: the remaining articles were read in full and only the studies closely related to this study's scope were kept, the non-related articles were ruled out.

Therefore, 17 articles remained to be fully analysed in this article.

2.1. Visual Maps

The Microsoft Power Business Intelligence (BI) and the VOSviewer software tools [4] were used to create visual maps. Using Microsoft Power BI, a graph showing publications per year and per journal (Figure 1) was created. Using VOSviewer, the co-occurrence of terms map (Figure 2) was created based on text data, using the following settings: source: bibliographic data; type of analysis: co-occurrence; unit of analysis: keywords; counting method: full counting; threshold: 3.

3. Major Discussions on Life Cycle Cost of Electricity Generation Systems

The Sustainable Development Goals (SDG) bring in their goal nº 7 "Ensure access affordable, reliable, sustainable and modern energy for all". This requires efforts in different areas such as technology, private and public investments and an increase in the access to renewable energy [5]. Therefore, without studies about the ways to produce clean energy, achieving this objective becomes difficult.

The studies addressing LCC mainly conduct comparisons between LCCs of different electricity generation sources. The sources include wind, biomass, diesel, and others. The studies were developed in many countries, thus, under different economic, environmental and social scenarios.

With two studies, Spain stands out. Research in the country includes investigating costs of melting aluminium for energy generation [6], comparing the potential impacts of three scenarios in the furnace: natural gas and alternating current power, natural gas and direct current power, and direct current. In turn, [7] conducted an LCC and assessed the economic feasibility of floating offshore wind farms.

The remaining contributing countries produced one study each, showing a lack of studies in the American and Asian continents. Hence, it can be noted the need for further studies in different areas and locations in order to increase the knowledge around the costs involved with different electricity generation sources, to allow more solid decision-making.

In the Netherlands, [8] excluded tax from their calculations and showed that using ethanol is more economically feasible than using gasoline, due to oil prices. On top of it, both studies of [9, 10] defended considering the costs of capital, operations and maintenance in their calculations.

[11, 12] used financial indicators as Present Value (PV) and Return on Investment (ROI) to analyze the economic feasibility of energy production from wind, solar power, hydrogen and other sources. The authors detected some elevated costs to these renewable energy pathways. Indicators as Internal Rate of Return (IRR) and Payback were used in the study of [7] to compare wind in different areas of Galiza, Spain. Some authors combined these indicators to some boundaries to their studies as Well-to-Tank (WTT), Tank-to-Wheel (TTW) and Well-to-Wheel (WTW). An example can be found in [12] comparing wind-hydrogen systems to generate electricity.

The study [13] analyzed three configurations in order to compare net present cost, energy cost and operating cost with the sales price, to estimate the payback times. The study by [2] point out that electricity production from biomass is important, making it possible to reduce government costs with organic waste and job implementation. In India, [1] showed that even with a positive economic and environmental performance of solar panels in different locations on the planet, it is still necessary to study each case, as the climate variable differs considerably from region to region. The work by [14], in China, addressed an analysis of the economic viability of the Solar Dish/Stirling (SDS) that uses solar energy. On an island belonging to Italy, in the Mediterranean, [3], discussed the economic feasibility of establishing an area dedicated to solar panels with the purpose of desalinating sea water, making it drinkable.

It can be seen, thus, many ways to conduct an LCC, with each study considering the local conditions. However, capital, operation, maintenance and final disposal costs, in most cases, are considered by the pieces of research. The lifetimes of the alternatives are also taken into consideration, the values range from 10 to 12 years depending on the energy source.

LCC studies have addressed different issues depending on when they seemed relevant. Therefore, Figure 1 shows the number of studies published per year, per journal and their respective impact factor.

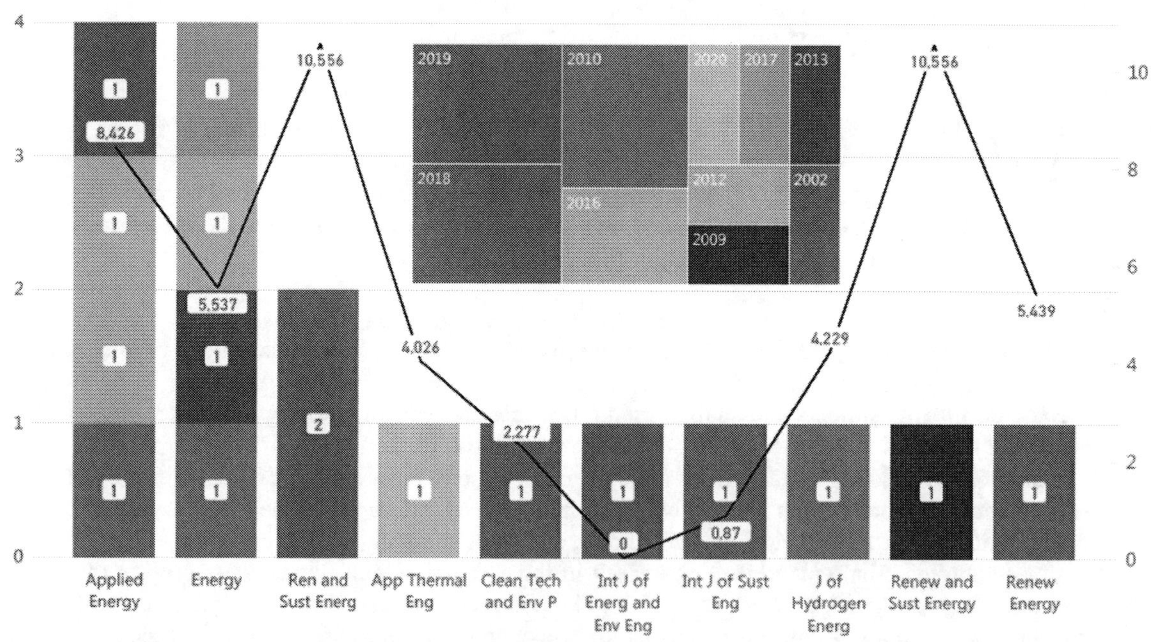

Figure 1. Number of LCC-related publications per year, per journal and their respective impact factors.

Applied Energy, which has the second greatest impact factor, showed four publications related to LCC. These are related to photovoltaic panel systems [15], combustion systems [16], thermal energy storage systems [17] and the economic feasibility of using solar panels to desalinate sea water [3]. Similarly to the journal Applied Energy, Energy also published four papers [10, 18, 7, 13].

The most recent studies analyzed transportation, installation, operation and maintenance costs comparing energy from diesel generators to energy from photovoltaic systems [9] and costs of melting aluminium for energy generation in European countries [6]. The study of [19] reinforces the awareness of the characteristics and economics of photovoltaic panels on roofs, being vital for the implementation of renewable energy. Nevertheless, there seems to be a shortage of recent studies regarding the LCC of electricity generation systems, which requires for further and greater contributions in order to enable better decision-making. Thus, a more specific analysis could offer further insights on this literature body; therefore, Figure 2 shows a term co-occurrence map of the LCC studies in the final portfolio.

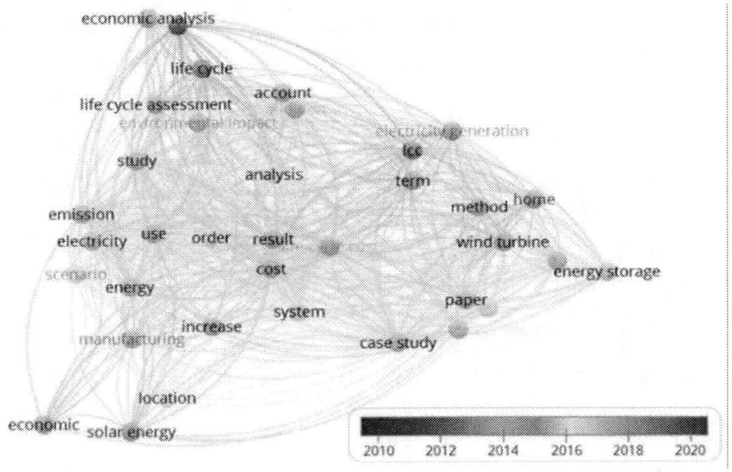

Figure 2. Co-occurrence of keywords in LCC-related studies.

Analyzing the term co-occurrence map and the afore discussion, the interest in studies of scenarios, technologies, electricity mixes, and economic feasibility have increased in the last few years. Concerns have switched from being only about the source for producing electricity to how to diversify the energy system including renewable energy in the energy mix. The methods to evaluate the feasibility have changed. The studies started combining economic indicators with LCA methods, but there is a lack of these studies in the literature yet. In the future, it is expected that analysis that evaluate the environmental and social impacts together with the economic dimension be conducted.

4. Final Considerations

This study aimed to discuss the use of Life Cycle Cost in the assessment of electricity generation systems, from which a few conclusions could be reached and some recommendations be drawn.

On the economic dimension, indicators are identified to evaluate a system of electricity production, as well as some boundaries were used by the authors to carry out their analyses. Depending on the scenario, in the long-term renewable energy sources are more economically feasible than the traditional ways of producing electricity. It is important to emphasize that each country has different characteristics according to where it is located.

The results in this study can be used to foment public policies towards more sustainable energy generation, as it could be seen that less environmentally impacting energy sources seem to also have positive economic results. On top of governmental implications, it enables practitioners and researchers to engage in partnerships towards the development of further research and technology by showing the main countries and institutions developing research on the subject worldwide.

This study contributes to knowledge generation on more sustainable practices towards electricity matrices/mixes, bringing out the knowledge of previous studies on the economic assessment of said systems. Such results present the scientific community with what has already been studied, as well as point out best practices on the economic dimension of electricity generation systems.

References

[1] Sherwani A F, Asjad, M, Haleem A 2019 Performance evaluation of solar photovoltaic electricity-generating systems: an Indian perspective. International Journal of Sustainable Engineering 12 1 70-75. https://doi.org/10.1080/19397038.2018.1491069

[2] Bagheri M, Shirzadi N, Bazdar E Kennedy C A 2018 Optimal planning of hybrid renewable energy infrastructure for urban sustainability: Green Vancouver. Renewable and Sustainable Energy Reviews 95 254-264 https://doi.org/10.1016/j.rser.2018.07.037

[3] Calise F, Cappiello F L, Vanoli R, Vicidomini M 2019 Economic assessment of renewable energy systems integrating photovoltaic panels, seawater desalination and water storage. Applied Energy, 253, 113575. https://doi.org/10.1016/j.apenergy.2019.113575

[4] Van Eck N J, Waltman L 2014 Visualizing bibliometric networks. In: Measuring scholarly impact. Springer Cham 285-320. https://doi.org/10.1007/978-3-319-10377-8_13

[5] UN (United Nations). 17 Goal to Transform our World, 2015. https://www.un.org/sustainabledevelopment/sustainable-consumption-production/. (accessed 29 January 2020).

[6] Royo P, Ferreira V J, López-Sabirón A M, García-Armingol T, Ferreira G 2018 Retrofitting strategies for improving the energy and environmental efficiency in industrial furnaces: A case study in the aluminium sector. Renewable and Sustainable Energy Reviews 82 1813-1822 https://doi.org/10.1016/j.rser.2017.06.113

[7] Castro-Santos L, Filgueira-Vizoso A, Carral-Couce L, Formoso J Á F 2016 Economic feasibility of floating offshore wind farms. Energy 112 868-882. https://doi.org/10.1016/j.energy.2016.06.135

[8] Luo L, Voet E V D, Huppes G 2009 Life cycle assessment and life cycle costing of bioethanol from sugarcane in Brazil. Renewable and Sustainable Energy Reviews 13 6-7 1613-1619. https://doi.org/10.1016/j.rser.2008.09.024

[9] Arceo A, Rosano M, Biswas W K 2018 Eco-efficiency analysis for remote area power supply selection in Western Australia. Clean Technologies and Environmental Policy 20(3) 463-475 https://doi.org/10.1007/s10098-017-1438-6

[10] Arslan O 2010 Technoeconomic analysis of electricity generation from wind energy in Kutahya, Turkey Energy 35(1) 120-131 http://dx.doi.org/10.1016/j.energy.2009.09.002

[11] El-Kordy M N, Badr M A, Abed K A, Ibrahim S M 2002 Economical evaluation of electricity generation considering externalities. Renewable Energy 25 2 317-328 https://doi.org/10.1016/S0960-1481(01)00054-4

[12] Lee J Y, An S, Cha K, Hur T 2010. Life cycle environmental and economic analyses of a hydrogen station with wind energy. International Journal of Hydrogen Energy 35 6 2213-2225. https://doi.org/10.1016/j.ijhydene.2009.12.082

[13] Islam M S, Akhter R, Rahman M A 2018 A thorough investigation on hybrid application of biomass gasifier and PV resources to meet energy needs for a northern rural off-grid region of Bangladesh: A potential solution to replicate in rural off-grid areas or not? Energy 145 338-355 https://doi.org/10.1016/j.energy.2017.12.125

[14] Zayed M E, Zhao J, Li W, Elsheikh A H, Zhao Z, Khalil A, Li H 2020 Performance prediction and techno-economic analysis of solar dish/stirling system for electricity generation. Applied Thermal Engineering 164 114427 https://doi.org/10.1016/j.applthermaleng.2019.114427

[15] Qoaider L, Steinbrecht D 2010 Photovoltaic systems: A cost competitive option to supply energy to off-grid agricultural communities in arid regions. Applied Energy 87 2 427-435. https://doi.org/10.1016/j.apenergy.2009.06.012

[16] Afrane G, Ntiamoah A 2012 Analysis of the life-cycle costs and environmental impacts of cooking fuels used in Ghana. *Applied Energy* 98 301-306 https://doi.org/10.1016/j.apenergy.2012.03.041

[17] Alimohammadisagvan B, Jokisalo J, Kilpeläinen S, Ali M, Sirén K 2016 Cost-optimal thermal energy storage system for a residential building with heat pump heating and demand response control. Applied Energy 174 275-287 https://doi.org/10.1016/j.apenergy.2016.04.013

[18] Caballero F, Sauma E, Yanine F 2013 Business optimal design of a grid-connected hybrid PV (photovoltaic)-wind energy system without energy storage for an Easter Island's block Energy 61 248-261 http://dx.doi.org/10.1016/j.energy.2013.08.030

[19] Kouhestani F M, Byrne J, Johnson D, Spencer L, Hazendonk P, Brown B 2019 Evaluating solar energy technical and economic potential on rooftops in an urban setting: the city of Lethbridge, Canada. International *Journal of Energy and Environmental Engineering* 10 **1** 13-32. https://doi.org/10.1007/s40095-018-0289-1

Acknowledgments
This study was financed in part by the Coordenação de Aperfeiçoamento de Pessoal de Nível Superior - Brasil (CAPES) - Finance Code 001, and the Conselho Nacional de Desenvolvimento Científico e Tecnológico - Brazil (CNPq) (Sponsored by CNPq 310259/2020-7 and 312285/2019-1)

Closing the material loop in additive manufacturing: A literature review on waste recycling

Yiran Yang[1,*], Fu Zhao[2,3]

[1]Department of Industrial, Manufacturing, & Systems Engineering, University of Texas at Arlington, Arlington, TX, 76019 USA
[2]School of Mechanical Engineering, Purdue University, West Lafayette, IN, 47907 USA
[3]Environmental and Ecological Engineering, Purdue University, West Lafayette, IN, 47907 USA
[*]Corresponding author. Email: yiran.yang@uta.edu

Abstract. Additive manufacturing (also referred to as 3D printing) technologies have found applications in a wide range of industries such as aerospace and automobile, due to their superior manufacturing capability and design freedom enabled by the layer-wise fabrication method. Over the past decade, the adoption of additive manufacturing has evolved from rapid prototyping and tooling to rapid manufacturing of end-use products, which, on the other hand, introduced new challenges for reducing the environmental impacts and enhancing resource sustainability of additive manufacturing from a lifecycle perspective. In current literature, some research efforts have been conducted on waste recycling aiming to close to material loop and relieve the environmental consequences caused by both pre- and post-consumer wastes generated from additive manufacturing. This article provides an overview of the state-of-the-art on additive manufacturing waste recycling and identifies critical gaps for future research in this field.

Keywords: Additive manufacturing; Recycling; Closed-loop material flow; Environmental impact; Resource sustainability.

1. Introduction

Additive manufacturing (AM), or 3-dimensional (3D) printing, refers to processes that fabricate products from 3D model usually layer-by-layer, and it is fundamentally different from traditional formative and subtractive manufacturing. Based on different processes, AM can be categorized into seven groups [6], i.e., binder jetting, directed energy deposition, material extrusion, material jetting, powder bed fusion, sheet lamination, and vat photopolymerization. In recent years, owing to technological advancements in AM processes and materials, the global AM market has been rapidly increasing. According to surveys conducted by the Wohlers Associates, in 2017, the AM industry consisting of all AM products and services worldwide, grew 21% from 2016 and reached to USD 7.34 billion [55]. The global AM market is expected to grow to USD 36.61 billion by 2027 from USD 8.44 billion in 2018, at a compound annual growth rate (CAGR) of 17.7% [32].

Compared to traditional manufacturing, AM can provide shortened product development cycle, additional design freedom, and increased manufacturing complexity. Additionally, enhanced environmental sustainability has been touted as one of the main advantages of AM with respect to high material efficiency, low scrap rate, and reduced or even eliminated need for tooling, cutting fluids and lubricants. Furthermore, AM can revolutionize the current supply chain structures by offering localized and/or regional strategies for production, transportation, as well as end-of-life (EOL) options. An example of AM supply chain is illustrated in Figure 1. In current literature, most research efforts are focused on developing new AM feedstock materials [26][33][46], improving printing technologies [28][30], and evaluating production performance and quality [11][13][29][57].

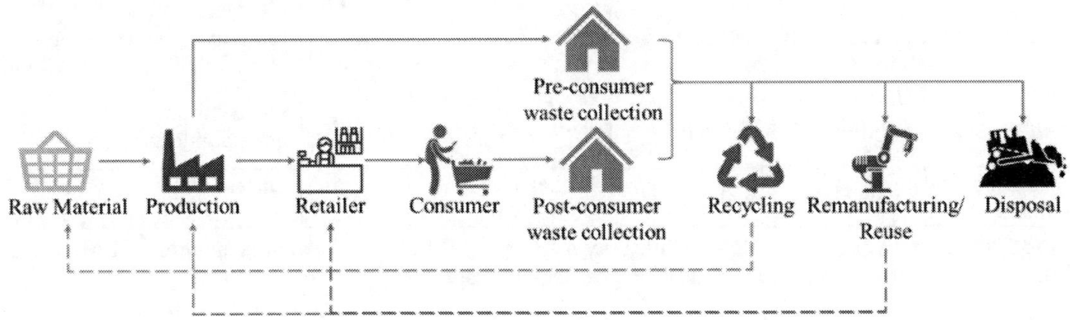

Figure 1. An example of additive manufacturing supply chain.

Comprehensive reviews exist for the state-of-the-art of AM in terms of design and modelling methodologies [9][56], production processes [8][23][45][58], feedstock materials [17][27][34][54], different applications [53][59], and environmental sustainability [1][2][35][39]. In this article, we aim to provide an overview on AM waste recycling and identify critical knowledge gaps in facilitating the closed-loop material flows in AM. The current literature on AM waste recycling is reviewed based on different feedstock materials, i.e., metal powder recycling as shown in Section 2 and plastic wastes recycling as shown in Section 3. In addition, conclusions and future research topics are discussed in Section 4 aiming to provide useful insights on AM waste recycling and facilitate multidisciplinary collaboration in the field.

2. The state-of-the art on AM metal powder recycling

The current literature on waste recycling of metal powders used in AM is summarized in Table 1. Most studies reviewed in this section are focused on evaluating the performance of recycled powders (micro-level performance in terms of porosity, particle size distribution, etc.) and printed parts using recycled powders (macro-level performance in terms of mechanical properties and surface quality). Meanwhile, the lifecycle environmental sustainability of recycling metal powders is also investigated. In addition to the literature shown in Table 1, research efforts have also been dedicated to improving the performance of recycled powders. As an example, thermal post-processing is adopted to enhance the recyclability of AlSi10Mg powders in SLM [31].

Table 1. A summary of current literature on metal powder recycling in AM.

Category	Citation	Material	AM	Main findings
Recycling performance evaluation	(Strondl, Lyckfeldt, Brodin, & Ackelid, 2015)	Ti-6Al-4V, Ni alloy	EBM, SLM	Differences in particle size and flow ability
	(Slotwinski et al., 2014)	Stainless steel & cobalt-chrome	DMLS	Increased powder size distribution
	(Asgari, Baxter, Hosseinkhani, & Mohammadi, 2017)	AlSi10Mg_200C	DMLS	Comparable powder characteristics, microstructure and mechanical behaviour
	(Ardila et al., 2014)	IN718	SLM	No significant change in powder and test part properties
	(Popov, Katz-Demyanetz, Garkun, & Bamberger, 2018)	Ti-6Al-4V	EBM	Recycling causes various defects

	(Renderos et al., 2017)	IN718	LMD	Increased porosity and decreased rupture strain
	(Gorji et al., 2019)	Stainless steel	PBF	More satellite and bonded particles & smaller roughness and higher hardness
	(Carroll et al., 2006)	Ni alloy	DMLS	No compromised part integrity, great variation in size distribution and composition
	(Hadadzadeh, Baxter, Amirkhiz, & Mohammadi, 2018)	AlSi10Mg	DMLS	Different microstructure, comparable strength
Sustainability evaluation	(Walachowicz et al., 2017)	Nickel alloy and stainless steel	LBM	Different recycling rates affect lifecycle footprint

(EBM: electron beam melting; SLM: selective laser melting; DMLS: direct metal laser sintering; LBM: laser beam melting; LMD: laser material deposition; PBF: powder bed fusion)

The following observations can be made from the current literature.

- Sieving is used as the sole recycling method in most metal powder recycling studies. New methods are required in order to further improve the quality of recycled powders.
- The comparison results on micro-level powder characteristics between virgin powders and recycled powders are not consistent for different materials and/or AM processes. As an example, no significant changes are observed in powders for IN718 in [4] while increased porosity is detected in [40].
- The macro-level properties of parts fabricated using recycled materials need to be further expanded to include the profiles and roughness of printed surfaces.
- Most existing studies indicate no significant change in mechanical properties within a certain range of recycling cycles.
- Almost all studies adopt empirical approaches, and therefore their results are only applicable to certain experimental conditions.

3. The state-of-the-art on AM plastic waste recycling

The current literature on waste recycling of plastic feedstock used in AM is summarized in Table 2. Two types of plastics are usually used in AM and investigated by the literature, i.e., thermosetting resin and thermoplastic polymers like ABS, HDPE and PLA. The current literature on recycling plastic wastes in AM is mainly focused on three aspects, 1) proposing recyclable materials to be used in AM; 2) recycling performance evaluation; and 3) sustainability evaluation, as categorized in Table 2.

Table 2. A summary of current literature on plastic recycling in AM.

Category	Citation	Material	AM	Recycling	Main findings
Proposing recyclable material	(Plummer, Vasquez, Majewski, & Hopkinson, 2012)	Thermoplastic PUR powder	LS	Sieving	No significant degradation in properties
	(Shi et al., 2017)	Thermosetting epoxy ink	DIW	Chemical dissolution	Good printability
	(Boparai, Singh, Fabbrocino, & Fraternali, 2016; Stoof & Pickering, 2017, 2018; Tian et al., 2017)	Composites	FDM	Fiber treatment	Enhanced mechanical properties & thermal stability
Recycling performance evaluation	(Baechler, DeVuono, & Pearce, 2013)	PLA	FDM	Mechanical (RecycleBot)	Technical evaluation (quality, time & energy consumption)

	(Fabio A. Cruz Sanchez, Boudaoud, Hoppe, & Camargo, 2017; Fabio Alberto Cruz Sanchez, Lanza, Boudaoud, Hoppe, & Camargo, 2015)	PLA	FDM	Mechanical (RecycleBot)	Slightly affected mechanical properties, reduced molecular weight & viscosity
	(Anderson, 2017)	PLA	FDM	Mechanical (Filabot EX2)	Similar mechanical properties, but more print failure
Sustainability evaluation	(Kreiger, Mulder, Glover, & Pearce, 2014)	HDPE	FDM	Mechanical (RecycleBot)	Great cost saving potentials
	(Hunt, Zhang, Anzalone, & Pearce, 2015)	Polymers	FDM	Mechanical (RecycleBot)	Recycling codes proposed

(FDM: fused deposition modeling; DIW: direct ink writing; LS: laser sintering, PLA: polylactic acid; PUR: polyurethane; HDPE: high density polyethylene)

The following observations can be made from the current literature.

- Most studies on thermoplastic polymer recycling are focused on PLA. Acrylonitrile butadiene styrene (ABS) needs to be studied as well.
- Almost all studies adopt empirical approaches, and therefore their results are only applicable to certain experimental conditions.
- The material properties of recycled plastics have not yet been comprehensively characterized, which is the root cause for potential decrease in print quality and mechanical properties.
- The sustainability of recycling needs to be further evaluated from both process-level and lifecycle perspective. For example, the cost-effectiveness of recycling wasted materials is currently unknown, and whether incorporating recycling into the AM lifecycle is environmentally sustainable has not yet been assessed.

4. Conclusions and Future Research Directions

In this article, the state-of-the-art on AM waste recycling is reviewed for both metal and plastic feedstock materials. To further advance the understandings of the recycling process, some future research directions are identified as follows.

- The relationships between waste material properties (for both metal powders or plastic scraps), recycling process design and planning, and achieved print performance using recycled materials, need to be characterized. This interrelation will enable a prior prediction on recycling performance and greatly aid the better control and optimization on the recycling process.
- The energy and material flow in the recycling process needs to be characterized and incorporated into a comprehensive life cycle assessment as inventory data, to evaluate the environmental sustainability of AM waste recycling.
- Different supply chain designs considering AM waste recycling need to be compared and evaluated jointly considering cost and carbon footprint.

References

[1] Ahn, D.-G. (2016). Direct metal additive manufacturing processes and their sustainable applications for green technology: A review. *International Journal of Precision Engineering and Manufacturing-Green Technology*, *3*(4), 381–395. https://doi.org/10.1007/s40684-016-0048-9

[2] Al-Meslemi, Y., Anwer, N., & Mathieu, L. (2018). Environmental Performance and Key Characteristics in Additive Manufacturing: A Literature Review. *Procedia CIRP*.

[3] Anderson, I. (2017). Mechanical Properties of Specimens 3D Printed with Virgin and Recycled Polylactic Acid. *3D Printing and Additive Manufacturing.* https://doi.org/10.1089/3dp.2016.0054

[4] Ardila, L. C., Garciandia, F., González-Díaz, J. B., Álvarez, P., Echeverria, A., Petite, M. M., … Ochoa, J. (2014). Effect of IN718 recycled powder reuse on properties of parts manufactured by means of Selective Laser Melting. *Physics Procedia.* https://doi.org/10.1016/j.phpro.2014.08.152

[5] Asgari, H., Baxter, C., Hosseinkhani, K., & Mohammadi, M. (2017). On microstructure and mechanical properties of additively manufactured AlSi10Mg_200C using recycled powder. *Materials Science and Engineering A.* https://doi.org/10.1016/j.msea.2017.09.041

[6] ASTM International. (2013). F2792-12a - Standard Terminology for Additive Manufacturing Technologies. In *Rapid Manufacturing Association.* https://doi.org/10.1520/F2792-12A.2

[7] Baechler, C., DeVuono, M., & Pearce, J. M. (2013). Distributed recycling of waste polymer into RepRap feedstock. *Rapid Prototyping Journal, 19*(2), 118–125. https://doi.org/10.1108/13552541311302978

[8] Bhavar, V., Kattire, P., Pawar, P., Khot, S., Gujar, K., & Singh, R. (2014). A Review on Powder Bed Fusion Technology of Additive Manufacturing. *Processes, Applications, and Performance of Materials in Additive Manufacturing, 4*(11), 991–995.

[9] Bikas, H., Stavropoulos, P., & Chryssolouris, G. (2016). Additive manufacturing methods and modelling approaches: a critical review. *The International Journal of Advanced Manufacturing Technology, 83*(1), 389–405. https://doi.org/10.1007/s00170-015-7576-2

[10] Boparai, K. S., Singh, R., Fabbrocino, F., & Fraternali, F. (2016). Thermal characterization of recycled polymer for additive manufacturing applications. *Composites Part B: Engineering.*

[11] Brandl, E., Leyens, C., & Palm, F. (2011). Mechanical properties of additive manufactured Ti-6Al-4V using wire and powder based processes. *IOP Conference Series: Materials Science and Engineering.* https://doi.org/10.1088/1757-899X/26/1/012004

[12] Carroll, P. A., Pinkerton, A. J., Allen, J., Syed, W. U. H., Sezer, H. K., Brown, P., … Li, L. (2006). The effect of powder recycling in direct metal laser deposition on powder and manufactured part characteristics. *Proceedings of AVT-139 Specialists Meeting on Cost Effective Manufacture via Net Shape Processing. NATO Research and Technology Organisation.*

[13] Christiyan, K. G. J., Chandrasekhar, U., & Venkateswarlu, K. (2016). A study on the influence of process parameters on the Mechanical Properties of 3D printed ABS composite. *IOP Conference Series: Materials Science and Engineering.* https://doi.org/10.1088/1757-899X/114/1/012109

[14] Cooke, A., & Slotwinski, J. (2015). Properties of metal powders for additive manufacturing: A review of the state of the art of metal powder property testing. In *Additive Manufacturing Materials: Standards, Testing and Applicability.* https://doi.org/10.6028/NIST.IR.7873

[15] Cruz Sanchez, Fabio A., Boudaoud, H., Hoppe, S., & Camargo, M. (2017). Polymer recycling in an open-source additive manufacturing context: Mechanical issues. *Additive Manufacturing.*

[16] Cruz Sanchez, Fabio Alberto, Lanza, S., Boudaoud, H., Hoppe, S., & Camargo, M. (2015). Polymer Recycling and Additive Manufacturing in an Open Source context : Optimization of processes and methods. *Proceedings of Solid Freeform Fabrication Symposium.*

[17] Frketic J, Dickens, T., & Ramakrishnan, S. (2017). Automated manufacturing and processing of fiber-reinforced polymer (FRP) composites: An additive review of contemporary and modern techniques for advanced materials manufacturing. *Additive Manufacturing*, Vol. 14, pp. 69–86. https://doi.org/10.1016/j.addma.2017.01.003

[18] Giannatsis J, & Dedoussis, V. (2009). Additive fabrication technologies applied to medicine and health care: A review. *International Journal of Advanced Manufacturing Technology, 40*(1–2), 116–127.

[19] Gorji, N. E., O'Connor, R., Mussatto, A., Snelgrove, M., González, P. G. M., & Brabazon, D.

(2019). Recyclability of stainless steel (316 L) powder within the additive manufacturing process. *Materialia*.
[20] Hadadzadeh, A., Baxter, C., Amirkhiz, B. S., & Mohammadi, M. (2018). Strengthening mechanisms in direct metal laser sintered AlSi10Mg: Comparison between virgin and recycled powders. *Additive Manufacturing*. https://doi.org/10.1016/j.addma.2018.07.014
[21] Haghighi, A., Yang, Y., & Li, L. (2017). Dimensional performance of as-built assemblies in polyjet additive manufacturing process. *Proceedings of the ASME 12TH International Manufacturing Science and Engineering Conference - 2017, VOL 2*, (27), 293–304. https://doi.org/10.1115/MSEC2017-2983
[22] Hunt, E. J., Zhang, C., Anzalone, N., & Pearce, J. M. (2015). Polymer recycling codes for distributed manufacturing with 3-D printers. *Resources, Conservation and Recycling*.
[23] Körner, C. (2016). Additive manufacturing of metallic components by selective electron beam melting - A review. *International Materials Reviews*, Vol. 61, pp. 361–377.
[24] Kreiger, M. A., Mulder, M. L., Glover, A. G., & Pearce, J. M. (2014). Life cycle analysis of distributed recycling of post-consumer high density polyethylene for 3-D printing filament. *Journal of Cleaner Production*, *70*, 90–96. https://doi.org/10.1016/j.jclepro.2014.02.009
[25] Kulkarni, P., Marsan, A., & Dutta, D. (2000). A review of process planning techniques in layered manufacturing. *Rapid Prototyping Journal*, *6*(1), 18–35. https://doi.org/10.1108/13552540010309859
[26] Leung, P. Y. V. (2017). Sugar 3D Printing: Additive Manufacturing with Molten Sugar for Investigating Molten Material Fed Printing. *3D Printing and Additive Manufacturing*, *4*(1).
[27] Lewandowski, J. J., & Seifi, M. (2016). Metal Additive Manufacturing: A Review of Mechanical Properties. *Annual Review of Materials Research*, *46*(1), 151–186. https://doi.org/10.1146/annurev-matsci-070115-032024
[28] Li, L., Haghighi, A., & Yang, Y. (2018a). A novel 6-axis hybrid additive-subtractive manufacturing process: Design and case studies. *Journal of Manufacturing Processes*, *33*, 150–160.
[29] Li, L., Haghighi, A., & Yang, Y. (2018b). Theretical Modeling and Prediction of Surface Roughness for Hybrid Additive-Subtractive Manufacturing Processes. *IISE Transactions*, *51*(2), 124–135.
[30] Lu, L., Guo, P., & Pan, Y. (2017). Magnetic-Field-Assisted Projection Stereolithography for Three-Dimensional Printing of Smart Structures. *Journal of Manufacturing Science and Engineering*, *139*(7), 071008.
[31] Maamoun, A. H., Elbestawi, M., Dosbaeva, G. K., & Veldhuis, S. C. (2018). Thermal post-processing of AlSi10Mg parts produced by Selective Laser Melting using recycled powder. *Additive Manufacturing*. https://doi.org/10.1016/j.addma.2018.03.014
[32] Market Study Report, L. (2019). *Additive Manufacturing Market*.
[33] Nordin, N. A. B., Johar, M. A. Bin, Ibrahim, M. H. I. Bin, & Marwah, O. M. F. Bin. (2017). Advances in High Temperature Materials for Additive Manufacturing. *IOP Conference Series: Materials Science and Engineering*. https://doi.org/10.1088/1757-899X/226/1/012176
[34] Parandoush, P., & Lin, D. (2017). A review on additive manufacturing of polymer-fiber composites. *Composite Structures*, Vol. 182, pp. 36–53. https://doi.org/10.1016/j.compstruct.2017.08.088
[35] Peng, T., Kellens, K., Tang, R., Chen, C., & Chen, G. (2018). Sustainability of additive manufacturing: An overview on its energy demand and environmental impact. *Additive Manufacturing*, Vol. 21, pp. 694–704.
[36] Plummer, K., Vasquez, M., Majewski, C., & Hopkinson, N. (2012). Study into the recyclability of a thermoplastic polyurethane powder for use in laser sintering. *Proceedings of the Institution of Mechanical Engineers, Part B: Journal of Engineering Manufacture*. https://doi.org/10.1177/0954405412440066
[37] Popov, V. V., Katz-Demyanetz, A., Garkun, A., & Bamberger, M. (2018). The effect of powder

recycling on the mechanical properties and microstructure of electron beam melted Ti-6Al-4 V specimens. *Additive Manufacturing*. https://doi.org/10.1016/j.addma.2018.06.003

[38] Raviv D, Zhao, W., McKnelly, C., Papadopoulou, A., Kadambi, A., Shi, B., … Tibbits, S. (2014). Active printed materials for complex self-evolving deformations. *Scientific Reports*, *4*, 1–8.

[39] Rejeski, D., Zhao, F., & Huang, Y. (2018). Research needs and recommendations on environmental implications of additive manufacturing. *Additive Manufacturing*. https://doi.org/10.1016/j.addma.2017.10.019

[40] Renderos, M., Torregaray, A., Gutierrez-Orrantia, M. E., Lamikiz, A., Saintier, N., & Girot, F. (2017). Microstructure characterization of recycled IN718 powder and resulting laser clad material. *Materials Characterization*. https://doi.org/10.1016/j.matchar.2017.09.029

[41] Sames, W. J., List, F. A., Pannala, S., Dehoff, R. R., & Babu, S. S. (2016). The metallurgy and processing science of metal additive manufacturing. *International Materials Reviews*, Vol. 61, pp. 315–360.

[42] Santos, A. R. C., Almeida, H. A., & Bártolo, P. J. (2013). Additive manufacturing techniques for scaffold-based cartilage tissue engineering. *Virtual and Physical Prototyping*, *8*(3), 175–186.

[43] Shi, Q., Yu, K., Kuang, X., Mu, X., Dunn, C. K., Dunn, M. L., … Jerry Qi, H. (2017). Recyclable 3D printing of vitrimer epoxy. *Materials Horizons*. https://doi.org/10.1039/c7mh00043j

[44] Slotwinski, J. A., Garboczi, E. J., Stutzman, P. E., Ferraris, C. F., Watson, S. S., & Peltz, M. A. (2014). Characterization of metal powders used for additive manufacturing. *Journal of Research of the National Institute of Standards and Technology*, *119*, 460.

[45] Song, B., Zhao, X., Li, S., Han, C., Wei, Q., Wen, S., … Shi, Y. (2015). Differences in microstructure and properties between selective laser melting and traditional manufacturing for fabrication of metal parts: A review. *Frontiers of Mechanical Engineering*, Vol. 10, pp. 111–125.

[46] Stoof, D., & Pickering, K. (2017). 3D Printing of Natural Fibre Reinforced Recycled Polypropylene. *Processing and Fabrication of Advanced Materials*.

[47] Stoof, D., & Pickering, K. (2018). Sustainable composite fused deposition modelling filament using recycled pre-consumer polypropylene. *Composites Part B: Engineering*.

[48] Strondl, A., Lyckfeldt, O., Brodin, H., & Ackelid, U. (2015). Characterization and Control of Powder Properties for Additive Manufacturing. *JOM*. https://doi.org/10.1007/s11837-015-1304-0

[49] Tian, X., Liu, T., Wang, Q., Dilmurat, A., Li, D., & Ziegmann, G. (2017). Recycling and remanufacturing of 3D printed continuous carbon fiber reinforced PLA composites. *Journal of Cleaner Production*.

[50] Turner, B. N., (Chemical and Materials Engineering, U. of D., Strong, R., Gold, S. A., N. Turner, B., Strong, R., & A. Gold, S. (2014). A review of melt extrusion additive manufacturing processes: I. Process design and modeling. *Rapid Prototyping Journal*, *20*(3), 192–204. https://doi.org/10.1108/RPJ-01-2013-0012

[51] Vayre, B., Vignat, F., & Villeneuve, F. (2013). Identification on some design key parameters for additive manufacturing: Application on Electron Beam Melting. *Procedia CIRP*, *7*, 264–269.

[52] Walachowicz, F., Bernsdorf, I., Papenfuss, U., Zeller, C., Graichen, A., Navrotsky, V., … Kiener, C. (2017). Comparative Energy, Resourceand Recycling Lifecycle Analysis of theIndustrial Repair Process of Gas TurbineBurners Using Conventional Machiningand Additive Manufacturing. *Journal of Industrial Ecology*, *21*(S1), S203–S215. https://doi.org/DOI: 10.1111/jiec.12637

[53] Wang, Xiaojian, Xu, S., Zhou, S., Xu, W., Leary, M., Choong, P., … Xie, Y. M. (2016). Topological design and additive manufacturing of porous metals for bone scaffolds and orthopaedic implants: A review. *Biomaterials*, Vol. 83, pp. 127–141. https://doi.org/10.1016/j.biomaterials.2016.01.012

[54] Wang, Xin, Jiang, M., Zhou, Z., Gou, J., & Hui, D. (2017). 3D printing of polymer matrix composites: A review and prospective. *Composites Part B: Engineering*, Vol. 110, pp. 442–

458.
[55] Wohlers Report 2018. (2018). In *Wohlers Associates*.
[56] Yang, S., & Zhao, Y. F. (2015). Additive manufacturing-enabled design theory and methodology: a critical review. *International Journal of Advanced Manufacturing Technology*, Vol. 80, pp. 327–342.
[57] Yang, Y., Li, L., & Zhao, J. (2019). Mechanical Property Modeling of Photosensitive Liquid Resin in Stereolithography Additive Manufacturing: Bridging Degree of Cure with Tensile Strength and Hardness. *Materials & Design*, *162*(15), 418–428.
[58] Yap, C. Y., Chua, C. K., Dong, Z. L., Liu, Z. H., Zhang, D. Q., Loh, L. E., & Sing, S. L. (2015). Review of selective laser melting: Materials and applications. *Applied Physics Reviews*, Vol. 2.
[59] Yap, Y. L., & Yeong, W. Y. (2014). Additive manufacture of fashion and jewellery products: a mini review. *Virtual and Physical Prototyping*, *9*(3), 195–201. https://doi.org/10.1080/17452759.2014.938993

Acknowledgements
This research is supported by the Research Experience for Undergraduates (REU) program, College of Engineering, University of Texas at Arlington (UTA). This research is also supported by the National Science Foundation under Grant No. 1605472. Any opinions, findings, and conclusions or recommendations expressed in this material are those of the authors and do not necessarily reflect the views of the National Science Foundation.

Mineral carbonation as a design project for green chemical engineering education

H Fantucci[1] and R M Santos[1,*]

[1] School of Engineering, University of Guelph, Guelph, ON, N1G 2W1, Canada

*Email: santosr@uoguelph.ca

Abstract. Accelerated mineral carbonation is a promising CO_2 sequestration technology that is strongly linked to concepts of sustainability and Green Chemistry, and its process requirements apply principles of reaction kinetics, transport phenomena, and materials characterization. The present work aimed to develop educational tools for including accelerated mineral carbonation in chemical engineering curricula. To this end, an experimental investigation laboratory procedure and a design project outline have been conceived. As a way to further engage students in this learning experience, the process conditions for the laboratory work are varied between groups of students, and the experimental data obtained are pooled to be used by every group for the subsequent design exercise. This is meant to give students motivation to generate accurate data that they knew would be useful for the entire class and, at the same time, provide students with the opportunity to use data generated by colleagues, much in the same way the design work is done in the industry. In the design project, students use the experimental data obtained by themselves and classmates on the accelerated mineral carbonation of wollastonite, to determine if this is a feasible process for industry to sequester carbon dioxide, in view of mitigating climate change. Also, they use the experimental data, acquired using a range of process conditions, to study the effect of the process variables (CO_2 pressure and mixing rate) on the carbonation kinetics and mass transfer rate. The focus of our previously published article was on the experimental investigation, while the focus of this conference paper is on the design project.

Keywords: Chemical engineering, green design, green chemistry, carbon dioxide

1. Introduction

One of the biggest concerns with the accelerated mineral carbonation technology is that if it is implemented as an industrial process, it will not deliver net CO_2 sequestration. The reason behind this is that all process equipment in the industry requires energy to function, and, in most cases, that energy is supplied from CO_2-intensive processes. Even so-called "green energy" has a CO_2 footprint. Hydroelectric fares the best, at 4 g, CO_2-eq/kWhe, while wind and solar emit (over the life-cycle) 12 and 22 g, CO_2-eq/kWhe, respectively (CO_2-eq stands for CO_2-equivalent, and kWhe is the rate of electrical energy generation) [1]. At the other end of the spectrum, the median CO_2 intensity of coal power generation, the worst performer, is a whopping 1001 g, CO_2-eq/kWhe. Thus, if too much energy is used for accelerating mineral carbonation, the process may end up emitting more CO_2 than sequestering it.

Due to the energetic challenges posed, it is vital to conduct an energetic analysis of any proposed accelerated mineral carbonation process. For example, Santos et al. [2] found that, for an ultrasound-enhanced mineral carbonation process (at lab-scale), for every 2.3 g of CO_2 sequestered, 79.2 g would be emitted (based on the energy-mix of Belgium). They proposed some strategies to reduce the CO_2 emissions further at the industrial-scale, such as securing a less CO_2-intensive energy source, increasing the solids loading in the reactor (i.e., more solids per volume of liquid), and improving the design of the reactor to maximize the utilization of sonication energy. Still, a 100-factor reduction in CO_2-intensity would be necessary to achieve net CO_2 sequestration, so it is not easy and perhaps not feasible.

More recently, for a more conventional carbonation process, utilizing steelmaking slag as the carbon sink, Costa et al. [3] estimated an energy requirement of 272−1750 kWh/tonne CO_2 sequestered. Based on the aforementioned energy-intensity of wind power generation, this would represent 38 kg, CO_2 emitted per 1000 kg, CO_2 sequestered. Hence, this would be a feasible process for net CO_2 sequestration. It should be noted, however, that steelmaking slags require less intense carbonation conditions compared to wollastonite (as the material is more reactive). Unfortunately, there is far less steelmaking slag being produced than CO_2 being emitted by steelmaking, so this material alone cannot substantially reduce emissions from this CO_2-intensive industry. Thus, natural materials, such as wollastonite, more abundantly available (though more expensive due to mining and transportation costs), are required to curtail industrial emissions.

So the question is, can accelerated carbonation of wollastonite become a feasible industrial process for net CO_2 sequestration? This can be addressed in an educational setting through a design project, herein proposed. This project is suitable for implementation in upper-year undergraduate (Bachelor or Advanced Diploma) engineering and engineering technology programs, and also for graduate courses at the Master's level.

2. Design project procedure

The following paragraphs highlight the main steps in the design procedure.

First, assemble a process flowsheet for sequestering CO_2 using wollastonite. The process starts with mined, coarse grade (80% passing 8 mm) wollastonite, and CO_2 that has been captured from a flue gas source. Capturing means concentrating the CO_2 from an industrial source (typically 10-30% CO_2 by volume) into a pure (>99 vol%) CO_2 stream). You must take the energy consumption of the capturing step into account (see table A1, in the Appendix), but you can ignore the energy consumption of wollastonite mining and transport. The process must use the following equipment (listed in arbitrary order):

a. pressurized, jacket heated, continuously stirred tank reactor.
b. CO_2 multi-stage compressor.
c. slurry pump(s).
d. heat exchanger(s).
e. solid/liquid separator (unspecified type).
f. size reduction grinder/mill (unspecified type).

Second, determine what process conditions you will use. Choose the combination of process conditions, from the experiments conducted by all students, that yielded the highest carbonation conversion, and conversely the highest CO_2 uptake.

The third step is to determine the scale of the process. That is, how much CO_2 is to be sequestered. For this, ideally, you should choose a rate of CO_2 sequestration that matches the rate of CO_2 emissions from a particular industrial facility (e.g., coal power plant, steelmaking mill, cement factory, pulp and paper plant, to name a few). Your flowsheet will have a single reactor, which should have a feasible size, so choose an emissions source that will fit the scale of the process you are able to design, or propose to capture only a fraction of the emissions with the single reactor if the scale is too large. It is your task to research emissions values and make an appropriate design choice. Assume that all CO2 that is compressed will be sequestered (i.e., there are no gaseous emissions). Based on the amount of CO_2 to be sequestered (tonnes/day), and the maximal CO_2 uptake (tonnes,CO_2/tonne, wollastonite) you are able to achieve (from the previous step), you can determine the amount of wollastonite (tonnes/day) that you will need to process.

Fourth is step is that once you know the amount of wollastonite that you will need to process, you can determine slurry volume and consequently the size of the reactor. Remember that the reactor is a batch process, so there is only so much wollastonite slurry that can be processed per batch, and only so many batches that can be processed per day (based on filling, heating and emptying times). Use your lab experiment experience, and your batch reactor theory and design knowledge, to make a suitable determination of the reactor size. You will not need to size any of the other equipment for this design exercise; it will be assumed that each equipment's energy consumption depends only on throughput and not on physical size.

In the fifth step, calculate the energy consumption of each process/equipment using the equations given in Table A. Table A presents the equations needed, along with definitions and units of each equation parameter, and suggestive values for some parameters. Other parameter values will come from the process conditions used in the lab experiment (that gave the best carbonation conversion), such as the mixing rate (G, in s^{-1}). Other parameters will have to be researched or assumed (e.g., the dynamic viscosity (μ) of the slurry); use justifiable assumptions. You can neglect the slurry pumping power. Note that the units for each energy term differ (kWh/tonne, Wh, J, etc.); these will have to be converted to a common energy unit (kWh/tonne CO_2 sequestered) before being summed to obtain the total energy consumption value.

In the sixth step, for the slurry heating power, take into account how much exothermic energy is produced in the reactor during the reaction, how much energy is lost to the environment by natural convection, and how much waste heat may be re-used to pre-heat the feed using a heat-exchanger. Some additional assumptions will be required (e.g., heat-exchange efficiency). Based on this analysis, you only need to take into account in the overall energy balance what extra heating energy is required that will consume external power.

For the seventh and final step, once you have the total energy consumption value, per tonne CO_2 sequestered, determine if the process achieves net CO_2 sequestration. That is, convert the energy consumption value into a carbon emission value, using an assumed CO_2-intensity value of power generation (find this value for Ontario), and compare the rate of emission to the rate of sequestration. Assume that all energy requirement for the process comes from electrical power obtained from the provincial grid.

2.1. Reaction rate law

In the energetic analysis, you will use experimental data to decide what the maximal CO_2 uptake of wollastonite is. However, during experimentation, only a limited number of experiments were performed. So it may be that the maximal CO_2 uptake occurs with a CO_2 pressure different than the values tested (maybe even higher or lower than the range covered). It is thus helpful for design purposes to obtain a rate law that describes how CO_2 pressure affects the carbonation rate.

In the Experimental Investigation [4], a rate law proposed by Ptáček et al. [5] was introduced, which accounted for the effect of temperature on the rate of carbonation. That law, which is based on the well-known Arrhenius equation, required experimental determination of two empirical parameters: the pre-exponential factor (A), and the activation energy (EA). Values for these two parameters, suggested by Ptáček et al., were cited: $1.8 \pm 0.9 \cdot 103\ s^{-1}$ for A in the case of wollastonite dissolution in acetic acid, and 47 ± 1 kJ/mol for EA. It should be noted, however, that these values were obtained from experimental studies on the dissolution of wollastonite in acetic acid, rather than its dissolution and immediate precipitation, as it occurs when wollastonite is exposed to carbonic acid (i.e., CO_2 at elevated pressure in contact with an aqueous slurry of wollastonite). It should also be noted that the value of EA obtained experimentally is "apparent", meaning that it is a measure value based on a specific set of process conditions, rather than the true activation energy based on fundamental principles of chemistry. Thus, it is plausible that it, in addition to A, would differ for substantially different process conditions (such as the dissolution of wollastonite in carbonic acid).

To take into account the effect of CO_2 pressure, in addition to temperature, a possibly suitable rate law is an extended version of the Arrhenius equation, as follows:

$$r_{mincarb} = A \cdot e^{-\frac{E_A}{R \cdot T}} \cdot (P_{CO2})^y \quad (1)$$

where $r_{mincarb}$ is the mineral carbonation rate (mol/(m^2·s)), A is the pre-exponential empirical factor (mol/(m^2·s)), E_A is the apparent activation energy of carbonation (kJ/mol), R is the universal gas constant (kJ/(mol·K)), T is the temperature (K), P_{CO2} is the partial pressure of CO_2 (bar), and y is the order of reaction.

Also recall that the rate of carbonation (R_{carb} (mol/s)) is proportional to the available mineral surface area ($A_{min,(t)}$ (m^2)) (Eq. 2), which itself is a function of time (Eq. 3). $\xi_{(t)}$ is the fractional extent of carbonation conversion of the mineral at the reaction time and $m_{wo,init}$ is the initial mass of wollastonite being carbonated (g), and SSA is the initial specific surface area (m^2/g).

$$\mathrm{R_{carb}} = \mathrm{r_{mincarb}} \cdot \mathrm{A_{min,(t)}} \quad (2)$$

$$\mathrm{A_{min,(t)}} = \mathrm{SSA} \cdot (1 - \xi_{(t)}) \cdot \mathrm{m_{wo,init}} \quad (3)$$

Hence, the first step in this analysis will be to convert the experimental carbonation conversions ($\xi_{(30min)}$) and rates of carbonation (R_{carb}) into reaction rates ($r_{mincarb}$). The value of R_{carb} will be calculated from the total CO_2 uptake value, and it will be assumed that the rate of carbonation is linear in the first 30 minutes (this is in agreement with the work of Huijgen et al. [6]).

Once values of $r_{mincarb}$ are obtained for every experiment (with varying CO_2 pressures and mixing rates), the rate law will be fitted to the data to account for the effect of pressure. For this, it is best to divide the data into sets of constant mixing rate, and for each set, fit the data based on variation of pressure alone. Fitting of the data should be done by plotting experimental and calculated values of rmincarb versus P_{CO2}, and adjusting parameters to minimize the difference between the lines. Having three unknowns (A, E_A and y), and no data on the effect of temperature, you will need to use some discretion in adjusting these parameters to fit the available data. For example, you may wish to use the values of Ptáček et al. [5] for A and E_A at first, solve for y, and then adjust A or E_A to better fit the experimental data trend. Repeat this for each mixing rate set to find out if the mixing rate has a large effect on the rate law, and if so on which rate parameter.

2.2. Mass transfer analysis

Mass transfer of ions in the aqueous mineral carbonation system occurs in two mediums: the solid phase and the liquid phase. In the solid phase, ion transport is governed by the diffusivity of the ions through the mineral matrix, with porosity and ion-solid interactions playing a role in aiding or restricting ion transport compared to the simple diffusion mechanism. In the liquid phase, mixing plays an important role in inducing convection in the bulk phase, in aiding the dissolution of CO_2 from the gaseous phase (which is over the liquid surface), and in reducing the thickness of the stagnant liquid film at the surface of solid particles. This increases the driving force for the transport of ions from and to the solid surface.

In this mass transfer analysis, you will focus on the effect of mixing rate on the rate of carbonation using the experimental data set you are given. In the experimental study, the mixing rate was varied along with CO_2 partial pressure. By dividing the data set into sub-sets, you can isolate the effect of mixing rate for a given CO_2 partial pressure. One question to address is if mixing can compensate for a lower CO_2 partial pressure, as the energy consumption of mixing is expected to be lower than the energy consumption of compressing CO_2 (is this what the energetic analysis indicated?). Another question to address is what the rate limiting step of the reaction is: mineral dissolution, ion diffusion

IOP Conf. Series: Materials Science and Engineering 1196 (2021) 012009 doi:10.1088/1757-899X/1196/1/012009

through the solid phase, or ion diffusion in the liquid phase (either from the solid phase or from the gaseous phase)?

To investigate the effect of the mixing rate, you will look at the Sherwood Number (*Sh*), which indicates if diffusion or convection dominates the mass transfer in the liquid phase. A high value of *Sh* implies that there is sufficient mixing to overcome diffusion limitations. But if the carbonation rate does not improve markedly with higher rate of mixing, it will indicate that the rate limiting step lies elsewhere. In that case, an assessment of the rate of diffusion through the solid phase, and comparison of that rate with the reaction rate, would further indicate if mineral dissolution or solid-phase mass transfer is rate-limiting.

The correlation between mass transfer and agitation is made by the Sherwood Number (*Sh*), which is a dimensionless number that represents the ratio of the convective rate of mass transfer to the rate of diffusive mass transport (Eq. 4) [7]. The value of *Sh* has been found to be a function of two additional dimensionless numbers common in the field of transport phenomena (Eq. 5): the Reynolds Number (*Re*), defined in Eq. 6 for an impeller-stirred tank; and the Schmidt Number (*Sc*), defined in Eq. 7.

$$Sh = \frac{K}{L/D} \tag{4}$$

$$Sh = f(\mathrm{Re}, Sc) \tag{5}$$

$$\mathrm{Re} = \frac{\rho_L \cdot N \cdot D_I^2}{\mu_L} \tag{6}$$

$$Sc = \frac{\mu_L}{\rho_L \cdot D_L} \tag{7}$$

where L is a characteristic length (m), which in the case of particles is the particle diameter [7], D is mass diffusivity ($m^2 \cdot s^{-1}$), K is the convective mass transfer film coefficient ($m \cdot s^{-1}$), ρ_L is the density of the liquid (kg/m^3), N is the rotational speed of the impeller (s^{-1}), D_I is the diameter of the impeller (m), μ_L is the dynamic viscosity of the liquid (Pa·s), and D_L is the diffusion coefficient of the chosen species in the liquid phase (m^2/s).

The actual value of *Sh* must be determined using empirical correlations. Many experimental studies of mass transfer from a particle to a fluid exist, some of which concern the dissolution of solid particles. The choice of a suitable equation should be based on the one that best fits experimental observations for the system in question. In the absence of prior experience, an equation that has been found to fit a variety of data reasonable well is the Frössling Equation (Eq. 8), which is valid for a wide range of *Sc* and *Re* up to 1000 [8].

$$Sh = 2 + 0.6 \cdot \mathrm{Re}^{1/2} \cdot Sc^{1/3} \tag{8}$$

To calculate *Sh*, you may start with the Frössling Equation, which requires calculation of *Re* and *Sc*. Confirm if the range of *Re*'s used matches the validity range of the equation. If not, search the literature for a more appropriate equation. Once you have determined values of *Sh* for the different mixing rates, and for the difference CO_2 partial pressure series, comment on the results. Can you tell what the rate limiting step is, or is further investigation needed? Does the effect of mixing differ at different CO_2 partial pressures, and if so why? How can this mass transfer analysis aid in process design to reduce the energy demand of the process? Are there other benefits that mixing may bring that were not assessed? Are there other mixing methods or reactor designs that could improve mixing or reduce the energy demand of mixing?

3. Report requirements

1. The report should contain the following sections: cover page, table of contents, table of figures, table of tables, energetic analysis (as per below), reaction rate law (as per below), mass transfer analysis (as per below), conclusions and recommendations (as per below), references, appendix (as per below).
2. Energetic analysis: (a) explain how you adapted the proposed procedure to reach the energy consumption and net CO_2 sequestration results (what assumptions had to be made, what changes you made, if any, to the proposed procedure, etc.); (b) show values (and sources) of all parameters used in the equations; (c) show the energy consumption value of each individual process/equipment; (d) show the calculation procedure of converting energy consumption to CO_2 emissions (footprint) and the compare the CO_2 emissions rate to the CO_2 sequestration rate to determine if net CO_2 sequestration is achieved; (e) discuss what process changes you may make, if more experimental or modelling information, to improve the efficiency of the process; (f) write conclusions and recommendations on whether accelerated mineral carbonation can become an industrially feasible process for CO_2 sequestration.
3. Reaction rate law: (a) explain the procedure used to develop the rate law(s), including data and graphs used; (b) explain assumptions taken and any limitation of the applicability of the rate law(s); (c) comment on what you learned about the mineral carbonation reaction based on the rate law(s) obtained; (d) comment on how the rate law(s) can aid in improving the process design and energy consumption presented earlier.
4. Mass transfer analysis: (a) explain the procedure used in the analysis, including data and graphs used; (b) explain assumptions taken and any limitation on validity of the analysis results; (c) comment on what you learned about the mineral carbonation reaction based on the results obtained; (d) comment on how the analysis can aid in improving the process design and energy consumption presented earlier.
5. Appendix: include any graph, figure or table that supports the design and analysis results, but that is not central to the report discussions. Add captions for each appendix entry.

4. Conclusions

The ultimate objective of the proposed design project is to better train chemical engineering students given their future careers. Graduates with experience in mineral carbonation investigation and design would potentially be better trained to develop engineering solutions that take into account sustainability criteria. Also, having gone through a memorable undergraduate exercise in mineral carbonation investigation and design would potentially reinforce learned concepts of chemical engineering fundamentals, and make it more likely that graduate would recall those fundamental concepts in their future careers.

5. References

[1] Moomaw W, Burgherr P, Heath G, Lenzen M, Nyboer J and Verbruggen A 2011 *Annex II: Methodology (IPCC Special Report on Renewable Energy Sources and Climate Change Mitigation)* ed Edenhofer O, Pichs-Madruga R, Sokona Y, Seyboth K, Matschoss P, Kadner S, Zwickel T, Eickemeier P, Hansen G, Schlömer S, *et al.* (Cambridge: Cambridge University Press) **pp** 973-1000

[2] Santos R M, Francois D, Mertens G, Elsen J and Van Gerven T 2013 *Appl. Therm. Eng.* **57**, 154-163

[3] Costa G, Polettini A, Pomi R, Stramazzo A and Zingaretti D 2017 *Greenh. Gases Sci. Technol.* **7**, 530-541

[4] Fantucci H, Sidhu J S and Santos R M 2019 *Sustainability* **11**, 4156

[5] Ptáček P, Nosková M, Brandštetr J, Šoukal F and Opravil T 2011 *Powder Technol.* **206**, 338-344
[6] Huijgen W J J, Witkamp G J and Comans R N J 2006 *Chem. Eng. Sci.* **61**, 4242-4251
[7] Heldman D R 2003 *Encyclopedia of Agricultural, Food, and Biological Engineering* (New York: Marcel Dekker) **p** 1184
[8] Rousseau R W 1987 *Handbook of Separation Process Technology* (New York:Wiley-Interscience) **p** 1024
[9] Perry R and Green D 2006 *Perry's Chemical Engineers' Handbook* (New York: McGraw Hill)
[10] Zingaretti D, Costa G and Baciocchi R 2014 *Ind. Eng. Chem. Res.* **53**, 9311-9324
[11] McCollum D L and Ogden J M 2006 *Techno-economic models for carbon dioxide compression, transport, and storage & correlations for estimating carbon dioxide density and viscosity* (Davis: UC Davis) **p** 88
[12] Air Liquide 2021 *Gas Encyclopedia*. Available online: http://encyclopedia.airliquide.com/encyclopedia.asp?GasID=26 (accessed on 02 August 2021)
[13] Robie R A and Hemingway B S 1995 *Bulletin* vol 2131 (Washington: U.S. G.P.O) **p** 461
[14] Intergovernmental Panel on Climate Change (IPCC) 2005 Mineral carbonation and industrial uses of carbon dioxide *Special Report on Carbon Dioxide Capture and Storage* (Cambridge: Cambridge University Press) chapter 7 **pp** 319-338

6. Appendix

Table A. Assumptions, mathematical equations and values of parameters adopted for the energetic evaluation (adapted from Costa et al. [3]). Re-used from Fantucci et al. [4], CC BY.

Unit operation	Assumptions and equations adopted	Definition and units of symbols and some numerical value suggestions
Size reduction	$E_{SR} = 10 \cdot W_i \cdot \left(\frac{1}{\sqrt{d_{fin}}} - \frac{1}{\sqrt{d_{in}}}\right)\frac{10.3 + d_{fin}}{1.145 \cdot d_{fin}}$	E_{SR} = milling energy (kWh/tonne,woll) W_i = Bond's work index (= 12.16 kWh/tonne) [9] d_{in} = initial particle size (80% passing) (μm) d_{fin} = final particle size (80% passing) (μm)
Reactor mixing	$E_{RM} = G^2 \times m \times V_{slurry} \times J_H$	E_{RM} = mixing energy (Wh) V_{slurry} = slurry volume (m^3) G = velocity gradient (s^{-1}) m = dynamic viscosity of the slurry (Pa·s) J_H = hydraulic residence time (h)
Slurry heating	$E_{HE} = V_{slurry} \times r_{slurry} \times c_{p,slurry} \times DT$	E_{HE} = heating energy (J) ρ_{slurry} = slurry density (kg/m^3) $c_{p,slurry}$ = specific heat of the slurry ($J \cdot kg^{-1} \cdot K^{-1}$) (for wollastonite $c_{p,woll} = 1.1125 \cdot 10^2 + 1.4373 \cdot 10^{-2} \cdot T + 16.936 \cdot T^{-0.5} - 2.7779 \cdot 10^6 \cdot T^{-2}$ ($J \cdot mol^{-1} \cdot K^{-1}$) [10] DT = temperature difference (assumed to be 30 K [10], the remaining heat to be supplied being recovered from the slurry recycled from the carbonation reactor)
CO_2 compression	Multi-stage compressor used (maximum number of stages = 5) [11]. For each stage: $E_{c,i} = \frac{Z \cdot R \cdot T_{in}}{M \cdot \eta_{is}} \cdot \frac{k_s}{k_s - 1} \cdot \left(CR^{\frac{k_s - 1}{k_s}} - 1\right)$ $E_c = \sum_i E_{c,i}$ (note: $P < P_{crit,CO2} = 7.38$ MPa)	$E_{c,i}$ = compression energy per stage (J/kg,CO_2) E_c = total compression energy (J/kg,CO_2) CR = compression ratio = $(P_{fin}/P_{in})^{1/N}$ P_{fin} = final pressure (end of stage) (Pa) P_{in} = initial pressure (start of stage) (Pa) N = number of compression stages Z = average CO_2 compressibility factor for each stage (e.g. Z = 0.99435 at 1.013 bar and 15 °C) [12] R = universal gas constant (8.314 $J \cdot mol^{-1} \cdot K^{-1}$) Tin = CO_2 temperature at the compressor inlet (K) M = molecular weight of CO_2 (0.044 $kg \cdot mol^{-1}$)) η_{is} = isentropic efficiency of the compressor k_S = (Cp/Cv) = average ratio of the specific heats of CO_2 for each stage.
Heat released by the reaction	Calculated from the enthalpy of reaction of wollastonite with CO_2 [13].	DH_{carb} = −86.99 kJ/mol,CO_2
Heat loss from the carbonation reactor	Heat transfer assumed by natural convection on the exterior ambient-air side. $\dot{Q}_{conv} = h \cdot A_s \cdot (T_s - T_\infty)$	$\dot{Q}_{conv}$ = heat flux from the reactor (W) h = average heat transfer coefficient ($W \cdot m^{-1} \cdot k^{-1}$) (calculated from Nusselt number expression for natural convection from reactor surface) A_s = surface area of the reactor (m^2) T_s = temperature at the outer surface of reactor (K) T_∞ = ambient temperature (K)
Solid/liquid separation	$E_{RM} = G^2 \times m \times V_{slurry} \times J_H$	Same as for rapid mixing, except assume $G = 60\ s^{-1}$
CO_2 capture [14]	$E_{Capt} = 4000\ kJ/kg\ CO_2\ captured$	Amount of energy required to capture (concentrate) CO_2 from emission source.

IOP Conf. Series: Materials Science and Engineering 1196 (2021) 012010 doi:10.1088/1757-899X/1196/1/012010

How to achieve the sustainability of the seafood sector in the European Atlantic Area?

I Ruiz-Salmón*, J Laso, C Campos, A Fernández-Ríos, D Hoehn, M Margallo, A Irabien and R Aldaco

[1] Department of Chemical and Biomolecular Engineering, University of Cantabria, Avda. de los Castros s/n, 39005, Santander, Spain.

* **E-mail:** Israel.ruizsalmon@unican.es

Abstract. Climate change, globalization or marine debris are on the spot of concerns for the most society. Particularly, fisheries are impacted by these and other issues. On the framework of the European Atlantic area, NEPTUNUS project (EAPA_576_2018) tries to provide opportunities for the transition to the circular economy of the seafood and aquaculture sectors by means of a consistent methodology for products eco-labelling and defining eco-innovation strategies. Furthermore, this project will provide key actions for resource efficiency based on life cycle thinking and the nexus water-energy-food, incorporating producers, policy makers and consumers in the decision-making process. This review addresses, therefore, the threats and challenges of the current Atlantic fisheries, the methodologies and actions to be face them and the expected results of the NEPTUNUS project

Keywords: sustainability, life cycle assessment, eco-label, seafood, climate change.

1. Introduction

Atlantic countries share commercial alliances and common interests in the production and consumption of food and product safety, transparency and the protection of consumer health must be guaranteed [1]. Particularly, in the case of seafood, the Atlantic area has abundant natural resources, which places it among the main exporters of seafood resources serving the European market. Indeed, 'Blue economy', in which fisheries and related activities are included, was directly employing over 4 million people in the European Union (EU) in 2017, generating €180 billion of gross value added [2]. The enormous contribution of this sector to the EU and global economy implies a high responsibility that needs to be articulated through tangible midland long-term actions. It jointly addresses a global concern and interest in terms of policies and strategies aimed at climate change mitigation, energy or food security. Recently, global institutions, such as the United Nations, highlighted the responsible consumption and production as one of the seventeen Sustainable Development Goal (SDG) [3]. Furthermore, the FAO is largely concerned about the state of the world seafood sector [4] or the specific impacts of climate change in Atlantic fisheries [5]. At European scale, the EU is working on the implementation of the Circular Economy [6] by means of specific regulation in fisheries [7]) or facilitating better information on the environmental performance of products and organisations [8], among others.

Although the vast majority of national and regional studies in these domains generate significant technical data, yet they fail to define weaknesses and strengths of transnational synergies (nexus or model) in infrastructure, innovation and skills that promote and maximise regional development and relevance to commensurate cross-jurisdiction needs from policy maker, industry and consumer approaches.

To address the challenges, sustainable and multilateral research cooperation is needed to define integrated methodologies and strategies. As part of the European Union's Cohesion Policy, some funding programmes, such as INTERREG Atlantic Area [9], promotes transnational cooperation among almost forty European Atlantic regions leading to support the development of renewable sources of energy, especially those deriving from the sea and fostering the transition to a resource-efficient society. Under the umbrella of this programme, NEPTUNUS project (EAPA_576/2018) [10] focuses on the idea of providing opportunities for a transition to the circular economy of the seafood sector in the Atlantic Area.

This review addresses the scope and the methodological framework of NEPTUNUS to face the current status of the seafood and aquaculture sectors in the European Atlantic region: environmental threats, e.g. marine litter [11] or climate change [12]; social aspects, e.g. people's awareness about public services and the planet [13] or worker`s rights protection [14]; and economic perspectives, e.g. globalization [15] or quality of products [16].

2. Research methodology

Clustering and knowledge transferring require specific and coordinated actions in order to obtain applicable and useful outputs in terms of sustainability. To make this, "nexus thinking" is applied to fisheries and activities related to transitioning from a linear economy to a circular one. The methodological framework implies, therefore, (i) the integration of environmental, nutritional and economic variables that meet regional needs through transnational strategies, and (ii) synergies in knowledge and experiences at the local level to help overcome challenges at a global level. Figure 1 illustrates the five interconnected research lines to be developed in NEPTUNUS:

IOP Conf. Series: Materials Science and Engineering 1196 (2021) 012010 doi:10.1088/1757-899X/1196/1/012010

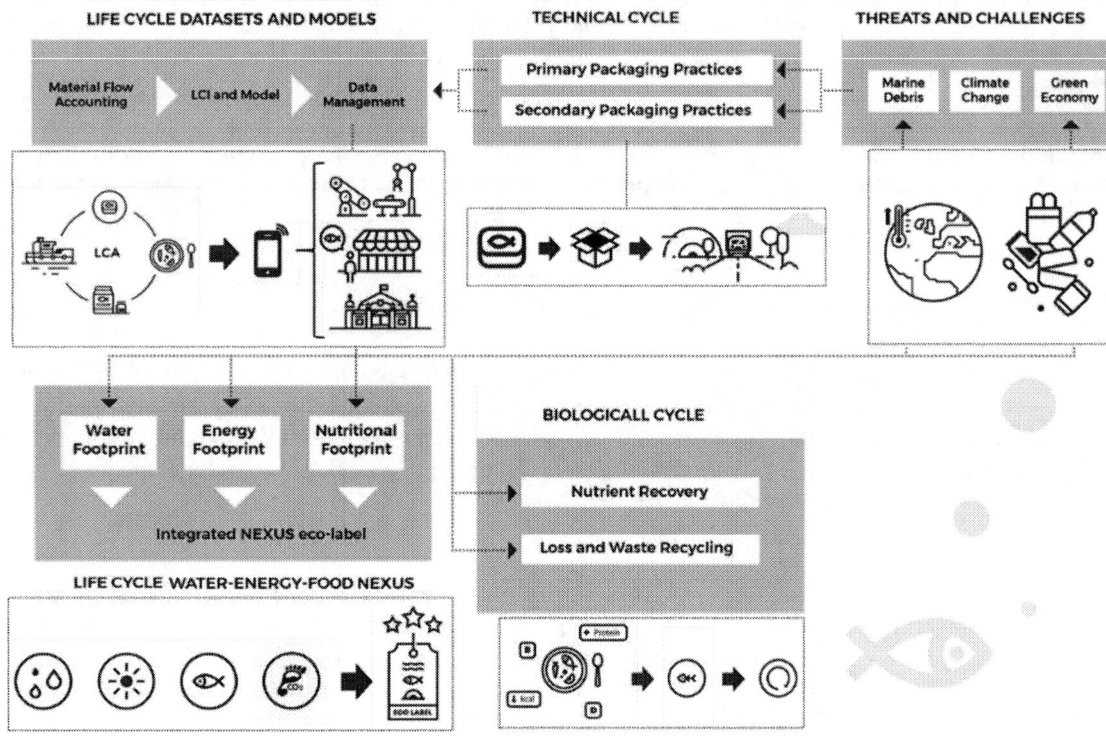

Figure 1. Methodological framework of NEPTUNUS project.

2.1. *Life cycle inventory*

Life cycle assessment (LCA) [17, 18] will be applied to the seafood sector to quantify main inputs and outputs of the product system in a life cycle inventory (LCI) according to European Life Cycle Database [8]. The model has to be as flexible as possible in order to create a user-friendly tool for assisting seafood produced, municipalities, communities and regions of the Atlantic area to easily obtaining LCA results on seafood production and consumption. Moreover, the tool has to integrate the imports/exports of seafood from/to other areas that will have an important environmental impact. The analysis of this variable can be conducted by material flow accounting (MFA) [19]. MFA will allow to address interactions between socioeconomics and biophysical systems. Thus, materials and energy inputs can be accounted in order to determine sustainability from an urban metabolic approach, as well as waste flows can be assessed, and therefore, sustainability by means of closing materials flows. Regarding the species, both marine captures –demersal fish, pelagic fish and molluscs- and aquaculture production –diadromous and mollusks- of the Atlantic region are considered.

2.2. *Lifecycle Water-Energy-Food Nexus*

Demanding transformations and the policy tools required, including behavioural change interventions that go beyond mere education to influence how people make decisions about buying and consuming marine products are also part of the research. A new approach to assess seafood lifecycle is proposed based on the nexus of water-energy-food systems. "Nexus" means that the action in one of the systems has impacts on the others and, therefore, any strategy that focuses on one system without considering its connections with other systems may lead to acute unpremeditated consequences. This concept allows assessing the life cycle of seafood products under a holistic manner considering the whole supply chain. Currently, there is no universally recognised methodology for nexus analysis. However, LCA is

IOP Conf. Series: Materials Science and Engineering 1196 (2021) 012010 doi:10.1088/1757-899X/1196/1/012010

particularly important for understanding the interconnections in the nexus, as it enables the consideration of entire supply chains.

Four footprints [20] and an eco-label will be calculated and designed. Firstly, the impact of the water use, based on the Footprint Category Rules Guidance [21], covering the blue water, i.e. the amount of surface and groundwater required to produce seafood products, green water, i.e. the amount of rainwater, and grey water, i.e. the amount of fresh water required diluting the wastewater generated in manufacturing, in order to maintain water quality determined by European standards. Secondly, the energy is the largest overhead cost associated with seafood processing. In this sense, FAO has recently recognised the need to reduce energy consumption and efficiency in food processes with a large demand of fossil fuels for use, transportation, storage and refrigeration [22]. The challenge is to identify bottlenecks to target actions for energy improvements, which can result in huge cost savings. Thirdly, the Carbon Footprint can be used to target hotspots in seafood supply chain establishing a link among fossil fuel consumption, carbon sequestration and greenhouse gas releases. Fourthly, the nutritional aspects. This action will conduct a global nutritional assessment, estimating the Protein Footprint based on the protein content of each ingredient involved in the seafood supply chain. Finally, the definition of an integrated nexus eco-label is considered, leading to facilitate the decision-making process to stakeholders and to allow the development of strategies based on the circular economy concept.

2.3. Eco-design and packaging

Better product design is key to facilitate recycling and to increase products' lifetime, while saving resources. Source reduction, recycling and reuse are the ways to implement practically the eco-design of packaging without sacrificing product safety [23, 24]. In this sense, the importance of applying a holistic approach to packaging eco-design clearly has been emerged. Recyclability is highly desirable, but other packaging features should also be considered. One of the most important of these is ensuring that products reach consumers undamaged and in good condition, and hence that the raw materials and energy required for their production are not wasted. Moreover, the packaging of a product is often the best way to visually engage and attract potential consumers. Consequently, by infusing eco-friendly methods into the design, retailers are able to creatively distinguish themselves apart from other less eco-conscious brands and promote innovative products from the Atlantic region. Since seafood is highly perishable and sensitive to harmful microbial growth, effective primary and secondary packaging is critical. Therefore, eco-design applied to the packaging is a source of added value for the Atlantic seafood food products.

The need and development of packaging is caused by the delocalisation between production and the consumption, and the required distribution and transportation of the produced goods. In the future, innovation and new packaging technologies will play a key role in the prevention of food waste [25]. Primary packaging eco-design strategies will be defined to mitigate climate change and to adapt the current type of containers validating their effectiveness preserving seafood. In the current level of global market, food distribution is crucial and the role of packaging is enormous. There are a number of factors to consider in choosing the appropriate secondary packaging solution. In addition to protective features, economic feasibility and ease of transport, environmental variable should be balancing these interests.

2.4. Circular economy strategies

A circular economy approach must be managed from a holistic point of view, i.e. from natural resources consumption to waste management. For the transition to a circular economy, the reuse of natural resources should be favoured. An approach in which residue streams are prevented as much as possible, and where unavoidable residues are viewed as useful resources, could provide a new impulse for tackling the old issue of seafood waste.

In this regard, it is necessary to link the potential high value use of residue streams with food safety. Therefore, in addition to nutrients identification, a full nutritional characterisation should be introduced. Based on LCA, techniques that produce several nutrient rich products (used to replace synthetic fertilisers) reducing CO_2 and N_2O emissions and increasing the economic revenue will be assessed. For instance, anaerobic digestion has further benefits in terms of providing by-products such as biogas [26]. The objective is to complement the reference European Best Available Techniques document [27] with recommendations based on food circular economy and the analysis will include the prevention and reuse options.

On the other hand, the European Energy Policy [28] aims at encouraging sustainable energy use and, consequently, waste to energy options should be considered. These residue streams, however, could be used more profitably as food, animal feed [29] or as resources for biorefining [30], depending on the raw material that is fermented. Waste management is often overlooked in seafood processing as a potential area for resources savings. The simplest way to foster an efficient use of resources is to reduce the amount of waste. Innovative reuse and recycling can transform wastes into valuable by-products. Three categories including biological digestion, non-biological volume/weight reduction, and thermal processing, are considered.

2.5. Threats and challenges fisheries in the Atlantic area

The main challenges of the fisheries in the Atlantic area are related to European circular economy transition, the adaptation to the climate change, and the growing threats of the marine debris. Climate change impacts in general on fisheries, including operational aspects at the harvesting stage [31], the likely impact along the supply change and particularly at the production stage [5, 32] but also at structure levels and policy development. Therefore, to mitigate climate effects on fisheries, actions are needed along the supply chain. The Atlantic area could be an adequate framework to apply this approach in order to propose regional strategies of climate change adaptation.

Furthermore, the seafood sector has to lead with the threat of marine debris. From interfering with food webs, ghost fishing and transferring toxins up the food chain, this marine debris significantly interferes in ecosystems and human health [33], and it has many socioeconomic impacts. The eradication of marine debris is essential to the wellbeing of our current and future generations, and the characteristics of marine debris as an environmental problem require an innovative approach to its eradication. Derelict fishing gear is particularly harmful as it often results in ghost fishing [34]. Therefore, a number of strategies or policies for tackling the problem of ghost nets are necessary to introduce in the Atlantic area.

3. Expected outputs

Expected results of NEPTUNUS project will be a reality if the nexus approach and the specific methodology and actions are correctly carried out, opening the opportunity to contribute from an environmental, economic and social point of view to a Europe Circular Economy Strategy. With the upcoming "circularity revolution" in the European and global arena, it is of vital importance to establish new methodologies and strategies, which will promote sustainable seafood production and consumption:

(i) Environmental point of view: The integrated approach of the project as a bidirectional challenge, considering production and consumption, will allow a better understanding of flows, obstacles and innovations that would seem to be necessary to promote the minimisation and recovery of food waste, contributing to the sustainable use of natural resources. The achievement of reduction targets in Atlantic Area needs an integrated approach based on circular economy. This perspective, in which residue streams are prevented as much as possible and where unavoidable residues are viewed as useful resources, will provide a new impulse for tackling the old issue of waste. Some expected effects from the environmental point of view include: an eco-efficiency gain, and the mitigation of climate change.

(ii) Economical point of view: The seafood value chains are of significant importance in many areas of the European Atlantic Area. A noteworthy socioeconomic benefit of a sound seafood value chain ensures employment security in less populated maritime areas. The focus in the present proposal could contribute to novel interests for new fishery production, important employment opportunities and value creation. The «closing the loop» initiatives of seafood will enable to address the economic needs of some specific seafood sectors, contributing to strengthen the fishery industry in the Atlantic Area. The consideration of the economic perspective will make possible to trigger investments; to increase business activity/capacity (new products, processes or services based on the pilots); to save cost and improve services (% savings and increasing users' satisfaction); and to rise jobs/employability (jobs and employment opportunities created).

(iii) Social point of view: Within a global context of increase population growth, reducing food losses and waste is necessary to meet forecasted nutritional needs. Therefore, the transition to a food circular economy will have significant societal relevance, being the central drivers: 1) the need for an environmentally, economically and societally sustainable development, in which foods and food waste take a central place; 2) the global issue linked to climate change and the need to reduce atmospheric greenhouse gases emissions; 3) the need to stimulate small regional maritime areas; and, 4) the main challenge of addressing food security on a global scale. This is especially noticeable under the seafood perspective, and taking into account healthy considerations. Specifically, throughout the NEPTUNUS project is expected a cluster development (increasing the durability of cluster connections, increasing the cluster size, and increasing the integration of activities among the partners).

4. Conclusions

The success of a cluster network under the umbrella of NEPTUNUS is expected but it strongly depends on the increment of the awareness and public acceptance based on the undoubtable positive aspects of collaboration and common development, among others. Nevertheless, there are some limitations to take into account for near future.

As a reminder, individual regions or countries working on their own cannot efficiently solve the environmental problems of the seafood supply chain in the Atlantic area. This could be either an issue affecting a clearly defined transnational geographical area or a common issue of interest, which transnational cooperation leads to more innovative and efficient solutions. Thus, in a global food economy, investments and new alliances between regions will be necessary, as well as balancing environmental influences with social and economic gain in policy and decision making for regional and global food flow. Therefore, the association requires including stakeholders from the whole society: from policy-makers and public bodies to citizens and not-for-profit organizations, as well as private companies.

However, the clustering beyond frontiers still results insufficient because only gives an answer from an administrative or technical point of view. But, what about the content and how to effectively deal with the environmental threats? Transitioning to a circular economy in a seafood context requires the "nexus thinking" aforementioned. This is essentially a transformative approach to governance, and also requires substantial changes in the way people behave. Of course, government policy informed by stakeholder engagement will be necessary to overcome obstacles that includes influencing consumer behaviour and change to embrace the importance of natural resources.

Accordingly, NEPTUNUS project addresses several overarching yet interrelated innovation actions, including the seafood flows, methodologies and tasks for a circular economy on a larger transnational scale that has not be pursed previously, this is, a robust framework that unites stakeholders and addresses gaps and barriers in terms of managing important interacting forces regionally and transnationally. Nevertheless, despite the pioneering scope of the project and the opportunity to extrapolate future results to other regions, seafood flows and related actions are limited from a territorial point of view that should be overcome in further researches.

References

[1] Schröder U 2008. Challenges in the traceability of seafood. *Journal für Verbraucherschutz und Lebensmittelsicherheit*, **3**(1), 45-48.

[2] European Commission 2019. *The EU Blue Economy Report 2019*. Luxembourg: European Commission.

[3] United Nations 2019. *Goal 12: Ensure sustainable consumption and production patterns*. Retrieved from https://www.un.org/sustainabledevelopment/sustainable-consumption-production/

[4] FAO 2018. *The State of World Fisheries and Aquaculture 2018* - Meeting the sustainable development goals.

[5] Peck M and Pinnegar J K 2019. Climate change impacts, vulnerabilities and adaptations: North Atlantic and Atlantic Arctic marine fisheries. *Impacts of climate change on fisheries and aquaculture*, **87**.

[6] European Commission 2019 *Communication from the Commission to the European Parliament, the Council, the European Economic and Social Committee and the Committee of the Regions: On the implementation of the Circular Economy Action Plan, COM/2019/0090*. Retrieve from: https://ec.europa.eu/commission/sites/beta-political/files/report_implementation_circular_economy_action_plan.pdf

[7] European Commission 2018 *Communication from the Commission to the Council and the European Parliament: delegation of powers referred to in Article 11(2), Article 15(2), (3), (6), (7) and Article 45(4) of Regulation (EU) No 1380/2013 of the European Parliament and of the Council of 11 December 2013 on the Common Fisheries Policy. COM/2018/0079*. Retrieve from: https://eur-lex.europa.eu/legal-content/EN/TXT/PDF/?uri=CELEX:52018DC0079&from=EN

[8] European Commission 2013 *Communication from the Commission to the European Parliament and the Council: Building the Single Market for Green Products. Facilitating better information on the environmental performance of products and organisations, COM/2013/0196*. Retrieve from: https://eur-lex.europa.eu/LexUriServ/LexUriServ.do?uri=COM:2013:0196:FIN:EN:PDF

[9] Interreg AA 2019. What is Interreg Atlantic Area? Retrieve from: https://www.atlanticarea.eu/page/2

[10] Neptunus project 2019 Overall objective of Neptunus project. Retrieve from: https://neptunus-project.eu/

[11] Sonnemann G and Valdivia S 2017 Medellin Declaration on Marine Litter in Life Cycle Assessment and Management. *The International Journal of Life Cycle Assessment*, **22**(10), 1637-1639.

[12] Hollowed A B, Barange M, Garçon V, Ito S I, Link J S, Aricò S, Batchelder H, Brown R, Griffis R and Wawrzynski W. 2019. Recent advances in understanding the effects of climate change on the world's oceans. *ICES Journal of Marine Science*, **76**(5), 1215–1220.

[13] Steffen W *et al.* 2015. Planetary boundaries: Guiding human development on a changing planet. *Science*, **347**(6223), 1259855.

[14] Lillie N 2008 *The ILO Maritime Labour Convention, 2006: A new paradigm for global labour rights implementation. Cross border social dialogue and agreements: An emerging global industrial relations framework?* Papadakis K. Geneva, International Labour Office, United Nations **191**.

[15] Anderson J L, Asche F and Garlock T 2018. Globalization and commoditization: The transformation of the seafood market. *Journal of Commodity Markets* **12** 2-8.

[16] Mansfield B 2003 Spatializing globalization: a "geography of quality" in the seafood industry. *Economic Geography*, **79**(1), 1-16.

[17] International Organization for Standardization 2006 Environmental Management: Life Cycle Assessment; Principles and Framework (No. 2006). *ISO*.

[18] International Organization for Standardization 2006. Environmental Management: Life Cycle Assessment: Requirements and Guidelines (No. 2006). *ISO*.

[19] Hinterberger F, Giljum S and Hammer M 2003. Material flow accounting and analysis (MFA). A valuable tool for analyses of nature-society interrelationships. *SERI Background Paper* **2**.

[20] European Commission 2019 *Suggestions for updating the product environmental footprint (PEF) method.* Joint Research Centre. Retrieve from: https://publications.jrc.ec.europa.eu/repository/bitstream/JRC115959/jrc115959_pef_method-online.pdf

[21] European Commission 2018 *Product Environmental Footprint Category Rules Guidance (Version 6.3 - May 2018).* Retrieve from: https://eplca.jrc.ec.europa.eu/permalink/PEFCR_guidance_v6.3-2.pdf
[22] FAO 2012 *Energy-Smart Food at FAO: An Overview.* Retrieve from: http://www.fao.org/3/an913e/an913e.pdf
[23] de la Caba K *et al.* 2019 From seafood waste to active seafood packaging: An emerging opportunity of the circular economy. *Journal of cleaner production* **208** 86-98.
[24] Heller M C, Selke S E and Keoleian G A 2019 Mapping the influence of food waste in food packaging environmental performance assessments. *Journal of Industrial Ecology* **23**(2) 480-495.
[25] Pauer E, Wohner B, Heinrich V and Tacker M 2019. Assessing the Environmental Sustainability of Food Packaging: An Extended Life Cycle Assessment including Packaging-Related Food Losses and Waste and Circularity Assessment. *Sustainability* **11**(3) 925.
[26] Panpong K, Srisuwan G, Sompong O and Kongjan P 2014. Anaerobic co-digestion of canned seafood wastewater with glycerol waste for enhanced biogas production. *Energy Procedia* **52** 328-336.
[27] Laso J, Margallo M, Fullana P, Bala A, Gazulla C, Irabien A and Aldaco R 2017 Introducing life cycle thinking to define best available techniques for products: application to the anchovy canning industry. Journal of Cleaner Production **155** 139-150.
[28] Kanellakis M, Martinopoulos G and Zachariadis T 2013. European energy policy—A review. *Energy Policy* **62** 1020-1030.
[29] Le Feon S, Thévenot A, Maillard F, Macombe C, Forteau L and Aubin J 2019 Life Cycle Assessment of fish fed with insect meal: Case study of mealworm inclusion in trout feed, in France *Aquaculture* **500** 82-91.
[30] Barr W J and Landis A E 2018 Comparative life cycle assessment of a commercial algal multiproduct biorefinery and wild caught fishery for small pelagic fish *The International Journal of Life Cycle Assessment* **23**(5) 1141-1150.
[31] Sainsbury N C, Genner M J, Saville G R, Pinnegar J K, O'Neill C K, Simpson S D and Turner, R. A. (2018). Changing storminess and global capture fisheries. *Nature Climate Change* **8**(8) 655.
[32] Nicola G G, Elvira B, Jonsson B, Ayllón D and Almodóvar A. 2018 Local and global climatic drivers of Atlantic salmon decline in southern Europe *Fisheries Research* **198** 78-85.
[33] Lusher A, Hollman P and Mendoza-Hill J 2017 Microplastics in fisheries and aquaculture: status of knowledge on their occurrence and implications for aquatic organisms and food safety *FAO Fisheries and Aquaculture Technical Paper* **615**.
[34] Barnes D K, Galgani F, Thompson R C and Barlaz M 2009 Accumulation and fragmentation of plastic debris in global environments. *Philosophical Transactions of the Royal Society B: Biological Sciences* **364**(1526) 1985-1998.

Acknowledgments
This work was supported by the EAPA_576/2018 NEPTUNUS project. The authors would like to acknowledge the financial support of Interreg Atlantic Area.

Transport electrification at all environmental costs? The case of large passenger duty-vehicles

A Bouter[1*], E Hache[1,2] and G Sokhna Seck[1]

[1] IFP Énergies nouvelles, 1-4 av. de Bois Préau, F-92852 Rueil-Malmaison, France.
[2] EconomiX-CNRS, University of Paris Nanterre, France.
* **E-mail:** anne.bouter@ifpen.fr

Abstract. Electrification is often highlighted as the future for transportation sector as it is considered as a zero-tailpipe emission. However, environmental impacts related to battery electric vehicles (BEVs) exist. The aim of this article is to identify the trade-off between batteries' weight and vehicle's autonomy from an environmental perspective in the BEVs' development as electrified mobility will consume more critical materials but reducing tailpipe emissions. Thus, a focus on climate change and resources has been done using a life cycle assessment (LCA) methodology. Large vehicles (D-segment) for BEVs and their fossil thermic counterparts have been assessed for two time horizons (2019, 2030) in Europe. Our linear programming world energy-transport model, TIAM-IFPEN, has been used in order to assess demand and import dependency on materials through to 2050. Results show that BEVs can generate higher impacts than their inherent conventional ones and that criticality assessment should be done in order to get a complete view of BEVs' deployment.

Keywords: life cycle assessment, IAM, electric vehicle, battery, resource depletion, criticality.

1. Introduction

The Paris Agreement sets out objectives to limit greenhouse gas (GHG) emissions in all sectors by fostering low carbon technologies. Transportation sector is brought to center stage as it accounts for almost a quarter of global CO_2 emissions in 2016 [16]. The market for BEVs has grown considerably and effective public policy incentives could reinforce this perception, as they are considered as zero tailpipe GHG emissions. However, other parameters shall be considered to properly assess BEVs' potential environmental benefits, such as the electricity production mix [25] and the battery production stage which accounts for an important part of total emissions related to the life cycle of BEVs [4].

Based on European Commission analysis [22], a substantial growth in the Lithium-ion (Li-ion) battery market is expected in the coming years. Indeed, consumer concerns about electric vehicle's operating range could impede the massive electrification of the transport sector worldwide [31] but should drive the production of larger batteries. Knowing that BEVs' autonomy depends on battery performance, such as capacity and energy density which in turn will impact their weight, it is likely that their environmental impacts highly depend on the battery technology and their size. In the last few years, significant improvements have been made to increase the energy density of battery [30].

The aim of this article is to identify the trade-off for BEVs' development between battery's weight and autonomy from a GHG perspective for large passenger duty-vehicle (D-segment) in Europe. To achieve this goal, large BEVs will be compared to their internal combustion engine (ICE) equivalents

IOP Conf. Series: Materials Science and Engineering 1196 (2021) 012011 doi:10.1088/1757-899X/1196/1/012011

and their future technological improvements by 2030. Based on LCA methodology [18, 19], the primary objective is to find out how much increasing BEV's operating range affects climate change and resource depletion indicators. The second objective is to consider the criticality issues of raw materials in a more global context as a complementary analysis. This criticality analysis will be provided by an integrated assessment model (IAM) called TIAM-IFPEN in order to develop a global three-dimensional approach taking into account geological, geopolitical and economic criteria with the transport evolution and global environmental constraints. Finally, our findings are discussed, and several insights are provided regarding resource dependency with climate constraints such as targets to limit global warming below 4°C or 2°C by 2100.

2. Materials and methods

The goal of this study is to perform a complete LCA for large passenger duty vehicles in Europe on a world-wide harmonized driving cycle (WLTC) [32] with a focus on climate change impact and resource depletion potential. The study aims at assessing the battery's specification for which BEV generate more environmental impacts than ICE based on these two midpoints, and at evaluating resources depletion through a criticality indicator also called supply risk method [33].

2.1. Life Cycle Assessment

LCA is performed according to ISO standards [18, 19] and follows the International Reference Life Cycle Data System [20]. The choice of LCA impact's methods complies with European Commission's recommendations [21, 33].

2.1.1. Goal definition

This paper is intended to forecast and analyse the environmental impacts on climate change and natural resources due to the roll-out of electric light duty vehicles. The aim is both to identify environmental performances of BEVs and to contribute to future transport policy decisions in a context where several European countries (France, Great Britain, Italy, Germany, etc.) are encouraging consumers to buy BEVs. The perimeter of the study is limited to Europe and the time coverage is 2019 and 2030. The target audience are governmental political decision makers and automotive sector.

2.1.2. Scope definition

Two larges (D-segment) BEVs: BEV and BEV+ (BEV with an extended battery capacity), and their fossil thermic counterparts are studied for 2019 and 2030 in a European context. Battery's technology is Li-ion battery. SimaPro® software and Ecoinvent 3.5 database [11] have been used. Vehicles were simulated over the homologated worldwide harmonized light vehicles test procedure (WLTP) as stipulated by ICCT [15]. The system boundaries comprise all stages of the energy pathway production till its use, and also the life cycle of the vehicle from cradle to grave.

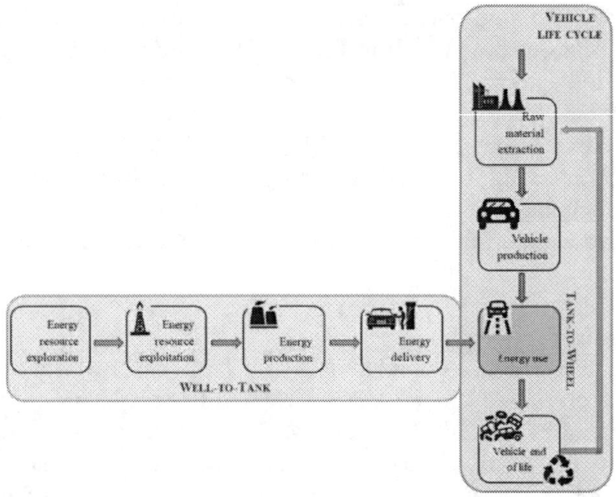

Figure 1. System boundaries of the study: well-to-wheel analysis including vehicle's life cycle assessment from cradle to grave.

The functional unit adopted is the movement of one person over 1 km within an overall European usage for a D-segment car expressed as person.km, for current time horizon (2019) and prospective (2030). The assumption of an average 1.3 persons per car is based on IFPEN expert discussions and Eurostat data.

2.1.3. Life cycle inventory (LCI)

According to the goal of the study, the LCI modelling is attributional and the Ecoinvent system model "cut-off" is selected for background data, i.e.: raw materials and manufacturing data, energy consumption, and end-of-life processes. Foreground data are based on several sources and represent the vehicle assembly [23, 24], the battery technology and sizing [29, 9, 25], the use stage (consumption and tailpipe emissions) [1, 2], the waste management (collection rate, sorting rate, etc.) [5, 6, 7], the parameters (lifetime, mileage, etc.), and the prospective choices.

Vehicles' and tyres' composition is modelled from a European study [23, 24] and components' weight is based on internal IFPEN assumptions of energy and power densities as follows: bare vehicle body with options and gearbox, ICE, electric engine and generator, battery. For 2030 horizon a reduction in vehicle's weight is considered [27] equivalent to the replacement of 30% of the vehicle's steel by an amount of aluminium weighing 65% of this 30% steel percentage. This vehicle downsizing and the improvement of engine efficiency leads to prospective energy pathway consumption for 2030.

Vehicles' lifetime is assumed to be 10 years with a 15,000 km/year mileage and 40,000 km for tyres, according to Notter et al. [26], Hawkins et al. [14], Ellingsen et al. [10], and others.

Batteries' chemistry is lithium-ion nickel manganese cobalt (LiNMC) technology modelled from confidential data from ADEME, the French environmental management agency. Batteries' energy densities and masses are based on IFPEN experts, manufacturers' feedbacks and battery's technology improvement and existing electric fleet. Table 1 presents the characteristics data for large BEVs updated from a former LCI [4].

Expected batteries' lifetime is 10 years after automotive sector's feedback and current guaranties. Pollutant emissions occurring during tank-to-wheel stage are recorded according to the Euro6b norms for CO, NOx, PM, and hydrocarbon compounds (HC) for gasoline. The Euro6b HC emissions norms for diesel vehicles were extrapolated based on the Euro6c norm and future Euro7 norm (IFPEN experts) as only the total of HC and NOx was provided in the standards. CO_2 correlates with fuel consumption, unlike other pollutant emissions and the low heating values (LHV) use the JRC's values [8]. An additional emission coefficient for NOx for diesel passenger cars has been added based on the requirements of several European member states to thereby properly assess real driving emissions. For 2030 horizon, the additional coefficient for NOx is removed and future Euro7 norm is taken.

Table 1. BEVs' characteristics for D-segment.

		Mass test [kg]	Battery mass [kg]	Battery capacity [kWh]	Battery energy density [Wh/kg]	WLTC consumption[1] [kWh/100 km]	Driving range[2] [km]
2019	BEV	2119	400	60	150	21.2	240
	BEV+	2252	533	80	150	21.9	310
2030	BEV	2059	400	80	200	14.6	493
	BEV+	2159	500	100	200	15.0	600

Vehicle's electricity and fuel consumptions are estimated using an IFPEN vehicle simulator based on Simcenter Amesim™ software [1, 2]. BEVs' use states in a European context assuming that the average electricity mix results in 460.7 g CO_2 eq./kWh [17]. No projection for electricity mix has been made for 2030. Biofuels' incorporation rates at filling stations have been also excluded.

Infrastructure, roads, charging stations, consumption by auxiliary functions, emissions related to tire wear and braking are not part of the scope but should be at fleet deployment scale.

2.1.4. Life cycle impact assessment (LCIA) methods

As BEVs are highlighted for their low GHG emissions, climate change is the first impact category assessed, based on the recommended methodology of the JRC-European Commission [21].

Furthermore, as BEVs are critical materials' users, resource depletion is also quantified based on the abiotic depletion indicator (ADP) from van Oers et Guinée [34] methodology for the LCA perspective, as recommended in UNEP/SETAC LCI [33].

A criticality indicator also named "supply risk method" in UNEP/SETAC LCI [33] is also considered as geopolitical and socio-economic aspects are included (section 2.2).

2.2. Raw material representation in an Integrated Assessment Model (IAM)

In addition to the LCA, metals availability contained in BEV's batteries (copper, lithium, nickel and cobalt) are assessed with the transport sector development through to 2050 and their geopolitical consequences. We rely on our linear programming world energy-transport model, TIAM-IFPEN which has been already presented in details in Hache et al. [13] The model allows examining two climate scenarios (2°C and 4°C) in order to estimate the consequences of the fast roll-out of low-carbon technologies in the transport sector and how it is likely to significantly increase metals demand by 2050. The import dependency on metals of the European Union in its attempt to electrify its transport sector was evaluated in order to estimate how this dynamic could increase metals criticality for this area. This notion of criticality is often described as all the risks related to the production, use or end-of-life management of a raw material [12] including geopolitical risks, economic risks, production risks and environmental or social risks.

3. Results

3.1. Climate change

[1] Energy consumption from the grid, including losses and efficiency. It is different from the battery consumption.

[2] Driving range corresponds to the battery capacity [kWh] weighted by the battery efficiency rate (efficiency between the electricity from the grid until the actual delivery of energy to the wheels: 0,85 in 2019 and 0.9 in 2030, based on IFPEN experts) divided by battery capacity [kWh].

Figure 2 presents the results for D-segment for BEVs and their thermic counterparts for current time and 2030. Technological progress in 2030 for gasoline vehicles can be mostly observed during the well-to-tank stage, based on consumption's reduction combined to a slightly decrease of vehicle's mass. Regarding BEVs, technological progress is quickly moving. Based on our assumptions, BEVs in 2030 will still remain slightly below the ICE regarding their impacts on climate change but frontier is small.

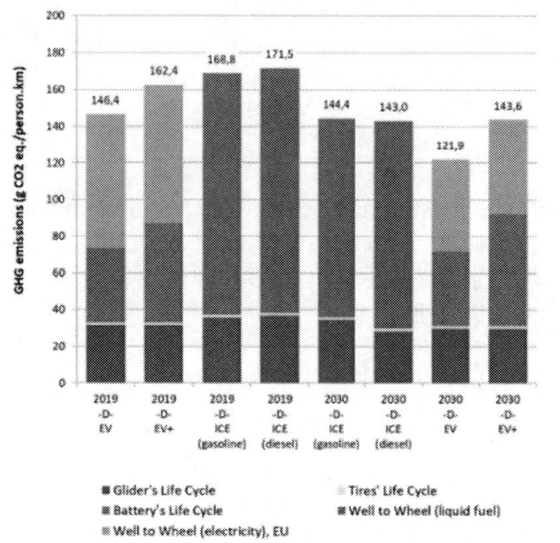

Figure 2. Potential impacts on climate change of one person transportation over one km for WLTP cycle for large cars, present time and 2030 projection, European electricity mix.

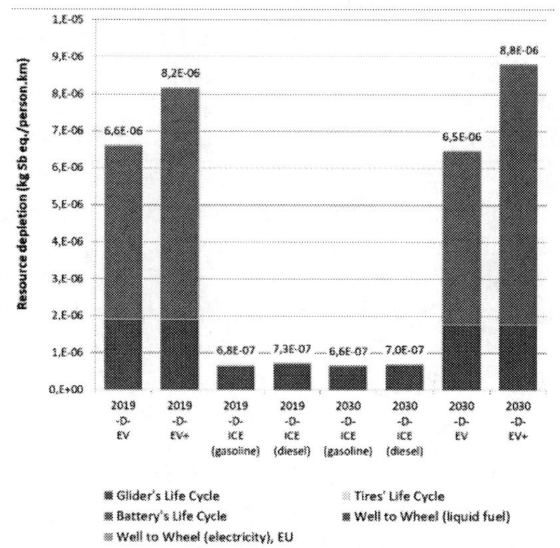

Figure 3. Potential impacts on resource depletion indicator (CML 2016) of one person transportation over one km for WLTP cycle for large cars, present time and 2030 projection, European electricity mix.

3.2. Trade-off identification for mass battery of EV in 2030

From climate change results, a methodology has been developed in order to assess the mass of battery required to exceed climate change impacts of ICE engines. The following assumptions have been considered:

$$I^{3}{}_{BEV,D} = I_{glider,BEV,D} + I_{tyres,BEV,D} + I_{battery,BEV,D} + I_{WTW,BEV,D} \quad (1)$$

With:

$I_{glider,BEV,D,2030} + I_{tyres,B\,EV,D,2030} = 31.10\ g\ CO_2\ eq./person.km$, based on Bouter et al. [4].

This parameter is assumed constant for D-segment BEVs through to 2030.

For the use phase, i.e. $I_{WTW,BEV,D}$, batteries' technologies' improvements are tough to project. However, the in-house vehicle simulator allows to assess electricity consumption which depends on mass test, also function of the mass of battery.

The same relation between the battery's environmental impact and the battery's mass can be made as battery's impact relies on parameters such as BEV's and battery's lifetimes, BEV's total mileage and the battery's emission factor (EF), considered constant on a kg-basis for both time horizon.

[3] I states for impact

Thus, the BEV's impact is a linear function of the battery's mass. Within the framework of the abovementioned hypothesis, the results show that the mass of battery should not exceed 457.5 kg, so that its performance does not fall below that of its thermal equivalent. A literature review of existing vehicles on market allowed us to make hypothesis regarding the range of BEVs' performances [28]. Lower boundary for capacity is 20 kWh and upper boundary is set at 200 kWh. Density is comprised between 75 $Wh.kg^{-1}$ and 500 $Wh.kg^{-1}$. Table 2 presents the possible couples related to the calculated battery mass.

Table 2. Possible couples associated with a battery mass within the defined range.

Mass battery [kg]	457	457	457	457	458	458	458	458	458
Battery size [kWh]	80	105	160	185	55	110	135	165	190
Energy density [$Wh.kg^{-1}$]	175	230	350	405	120	240	295	360	415

The limit of electrification for large cars (D-segment and above) is pinpointing as these batteries' parameters are fully possible to electrify the large vehicles fleet.

However, environmental impacts of BEVs are also located on resource depletion as batteries' production needs critical material. In order to assess this issue, two resource indicators are calculated: an LCA resource depletion indicator in section 3.3 and a criticality indicator in section 3.4.

3.3. Resource depletion

Based on UNEP/SETAC LCI [33] recommendations, the first approach chosen to characterize resource depletion for electric mobility is the Abiotic Depletion Potential (ADP) method [34] presented in Figure 3.

The process contribution analysis for battery's life cycle shows that 72.3% of the impact is related to the gold. This is not surprising as CML indicator identified gold with the highest characterization factor. Gold comes from the use of printed wiring board to produce battery. Gold is followed by copper (11%), silver (4.1%), zinc (1.4%), nickel (1.7%), and tin (0.8%). Lithium contribution to CML resource depletion indicator is less than 0.003%. Cobalt has almost no contribution to this indicator. However, it is important to consider the depletion of resources throughout a geopolitical risk. For this reason, a criticality indicator, also known as supply risk indicator, is evaluated.

3.4. Criticality

Figure 4 depicts the evolution of the total cumulative demand of copper, lithium, nickel and cobalt with climate constraints. It can be observed that 96.1%, 26%, 56.9% and 93.6%, respectively for copper, lithium, nickel and cobalt, would be consumed by 2050 in the most stringent climate scenario.

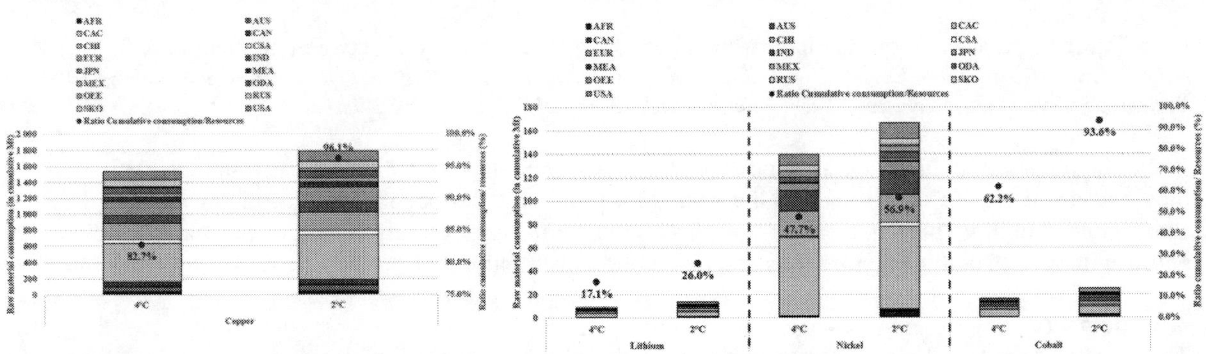

Figure 4: Comparison of cumulative consumption of raw materials by regions by 2050 and their known resources in 2010.

IOP Conf. Series: Materials Science and Engineering 1196 (2021) 012011 doi:10.1088/1757-899X/1196/1/012011

In this case, the transport sector accounts for 26.9%, 85.6%, 40.6% and 69.8% of copper, lithium, nickel, and cobalt respectively (Figure 5). Due to the fast roll-out of electric vehicles, the D-segment represents around 20% of the transport cumulative consumption of any raw material considered while it is less than 15% of the global fleet worldwide.

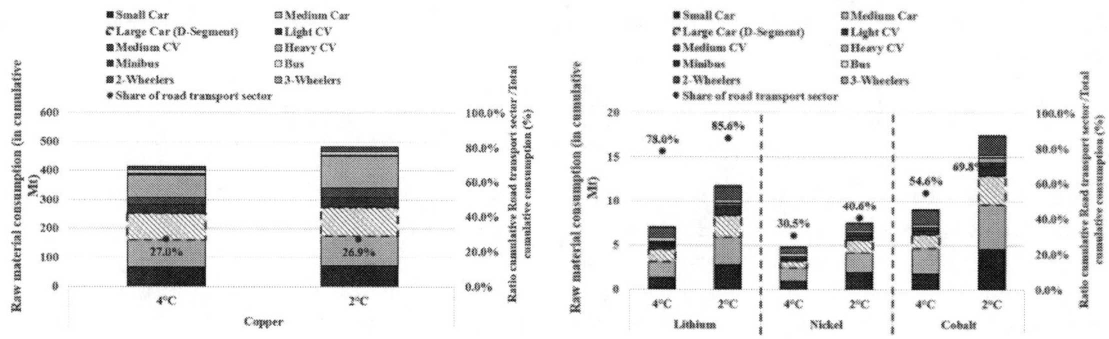

Figure 5: Evolution of the raw material consumption disaggregated by road transport segments.

This implies that this segment would certainly play an important role in raw material imports for Europe because they represent around 8% of the D-segment worldwide (Figure 6). In this context, European import dependency on raw materials for batteries is on stake. The European willingness to create a battery consortium in order to limit this future risk should be considered as the development of battery recycling sector.

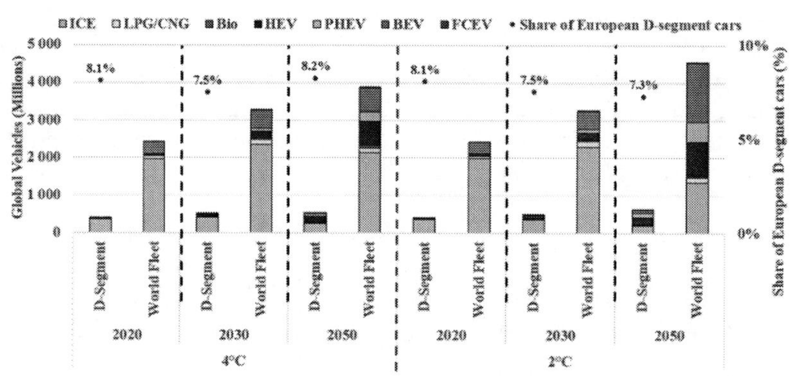

Figure 6: Evolution of the global vehicle stock and the D-segment between 2020 and 2050.
ICE: Internal Combustion Engine; HEV: Full hybrid vehicle; PHEV: Plug-In hybrid electric vehicle; BEV: Battery-powered electric vehicle; FCEV: Fuel cell electric vehicle

Public transport regulators in Europe should also integrate in their policies the resource depletion as they had implemented CO_2 issues in the past. More globally, European public policies dedicated to sustainable mobility (non-motorized transport, public transport) could constitute the most efficient pathway to reduce both CO_2 emissions and resources consumption.

4. Conclusion

This study presents a LCA analysis focused on the climate change impact and resource depletion potential of large (D-segment) BEVs and ICEs for two time horizons (2019, 2030) in Europe simulated over the WLTC cycle. Demand and import dependency on materials through to 2050 were assessed thanks to our linear programming world energy-transport model, TIAM-IFPEN.

Results obtained show that BEVs could be a game changer for GHG reduction according to batteries' weight and energetic sizing. Nevertheless, batteries' production consumes critical materials and the two resource indicators pinpoint the fact that these two methodologies should be performed in parallel in order to have a broad view of the environmental impacts of the future development of transport sector. Indeed, the resource depletion shows that raw materials such as gold or copper have more environmental impacts than others such as cobalt, nickel or lithium. However, their criticality assessment gives a more complete view by considering a tri-dimensional approach based on geological, economical, and geopolitical factors.

5. References

[1] Badin F, Le Berr F, Briki H, Dabadie J, Petit M, Magand Sand Condemine E 2013 *WEVJ* **6**.
[2] Badin F, Berr F, Castel G, Dabadie J, Briki H, Degeilh Pand Pasquier M 2015 *WEVJ* **7** 475.
[3] Bouter A, Ternel Cand Melgar J 2019 LCA Study of Vehicles Running on NGV and bioNGV.
[4] Bouter A, Hache E, Ternel Cand Beauchet S 2020 *Int J Life Cycle Assess* **22** 1441.
[5] *Off J Eur Communities* L 269 pp 34–43.
[6] *Off J Eur Union* L 310 pp 11–27.
[7] *Off J Eur Union* L 266 pp 1–13.
[8] European Commission 2014 Well-to-wheels analysis of future automotive fuels and powertrain in the European context. Description, results and input data per pathway. WTT_Appendix_4_v4a_APRIL2014.
[9] Ellingsen L A-W, Majeau-Bettez G, Singh B, Srivastava A K, Valøen L Oand Strømman A H 2014 *J Ind Ecol* **18** 113.
[10] Ellingsen L A-W, Singh Band Strømman A H 2016 *Environ. Res. Lett.* **11** 54010.
[11] Frischknecht R et al. 2005 *Int J Life Cycle Assess* **10** 3.
[12] Graedel T E, Harper E M, Nassar N T, Nuss Pand Reck B K 2015 *Proceedings of the National Academy of Sciences of the United States of America* **112** 4257. Available at https://www.pnas.org/content/pnas/112/14/4257.full.pdf.
[13] Hache E, Seck G, Simoen M, Bonnet Cand Carcanague S 2019 *Appl Energ* **240**.
[14] Hawkins T R, Singh B, Majeau-Bettez Gand Strømman A H 2013 *J Ind Ecol* **17** 53.
[15] ICCT ed 2014 The WLTP: How a new test procedure for cars will affect fuel consumption values in the EU. Working paper.
[16] IEA 2017 CO2 Emissions from Fuel Combustion Statistics (OECD).
[17] IEA 2017 World Energy Statistics and Balances.
[18] ISO 2006 Environmental management - life cycle assessment - principles and framework (Geneva: International Organization for Standardization) 13.020.60.
[19] ISO 2006 Environmental management - life cycle assessment - requirements and guidelines (Geneva: International Organization for Standardization) 13.020.10 ; 13.020.60.
[20] JRC - European Commission 2010 ILCD Handbook. General guide for LCA - Detailed guidance.
[21] JRC - European Commission 2012 Recommendations for Life Cycle Impact Assessment in the European context. - based on existing environmental impact assessment models and factors (Luxembourg: Publications office of the European Union).
[22] JRC - European Commission 2018. Available at http://www.avere-france.org/Uploads/Documents/1548933066974482d1b51c56632ad2f02a4078ec5f-kjna29440enn.pdf.
[23] Nemry F, Leduc G, Mongelli Iand Uihlein A 2008 Environmental Improvement of Passenger Cars (IMPRO-car) (Luxembourg: Publications Office).
[24] Nemry F et al. 2009 Feebate and Scrappage Policy Instruments. Environmental and Economic Impacts for the EU27 (Luxembourg: OPOCE).
[25] Nordelöf A, Messagie M, Tillman A, Ljunggren Söderman Mand van Mierlo J 2014 *Int J Life Cycle Assess* **19** 1866.

[26] Notter D A, Gauch M, Widmer R, Wäger P, Stamp A, Zah Rand Althaus H-J 2010 *Environ Sci Technol* **44** 6550.
[27] PE International AG and Gingko 21 2013 Elaboration selon les principes des ACV des bilans énergétiques, des émissions de GES et des autres impacts environnementaux induits par l'ensemble des filières de véhicules électriques et de véhicules thermiques, VP de segment B (citadine polyvalente) et VUL à l'horizon 2012 et 2020. Etude réalisée pour le compte de l'ADEME.
[28] Peters J F, Baumann M, Zimmermann B, Braun Jand Weil M 2017 *Renew Sust Energ Rev* **67** 491.
[29] Sullivan J L, Gaines Land Systems E 2010 A review of battery life-cycle analysis. State of knowledge and critical needs (United States).
[30] Thackeray M M, Wolverton Cand Isaacs E D 2012 *Energy Environ. Sci.* **5** 7854.
[31] Tran M, Banister D, Bishop J D Kand McCulloch M D 2012 *Nature Clim Change* **2** 328. Available at https://www.nature.com/articles/nclimate1429.pdf.
[32] Tutuianu M, Bonnel P, Ciuffo B, Haniu T, Ichikawa N, Marotta A, Pavlovic Jand Steven H 2015 *Transp Res D Transp Environ* **40** 61. Available at http://www.sciencedirect.com/science/article/pii/S1361920915001030.
[33] UNEP SETAC LCI 2019.
[34] L. van Oers et al. 2002 Abiotic resource depletion in LCA. Improving characterisation factors for abiotic resource depletion as recommended in the new Dutch LCA Handbook.

Recycled concrete for foundation structure: LCA case study

J Pešta[1,2,*], T Pavlů[2], K Fořtová[2] and J Řepka[2]

[1]Department of Environmental Chemistry, Faculty of Environmental Technology, University of Chemistry and Technology, Prague, the Czech Republic,
[2]University Centre for Energy Efficient Buildings of Czech Technical University in Prague; the Czech Republic

***Email:** jan.pesta@cvut.cz

Abstract. Concrete containing primary raw resources such as sand and gravel can be replaced by recycled aggregate concretes (RAC). But recycling processes are energy-consuming and thus energy resources are consumed. To investigate the resource consumption and overall environmental impact related to the life cycle of RAC, Life Cycle Assessment (LCA) was used to compare ten RAC foundation structures with two foundation structures containing just natural aggregate (NA). This paper presents the characteristics of two types of RAC for the foundation with various replacement rates of NA (30%, 50%, 100%). Various concrete mixtures were designed with the three replacement ratios of natural gravel, the two amount of cement and the two effective water-cement ratios. Unsurprisingly, the NA foundation structures have smaller crude oil consumption. The NAC I 16/20 foundation consumed 84 kg of crude oil. Contrarily, this foundation has the biggest consumption of natural aggregate (7.6 t). Moreover, the RAC foundation structures have beneficial impacts in the Metal depletion category. For example, the impact of the RAC I C30_V1 foundation is -31,7 kg Fe eq. Most of the RAC foundation structures demonstrated the overall environmental impact lower than NA foundation structures, suggesting that RAC could be suitable replacements for standard concretes.

1. Introduction

Concrete is the second most used material after water, which is dependent on primary natural resources such as virgin aggregate, cement and water. It is evident that, due to its physical properties, concrete will remain a key construction material in the future. According to this prediction, its global use will continue to increase in the future. For this reason, reusing recycled aggregate concrete (RAC) as an aggregate for a new concrete mix follows crucial importance. Simultaneously, there will be an increase in the amount of demolished concrete structures and the need for the recycling and reuse of RAC. The key hierarchy of sustainability goes through the three "R"s: the first is reduce, then reuse, and finally recycle. When the materials have been used in construction and there is no way to reuse them, recycling is the best possible way to reduce primary sources. The research in the field of RAC dates back to the 1940s [1].

Due to the suitable characteristics of concrete, it is the most used material for foundation structures. Furthermore, the foundation structures usually consume a large volume of materials, because it is necessary to carry the load from the building to the subsoil. Given these facts, the foundation structures have been one of the structures, which represented the highest environmental impact in nearly every impact category analyzed by LCA [2–4].

The environmental assessment of RAC for structural use and its comparison with natural aggregate concrete (NAC) has been published [5–17]. The LCA method is one possible way to compare the environmental impacts of concrete structural elements [18]. Few studies have been published on the comparison of different types of natural aggregate (NA) and recycled aggregate (RA) e.g., river NA, crushed NA, and coarse RA [5,9,19,12,20]. The RA with NA production for the production of lower-grade concrete products was studied [19]. The utilization of the RA as a replacement of NA in concrete mixtures mostly leads to decreasing the properties of concrete and its durability [6,21]. Due to this fact, in some studies [7,22] the decline of RAC properties was compensated with additional cement.

The main aim of this paper was to optimize the RAC mixture for structural utilization according to the examined properties and environmental impacts. This paper presents the characteristics of RA and RAC and their comparison with conventional materials. Two types of recycled aggregate prepared by different recycling processes were used to find the optimal way how to get the suitable quality of RA with the lowest environmental impact. The concrete was optimized according to the exposition class of concrete and future use as the structural element for foundations.

2. Materials and methods

To compare the environmental impact of several new concrete mixtures with recycled aggregates, a holistic approach taking into account a whole product system should be used. The Life Cycle Assessment (LCA) was selected as the most suitable method for this purpose. This method was performed following the ISO 14040:2006 [23]. According to these requirements, LCA consists of four main parts: the goal and scope definition, life cycle inventory (LCI), life cycle impact assessment (LCIA) and life cycle interpretation [24].

2.1. Materials

Ten RAC mixtures and two NAC mixtures were designed to optimize the use of RA for the same structural use. The different replacement ratios of aggregate with two different amounts of cement and the two different effective water-cement ratios were designed to examine the impact of the proposed technique. Mechanical and physical properties of manufactured mixtures are investigated. According to previous researches, higher water absorption and lower density of RA have been identified [11,25]. Water absorption of RCA after 24 h for fraction 4/8 mm was 7.0% for both types of RA and for fraction 8/16 mm varies from 6.0% (RA 2) to 9.0% (RA 1). The oven-dried density varied from 2380 kg/m3 (RA 1) to 2420 kg/m3 (RA 2) for fraction 4/8 mm and it was 2420 kg/m3 for fraction 8/16 mm of both types of RA. The results of the RA properties correspond with the results of previous studies. Twelve concrete mixtures with two types of RA, various replacement ratios, two different amounts of cement CEM I 42.5 R (OPC) and different water-cement ratios were used for experimentation. The considered exposition classes were X0 and XC1 according to the Czech European Standard [26]. The cement content was kept constant, i.e. 260 kg/m^3 (NAC I, RAC I) and 300 kg/m^3 (NAC II, RAC II). The effective water/cement (w/c) ratio was kept constant, i.e. 0.65 (NAC I, RAC I) and 0.55 (NAC II, RAC II). The replacement ratio of coarse aggregate for RAC I was 30%, 50% and 100% (RAC I C30, RAC I C50, RAC I C100) and for RAC II was 30% and 100% (RAC II C30, RAC II C100). The composition of concrete mixtures per cubic meter is shown in **Table 1**.

Table 1. Concrete mixtures of NAC and RAC, per cubic meter.

	NAC I	RAC I C30_V1	RAC I C30_V2	RAC I C50_V1	RAC I C50_V2	RAC I C100_V1	RAC I C100_V2
CEM 42.5 R	260	260	260	260	260	260	260
Water	169	201	184	206	206	179	211
NA (0-4 mm)	736	632	632	611	611	415	588
NA(4-8 mm)	533	0	0	0	0	0	0
NA (8-16 mm)	570	656	656	311	311	0	0
RCA (4-8 mm)	0	485	485	506	506	239	526

RCA (8-16 mm)	0	0	0	283	283	1134	538
w/c	0,65	0,77	0,71	0,79	0,79	0,69	0,81
Effective w/c	0,65	0,65	0,65	0,65	0,65	0,65	0,65

	NAC II	**RAC II C30_V1**	**RAC II C30_V2**	**RAC II C100_V1**	**RAC II C100_V2**
CEM 42.5 R	300	300	300	300	300
Water	165	200	183	175	208
NA (0-4 mm)	700	615	615	364	549
NA(4-8 mm)	538	0	0	0	0
NA (8-16 mm)	601	674	674	0	0
RCA (4-8 mm)	0	485	485	225	533
RCA (8-16 mm)	0	615	615	1198	564
w/c	0,55	0,67	0,65	0,58	0,69
Effective w/c	0,55	0,55	0,55	0,55	0,55

The results of the basic examined mechanical and physical properties are shown in Table 2. The target concrete strength class was examined due to the characteristic compressive cube strength according to the Eurocode and ISO 12491. The target strength class for the NAC mixture was C25/30 (characteristic compressive cube strength equal to 30 MPa) and C30/37 (characteristic compressive cube strength equal to 37 MPa). The target concrete strength classes for RAC mixtures ranged from C8/10 to C25/30. This method of determination of strength class of concrete is chosen due to the consideration of the number of samples to eliminate their influence.

Based on these results, the foundation structural elements were designed to carry an equal load and had the same effective loading area [m^2]. The element is designed from plain concrete without reinforcement. The element made of lower-strength-class concretes were designed to maintain the same utility properties which led to a greater element height. It followed that for the same utility properties the structural element with the lower strength class had a larger volume, which means a larger amount of concrete must be used to obtain the same utility properties. The volume of concrete elements for foundations is shown in Table 2. These volumes were used as referential flows for environmental analysis.

Table 2. Properties of NAC and RAC

Type of concrete	**Density** [kg/m^3]	**Water absorption** [%]	**Compressive strength** [MPa]	**Flexural strength** [MPa]	**Static elastic modulus** [GPa]	**Target strength class** [-]	**Volume of element** [m^3]
RAC I C30_V1	2143	7.6	21.9	4.1	23.6	C12/15	4.62
RAC I C30_V2	2199	5.6	32.4	5.6	28.9	C25/30	3.74
RAC I C50_V1	2023	13.4	22.1	3.8	18.0	C16/20	4.40
RAC I C50_V2	2168	6.1	33.5	5.0	25.4	C25/30	3.74
RAC I C100_V1	1977	15.0	15.2	3.6	14.2	C8/10	5.06
RAC I C100_V2	2054	11.8	22.3	3.3	18.0	C16/20	4.40

RAC C100_V1	II	2006	14.3	21.9	4.2	13.8	C12/15	4.62
RAC C100_V2	II	2109	11.0	32.0	4.0	24.5	C25/30	3.74
RAC C30_V1	II	2141	9.4	24.9	4.7	22.7	C16/20	4.40
RAC C30_V2	II	2200	6.7	32.4	5.4	28.5	C25/30	3.74
NAC I		2250	5.5	31.4	4.4	25.7	C25/30	3.74
NAC II		2277	5.4	46.3	7.3	33.2	C30/37	3.52

2.2. Goal and scope definition, functional unit and system boundaries

The primary goal of this study was to determine the concrete mixture with the lowest environmental impact in the comparison, where the mixtures are used for the same function in building foundation structure. The secondary goal is to describe the influence of the type of recycled aggregate on the environmental impact of concrete mixture, or concrete strength class classification, respectively.

In this study, compared foundation structures were designed to have the same function. Therefore, the compared functional unit was one foundation structural element with an equal load and the same effective loading area. As compensation for the lower strength class, the higher element heigh was designed and a larger volume of concrete had to be used for the same utility properties of the structure. The referential flows of concrete mixtures, which were needed to reach defined FU are different and they are described as the calculated volume in **Table 2**. The system boundaries for each concrete mixture include all life cycle phases: production of materials, production of concrete and construction of foundation structure, deconstruction and EoL. Production of primary raw materials includes excavation and production of RCA as secondary raw material starts with unloading demolition and construction waste in a recycling plant.

Two types of aggregates were considered: V1 from the one-step recycling process and V2 from the two-step recycling process. The first step is the same. The demolition and construction waste is crushed and sorted. Also, steel scrap is separated. Through this recycling process, three fractions are produced. Two of them are used as filling into bed under infrastructure. The third is V1, which can be used as aggregate for concrete or it can be crushed and sorted again. In this second step, another three fractions are produced and one of them is V2 aggregate. The two-step process helps with the separation of clay particles and therefore V2 type is more suitable for concrete.

2.3. Life cycle inventory (LCI)

The Gabi 9 software was used to conduct the data. As a priority, specific data for the Czech Republic were used and also generic data from GaBi 9 database and Ecoinvent were used [27]. Specific data for modelling concrete production in the Czech Republic was assumed according to Fiala [28]. The EoL of foundation structures includes deconstruction, transportation and landfilling as a typical type of removal for construction and demolition waste.

2.4. Life cycle impact assessment, normalization and weighing

Environmental impacts were evaluated according to the ReCiPe characterization method(Goedkoop M., 2013). In this study, Hierarchist (H) perspective for this method was chosen. Version 1.08 of this method was used. For this study, considered indicators and their results are in **Table 5.** To compare the overall impact of each foundation structure, the results of the impact indicators were normalized and weighted. Normalization means relating the results to the global impact. The results are multiplied by global factors for each category(Guinee, 2001). In this case, normalization was carried out with factors from ReCiPe 1.08 (H), Mid-point Normalization, World, incl. biogenic carbon (person equivalents)(Dr. Thilo Kupfer, 2019). The weighing means adding specific value to each category. The results after normalization are multiplied by factors, which are based on the opinion of the scientific community. In this study, the weighing was performed according to the thinkstep LCIA Survey 2012, Global, thinkstep LCIA Survey 2012, Global, ReCiPe 1.08 (H), incl. biogenic carbon (person Equiv. weighted)(Dr. Thilo Kupfer, 2019).

3. Results

3.1. *Life cycle inventory analysis*

Table 3. shows chosen primary resources, which were consumed in the life cycle of compared foundations. Negative flows in this table represent the consumption of primary resources, which were avoided by the production of secondary raw materials or by waste materials recycling, respectively. For example, the negative flow of iron is caused by iron scrap recycling. In our production system, the construction and demolition waste is sorted and iron scrap is used again for the production of new steel. Production of recycled aggregate has a similar effect on natural aggregate flow, which is negative or significantly reduced in the case of foundations with recycled concrete.

Table 3. The chosen resources demand of concrete foundation structure with V1 and V2 recycled aggregate.

	RAC I C30_V1	RAC I C50_V1	RAC I C100_V1	RAC II C30_V1	RAC II C100_V1	NAC I	NAC II
Non renewable energy resources (kg)							
Crude oil	103	97	116	128	109	84	81
Hard coal	16.1	0.5	-31.3	-6.9	-26.4	32.9	34.6
Lignite	89.9	85.3	98.1	99.1	101.1	73.0	77.4
Natural gas	55.9	52.8	53.7	60.0	48.1	48.0	44.8
Peat	0.0246	0.0189	0.0097	0.0203	0.0078	0.0268	0.0252
Uranium	0.000887	0.000734	0.000665	0.000874	0.000686	0.000855	0.000877
Non renewable elements (kg)							
Copper	0.0038	-0.0074	-0.0302	-0.0117	-0.0274	0.0174	0.0179
Iron	-12.7	-31.3	-77.8	-46.5	-72.5	15.6	16.1
Lead	-0.105	-0.165	-0.334	-0.231	-0.315	0.004	0.004
Zinc	-0.060	-0.109	-0.235	-0.152	-0.221	0.020	0.020
Non renewable resources (kg)							
Basalt	-0.002	-0.024	-0.065	-0.027	-0.061	0.026	0.025
Clay	493	458	537	598	505	405	393
Dolomite	-1.56	-2.52	-5.16	-3.53	-4.88	0.13	0.13
Limestone	1980	1887	2172	2180	2285	1602	1738
Natural Aggregate	3188	-737	-8135	-1067	-8021	7566	7131
Pyrite	0.00141	0.00142	0.00180	0.00169	0.00186	0.00104	0.00113
Quartz sand	302	280	329	372	305	249	238
Renewable resources (kg)							
Water	416163	311513	221936	384683	235016	440835	448728
	RAC I C30_V2	RAC I C50_V2	RAC I C100_V2	RAC II C30_V2	RAC II C100_V2		
Non renewable energy resources (kg)							
Crude oil	91	95	115	126	101		
Hard coal	-46.0	-95.6	-165.2	-139.8	-142.0		

Lignite	68.2	65.2	73.5	74.1	71.5
Natural gas	47.9	49.4	59.5	57.2	50.3
Peat	0.0044	-0.0089	-0.0249	-0.0176	-0.0227
Uranium	0.000699	0.000594	0.000587	0.000701	0.000564
Non renewable elements (kg)					
Copper	-0.0227	-0.0482	-0.0838	-0.0683	-0.0724
Iron	-86	-150	-245	-212	-214
Lead	-0.350	-0.571	-0.907	-0.797	-0.795
Zinc	-0.220	-0.370	-0.596	-0.516	-0.522
	RAC I C30_V2	RAC I C50_V2	RAC I C100_V2	RAC II C30_V2	RAC II C100_V2
Non renewable resources (kg)					
Basalt	-0.0016	-0.0202	-0.0436	-0.0226	-0.0382
Clay	385	372	420	481	368
Dolomite	-5.4	-8.9	-14.2	-12.5	-12.4
Limestone	1607	1611	1898	1862	1859
Natural Aggregate	-11793	-24006	-41256	-33503	-36380
Pyrite	0.00145	0.00171	0.00229	0.00213	0.00213
Quartz sand	234	225	252	295	217
Renewable resources (kg)					
Water	330046	254944	219054	312462	214074

Mostly, the water is consumed for cement production and this effect depends on cement demand. The second process with big water demand is construction waste landfilling. Besides these processes, the production of recycled aggregate is also connected with the production of other types of recycled aggregates, which are not suitable for concrete. By producing these other aggregates, the excavation of primary resources is avoided and so water demand for this excavation is reduced. Therefore, the consumption of water is lower for foundations structure with recycled aggregates. The demand for energy resources reflects the properties of the concrete mixture. Mixtures with higher strength classes are used in smaller volumes and so fewer materials are transported. The V2 mixtures with two-step aggregate have a smaller volume than mixtures with one-step aggregate hence V2 mixtures have a smaller demand for crude oil for transport. Similarly, the recycling process affects hard coal demand. The reduction of hard coal is caused by coal, which is avoided by steel scrap recycling. The more construction and demolition waste is recycled, the more steel scrap is separated and recycled and so more demand for hard coal is avoided. A significant flow is the construction and demolition waste, which is needed to produce enough aggregates to a foundation structure. The production of 1 t of V2 type of aggregate processes more construction and demolition waste than the production of 1 t of V1 type. **Table 4**. shows the amount of construction and demolition waste for each foundation structure as the main input into the production system.

Table 4. Demolition waste consumption for building concrete foundation structure (kg).

RAC I C30_V1	RAC I C30_V2	RAC I C50_V1	RAC I C50_V2	RAC I C100_V1	RAC I C100_V2	RAC II C30_V1	RAC II C30_V2	RAC II C100_V1	RAC II C100_V2
5602	18139	8679	29509	17368	46816	12100	41140	16436	41028

3.2. Life cycle impact assessment

The results of the impact category indicators are shown in **Table 5**. Considering Climate Change, the reference foundations have a smaller impact than foundation structures with V1 RAC, but foundation structures with V2 RAC have an even smaller impact in this category. The difference between V1 and V2 is caused mainly due to the difference between the impacts of 1 t of V1 and V2 aggregate. The best result in this category was reached with the RAC I C50_V2 mixture (865 kg CO_2 eq.) and the RAC II C100_V2 mixture (865 kg CO_2 eq.). The very similar results are caused by balancing two factors: the amount of cement and the amount of V2 aggregate. While RAC I C50_V2 have a smaller amount of cement, RAC II C100_V2 has a bigger amount of V2 aggregate. For this category, these factors were combined with the most appropriate way, which was considered in this study.

In our comparison, the reference foundation (NAC I and NAC II) reached the best result only in the category of Agricultural land occupation (26.0 m^2a). It is mainly the consequence of a smaller volume of the concrete mixture. In the comparison with recycled concrete, reference concretes are not influenced by the impact of the recycling process, where specifically in this category is the significant negative impact of recycling of iron steel scrap and additional transport. Interestingly, these reference foundations have the biggest impact in the Urban land occupation category, where the main contributor is gravel production. And because the CDW recycling process leads to avoiding gravel consumption, the foundation with recycled concrete reach a better result. Similarly, the bigger impact of reference concretes in the categories Ozone depletion, Freshwater eutrophication and Metal is caused due to the same reasons: no gravel avoided and no credits for steel recycling.

Table 5. shows, that foundation with mixture RAC I C100_V1 reached the best results in various impact categories such as Freshwater ecotoxicity, Ozone depletion, Urban land occupation. In these categories, the main influence is caused by credits for avoided gravel. In this foundation structure, the biggest mixture volume is used and also the biggest amount of V1 aggregate. The production of recycled aggregate is connected with the production of other fractions of recycled aggregates. These other aggregates can be used also to avoid primary gravel production. Therefore, more V1 aggregate is produced, more primary gravel is avoided. Different approach to design a mixture was used in the foundation structure with RAC II C100_V2. This foundation structure with a higher percentage of V2 aggregate in comparison with RAC I C50_V2 and smaller volume comparing to RAC I C100_V2 reached also some pretty good results in the various indicators as is shown in **Table 5.**

To summarize, the two mixtures are more suitable for considered function than others: RAC I C100_V1 and RAC II C100_V2. The first of them has the biggest volume of foundation among all considered options. The results are mainly influenced by the big amount of avoided gravel. The second mixture contains V2 aggregates. The better properties of the mixture with these V2 aggregates leads to better characteristics of foundation structure and its smaller volume. Likewise, the V2 aggregates are more beneficial in categories, in which the impact is affected by the recycling of CDW and specifically by the iron scrap recycling.

Table 5. Results of selected impact category indicators for foundation structures

Impact category indicator	RAC I C30_V1	RAC I C50_V1	RAC I C100_V1	RAC II C30_V1	RAC II C100_V1	NAC I
Agricultural land occupation [m2a]	32.7	31.2	37.6	39.4	36.4	26.0
Climate change, incl. biogenic carbon [kg CO2 eq.]	1379	1268	1369	1491	1395	1177
Fossil depletion [kg oil eq.]	197	176	179	214	170	175
Freshwater ecotoxicity [kg 1,4 DB eq.]	0.058	-0.030	-0.178	-0.025	-0.167	0.153

Freshwater eutrophication [kg P eq.]	0.00169	-0.00073	-0.00475	-0.00054	-0.00447	0.00426
Human toxicity [kg 1,4-DB eq.]	63.7	57.3	63.4	73.8	61.0	54.9
Ionising radiation [U235 eq.]	19.6	12.0	2.5	14.8	3.6	24.2
Metal depletion [kg Fe eq.]	-31.7	-63.6	-145.5	-91.7	-136.5	19.3
Natural land transformation [m2]	0.306	0.245	0.217	0.328	0.199	0.302
Ozone depletion [kg CFC-11 eq.]	-9.4E-08	-1.1E-06	-3.0E-06	-1.2E-06	-2.8E-06	1.2E-06
Particulate matter formation [kg PM10 eq.]	1.11	0.99	1.04	1.19	1.04	0.97
Photochemical oxidant formation [kg NMVOC eq.]	2.50	2.26	2.44	2.78	2.42	2.16
Terrestrial acidification [kg SO2 eq.]	2.27	2.05	2.20	2.51	2.19	1.97
Urban land occupation [m2a]	0.22	-1.42	-4.39	-1.54	-4.16	2.19
Water depletion [m3]	412	306	211	377	225	440

Indicator category indicator	RAC I C30_V2	RAC I C50_V2	RAC I C100_V2	RAC II C30_V2	RAC II C100_V2	NAC II
Agricultural land occupation [m2a]	26.9	27.5	32.8	34.8	29.6	26.0
Climate change, incl. biogenic carbon [kg CO2 eq.]	983	865	888	969	865	1224
Fossil depletion [kg oil eq.]	131	104	94	118	84	172
Freshwater ecotoxicity [kg 1,4 DB eq.]	0.041	-0.035	-0.123	-0.035	-0.106	0.150
Freshwater eutrophication [kg P eq.]	0.00123	-0.00083	-0.00318	-0.00075	-0.00276	0.00417
Human toxicity [kg 1,4-DB eq.]	51.3	48.7	54.2	62.4	48.6	54.3
Ionising radiation [U235 eq.]	16.19	10.74	6.82	13.34	7.04	24.5
Metal depletion [kg Fe eq.]	-155	-264	-426	-370	-372	19.7
Natural land transformation [m2]	0.195	0.126	0.072	0.162	0.059	0.290
Ozone depletion [kg CFC-11 eq.]	-7.6E-08	-9.3E-07	-2.0E-06	-1.0E-06	-1.8E-06	1.1E-06
Particulate matter formation [kg PM10 eq.]	0.553	0.293	0.064	0.241	0.109	0.99
Photochemical oxidant formation [kg NMVOC eq.]	1.95	1.81	1.97	2.20	1.83	2.19
Terrestrial acidification [kg SO2 eq.]	1.68	1.49	1.54	1.77	1.45	2.01
Urban land occupation [m2a]	-0.16	-1.74	-3.73	-2.06	-3.27	2.13
Water depletion [m3]	319	238	192	289	190	448

3.3. Normalization and weighting

Figure 1. shows the results of impact indicators after normalization and their summarization for each foundation structure. These summarizations are presented in columns, which have positive and negative parts. The positive part represents the damages to the environment, which were caused by the foundation structures in some categories. The negative part of the columns shows the impact, which was avoided during the compared life cycles.

In LCIA, the two mixtures reached the best result: RAC I C100_V1 and RAC II C100_V2. The second has more beneficial results in the Metal depletion category and it has a less damaging impact in the Human toxicity category. After summarization of normalized and weighted results in one result of the overall impact, RAC I C100_V2 can be proposed as the most suitable option in this study.

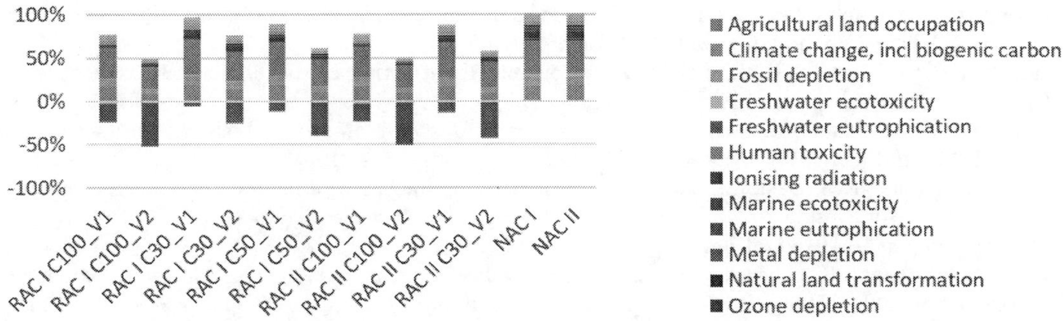

Figure 1. Results of impact indicators after normalization and weighting.

4. Conclusions

We have successfully demonstrated that RAC can be used for foundation structure and they have a smaller overall environmental impact than foundation structure containing NAC. Such utilization of RAC for the foundation structure will enable us to reduce primary resource consumption and construction and demolition waste production. For example, the RAC II C30_V2 foundation structure consumes almost 41 t of demolition waste. Furthermore, we identified the four key factors which affect the environmental impact of foundation structure: the amount of recycled aggregate, type of aggregate, amount of cement and volume of concrete. The RAC I C100_V1 foundation structure reached the best results in the Urban land occupation impact category (-4.39 m2a) and other categories, which are influenced by the amount of avoided natural gravel. However, the RAC I C100_V2 foundation structure has the biggest amount of avoided natural aggregate (41 t). Similarly, the RAC II C100_V2 foundation structure has the most beneficial impact in the Fossil depletion category (84 kg oil eq.) because of the smallest volume of the mixture. Among the considered mixtures, the best combination of the above-mentioned factors is represented by RAC I C100_V2. The sum of normalized results for this foundation structure is -0.31. Foundation with this mixture reached the best environmental result after normalization due to impact in category Metal depletion (-426 kg Fe eq.), which is affected by the amount of two-step recycled aggregate.

Although our results do not represent all possible utilization of recycled aggregate, we suggest concrete containing two-step recycled aggregate as a more environmentally beneficial alteration of natural aggregates.

To support the use of two-step recycled aggregate, future work is planned to designed reusable elements containing RAC. These elements will enable us to reach an even more circular building industry.

5. References

[1] Hansen T C. *Recycling of demolished concrete and masonry: report of Technical Committee 37-DRC, Demolition and Reuse of Concrete*. London; New York: E & FN Spon; 1992.

[2] Guardigli L, Monari F, Bragadin M A. Assessing Environmental Impact of Green Buildings through LCA Methods: Acomparison between Reinforced Concrete and Wood Structures in the European Context. *Procedia Engineering* 2011;**21**:1199–206. https://doi.org/10.1016/j.proeng.2011.11.2131.

[3] Hoxha E, Habert G, Lasvaux S, Chevalier J, Le Roy R. Influence of construction material uncertainties on residential building LCA reliability. *Journal of Cleaner Production* 2017;**144**:33–47. https://doi.org/10.1016/j.jclepro.2016.12.068.

[4] Thiel C L, Campion N, Landis A E, Jones A K, Schaefer L A, Bilec M M. A Materials Life Cycle Assessment of a Net-Zero Energy Building. *Energies* 2013;**6**:1125–41. https://doi.org/10.3390/en6021125.

[5] Marinković S B, Malešev M, Ignjatović I. 11 - Life cycle assessment (LCA) of concrete made using recycled concrete or natural aggregates. *Eco-efficient Construction and Building Materials, Woodhead Publishing*; 2014, **p**. 239–66. https://doi.org/10.1533/9780857097729.2.239.

[6] Marinković S B. Life cycle assessment (LCA) aspects of concrete. *Eco-Efficient Concrete, Elsevier*; 2013, **p**. 45–80. https://doi.org/10.1533/9780857098993.1.45.

[7] Braunschweig A, Kytzia S, Bischof S. Recycled concrete: Environmentally beneficial over virgin concrete? n.d.:12.

[8] Pacheco Torgal F, Jalali S. *Eco-efficient Construction and Building Materials*. London: Springer London; 2011. https://doi.org/10.1007/978-0-85729-892-8.

[9] Tošić N, Marinković S, Dašić T, Stanić M. Multicriteria optimization of natural and recycled aggregate concrete for structural use. *Journal of Cleaner Production* 2015;**87**:766–76. https://doi.org/10.1016/j.jclepro.2014.10.070.

[10] Serres N, Braymand S, Feugeas F. Environmental evaluation of concrete made from recycled concrete aggregate implementing life cycle assessment. *Journal of Building Engineering* 2016;**5**:24–33. https://doi.org/10.1016/j.jobe.2015.11.004.

[11] Tam V W Y, Soomro M, Evangelista A C J. A review of recycled aggregate in concrete applications (2000–2017). *Construction and Building Materials* 2018;**172**:272–92. https://doi.org/10.1016/j.conbuildmat.2018.03.240.

[12] Marinković S, Radonjanin V, Malešev M, Ignjatović I. Comparative environmental assessment of natural and recycled aggregate concrete. *Waste Management* 2010;**30**:2255–64. https://doi.org/10.1016/j.wasman.2010.04.012.

[13] Pacheco Torgal F, Cabeza L F, Labrincha J A. *Eco-efficient construction and building materials: life cycle assessment (lca), eco-labelling and case studies*. Philadelphia, PA: Woodhead Pub; 2014.

[14] Marinković S B, Ignjatović I, Radonjanin V. 23 - *Life-cycle assessment (LCA) of concrete with recycled aggregates (RAs)*. In: Pacheco-Torgal F, Tam VWY, Labrincha JA, Ding Y, Brito J de, editors. Handbook of Recycled Concrete and Demolition Waste, Woodhead Publishing; 2013, **p**. 569–604. https://doi.org/10.1533/9780857096906.4.569.

[15] Evangelista L, de Brito J. Environmental life cycle assessment of concrete made with fine recycled concrete aggregates n.d.:6.

[16] Marinković S, Dragaš J, Ignjatović I, Tošić N. Environmental assessment of green concretes for structural use. *Journal of Cleaner Production* 2017;**154**:633–49. https://doi.org/10.1016/j.jclepro.2017.04.015.

[17] Knoeri C, Sanyé-Mengual E, Althaus H-J. Comparative LCA of recycled and conventional concrete for structural applications. *Int J Life Cycle Assess* 2013;**18**:909–18. https://doi.org/10.1007/s11367-012-0544-2.

[18] Gámez-García D C, Gómez-Soberón J M, Corral-Higuera R, Almaral-Sánchez J L, Gómez-Soberón M C, Gómez-Soberón L A. LCA as comparative tool for concrete columns and glulam columns. *Journal of Sustainable Architecture and Civil Engineering* 2015;**11**:21–31. https://doi.org/10.5755/j01.sace.11.2.10291.

[19] Hossain M d U, Poon C S, Lo I M C, Cheng J C P. Comparative environmental evaluation of aggregate production from recycled waste materials and virgin sources by LCA. *Resources, Conservation and Recycling* 2016;**109**:67–77. https://doi.org/10.1016/j.resconrec.2016.02.009.

[20] Jullien A, Proust C, Martaud T, Rayssac E, Ropert C. Variability in the environmental impacts of aggregate production. *Resources, Conservation and Recycling* 2012;**62**:1–13. https://doi.org/10.1016/j.resconrec.2012.02.002.

[21] de Brito J, Saikia N. *Recycled Aggregate in Concrete*. London: Springer London; 2013. https://doi.org/10.1007/978-1-4471-4540-0.

[22] Weil M, Jeske U, Schebek L. Closed-loop recycling of construction and demolition waste in Germany in view of stricter environmental threshold values. *Waste Management & Research* 2006;**24**:197–206. https://doi.org/10.1177/0734242X06063686.

[23] CEN. *Environmental management - Life Cycle Assessment* - Principles and Framework (ISO 14040:2006). 2006.

[24] Guinee J. Handbook on life cycle assessment - Operational guide to the ISO standards. *Int J Life Cycle* Assess 2001;**6**:255–255. https://doi.org/10.1007/bf02978784.

[25] Verian K P, Ashraf W, Cao Y. Properties of recycled concrete aggregate and their influence in new concrete production. Resources, *Conservation and Recycling* 2018;**133**:30–49. https://doi.org/10.1016/j.resconrec.2018.02.005.

[26] CSN EN 206+A1 *Concrete: Specification, performance, production and conformity, (in czech),* Prague 2018.
[27] GaBi Professional. n.d.
[28] Fiala C. *Optimalizace betonových konstrukcí v environmentálních souvislostech.* 1st ed. Prague: Faculty of Civil Engineering, CTU in Prague; 2011.

Acknowledgement
This work is part of the project „ Design of performance based concrete using sand reclaimed from construction & demolition wastes," LTAIN19205, supported by the Ministry of Education, Youth and Sports- of the Czech Republic, through the program INTER-EXCELLENCE.

The state-of-the-art of methodologies related to biodigester location decisions in clustering situations

Rômulo Henrique Gomes de Jesus[1], Fábio Neves Puglieri[1], Antonio Carlos de Francisco[1], Cassiano Moro Piekarski[1] and Jovani Taveira de Souza[1]

[1]Universidade Tecnológica Federal do Paraná (UTFPR), Ponta Grossa Campus, Graduate Program in Industrial Engineering, Sustainable Production Systems Laboratory (LESP)

* Corresponding Author's E-mail: romulohgj@gmail.com

Abstract. Bioenergy is a renewable energy obtained from biomass and its main benefits are the reduction of greenhouse gases and the reduction of waste disposal. A common problem in feasibility analysis is the biodigester location due to biomass availability and its spatial dispersion. Therefore, decisions regarding to biodigester location is an important aspect of feasibility design when multiple biomass producers are involved in clusters. Studies in this research field of biodigester location address different methodologies, such as mathematical programming, multicriteria approach, scenario analysis, and Geographic Information System (GIS). This paper presents the state-of-the-art in the biodigester location decisions considering clustering situations. To accomplish that, a systematic literature review using Prisma approach was carried out. As results, a set of feasibility opportunities was arranged after reading 24 papers, allowing to identify that the main methodologies used for biodigester location decisions were mathematical programming (41,16%), followed by multicriteria approaches (29,16%) and GIS (12,50%).

Keywords: Biodigester, location decisions, biomass, GIS

1. Introduction

Renewable energy is a solution to problems such as pollution, climate change and energy crisis [1] fulfilling its important role in the global strategy for reducing greenhouse gases [2]. In the context of renewable energy appears as an alternative bioenergy [3]. It is obtained through biomass decomposition [4] that is, waste of vegetation, waste of animal, waste of agriculture in addition to waste urban and certain types of industrial waste [2].

The treatment of biomass can be done through aerobic decomposition (with the presence of air) or anaerobic decomposition (without the presence of air). Anaerobic decomposition is associated with the reduction of pollution and also the production of biogas (methane, carbon dioxide, water vapor, nitrogen and traces of hydrogen, ammonia and carbon monoxide concentration) and biofertilizers [5–7].

Where to locate the anaerobic digester is important because the biomass is geographically and spatially dispersed [8] and its poorly chosen location may interfere with the operational viability of the anaerobic digester project. In this context, studies that address the location of the anaerobic digester are being carried out in different countries.

K. Ioannou et al. [9] developed a study in Greece that aims locate anaerobic digester. Therefore, the authors combine spatial decision methodology, fuzzy logic and the Analytical Hierarchy Process (AHP). The main result of the research is the use of tools and statistics for identify the location where the anaerobic digester can be installed, maximize use and reduce environmental impact and costs. In Portugal, a survey by S. Silva, L. Alçada-Almeida and L.C. Dias [10] addresses the problem of anaerobic digester location in milk producers, thus, randomly defining the size of the plant and the amount of manure that will be processed for energy production. In the study it is proposed about a linear mixed multiple integer program (MMILP) model to optimize the location of the anaerobic digester. In Spain, J.S. Jeong and Á. Ramírez-Gómez [8], proposed a model that seeks to optimize the location of anaerobic digester through the combination of techniques such as Geographic Information System (GIS), Multi-Criteria Decision Analysis (MCDA) and Fuzzy-Decision-Making Trial and Evaluation Laboratory (FDEMATEL). The main criteria studied by the authors were divided between the environmental, geophysical and socio-economic groups. Finally, the most appropriate place for the construction of the anaerobic digester was determined by overlaying maps. In the United States, the authors D. Mukherjee et al. [11] carry out a research that aims to map sources of useful bio-waste for co-digestion in a region dominated by small dairy farms and seeks to find suitable places to install anaerobic biodigester plants. As it also selects the appropriate size and number of digesters in the study, the Mixed Integer Programming (MIP) technique is used in the optimization model.

The bioenergy generated through the biodigester becomes interesting to farms that have agricultural or livestock production because animal and crop residues are generated and if not treated can pollute the environment. Thus, the biomass's application as a raw material is an environmental solution because the waste is does not dispersed in the environment and also an economical solution when biomass is used to generate renewable energy [1].

In the presented context, it is necessary to verify in greater depth the techniques used to locate biodigester because in literature there are not enough studies about this topic due to a gap in the literature. Thus, this study seeks to answer the following question: Which optimization technique is most present in the literature to find the optimal location of biodigesters? To achieve the objective, a literature review was carried out.

This article was structured as follows: section 2 presents the materials and methods. In section 3 the result and discussion are presented and finally in section 4 contains the conclusion and suggestions for future studies.

2. Materials and methods

The intention of this research is to identify in the literature the most present optimization techniques for the location of the biodigester. To obtain the final portfolio of papers was conducted PRISMA approach proposed by A. Liberati et al. [12]. Thus, four steps were taken in PRISMA approach: Research Phase, Filtering Phase, Reading Phase and Inclusion Phase, as shown in figure 1. A systematic literature review was applied through the Methodi Ordinatio (MO) proposed by R. Pagani, J. Kovaleski and L. Resende [13] to ranking the existing knowledge regarding the location of a biodigester. This method considers the three most important factors in a publication: impact factor, year of publication and number of citations.

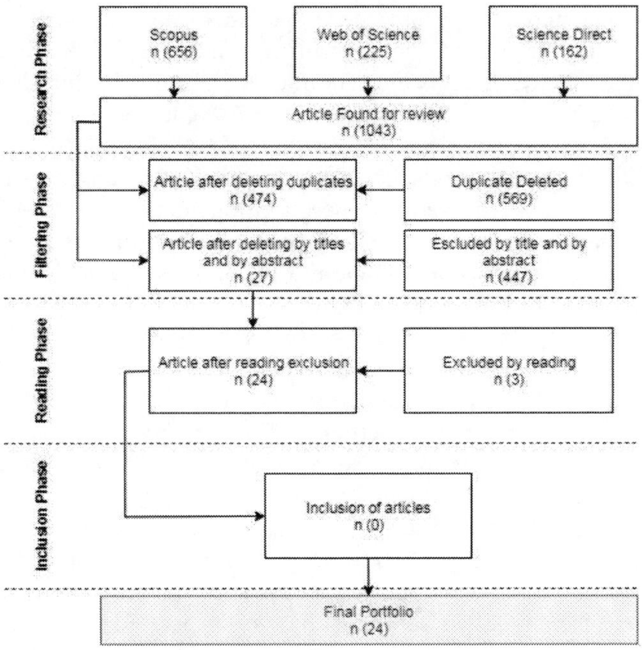

Figure 1. Prisma approach.

In the Research phase, the search through keywords was performed, as shown in table 1, in the Scopus, Web of Science and Science Direct databases.

Table 1. Keywords and databases.

Query (keywords)	Scopus	Web of science	Science direct
"anaerobic digestion" AND "location"	176	66	64
"anaerobic digestion" AND "spatial decision"	1	1	1
"bioreactor" AND "location"	459	138	88
"biomass" AND "spatial decision"	15	14	9
(("anaerobic digestion" AND "location") AND "multicriteria")	2	3	0
(("anaerobic digestion" AND "spatial decision") AND "multicriteria")	1	1	0
(("bioreactor" AND "location") AND "multicriteria")	1	1	0
(("anaerobic digestion" AND "location") AND "multiobjective")	1	1	0
Total per database	656	225	162
Total		1043	

In the research phase, 1043 articles were found. The next step was the filtering phase, 569 duplicate articles were excluded with the use of the Mendeley Desktop® software, leaving the portfolio with 474 articles. After it the reading of the titles and abstracts of the articles were made, eliminating 447 articles that did not fit the research proposal, leaving the final research portfolio with 27 articles.
In the reading phase were excluded 3 papers. It became the article portfolio with 24 papers that will be analysed. Thus, the JabRef® software was used to export the data to an Excel® software spreadsheet, in order to apply the MO to classify the most relevant articles for the study. Table 2 was prepared, and it contains inOrdinatio index and title of articles. The InOrdinatio rank the papers to the most relevant until the least relevant.

Table 2. InOrdinatio index and title of articles.

Title of articles	InOrdinatio
A network design model for biomass to energy supply chains with anaerobic digestion systems	105
Optimizing the location of a biomass plant with a fuzzy-DEcision-MAking Trial and Evaluation Laboratory (F-DEMATEL) and multi-criteria spatial decision assessment for renewable energy management and long-term sustainability	102
Evaluation of biogas potential from livestock manures and rural wastes using GIS in Iran	101
Multicriteria assessment in GIS environments for siting biomass plants	101
A spatial analysis of biogas potential from manure in Europe	97
Optimal location of biorefineries considering sustainable integration with the environment	97
A cyberGIS approach to uncertainty and sensitivity analysis in biomass supply chain optimization	96
A spatial decision support system framework for the evaluation of biomass energy production locations: Case study in the regional unit of drama, Greece	96
What can the location of biogas plants tell us about agricultural change? A case study from the Czech Republic	94
Multiscale scheme for the optimal use of residues for the production of biogas across Castile and Leon	93
Planning the optimal site, size, and feed of biogas plants in agricultural districts	93
Multiobjective programming for sizing and locating biogas plants: A model and an application in a region of Portugal	91
Fuzzy spatial decision tool to rank suitable sites for allocation of bioenergy plants based on crop residue	89
Assessment of the impact of incentives and of scale on the build order and location of biomethane facilities and the feedstock they utilise	84
A multicriteria GIS-based assessment to optimize biomass facility sites with parallel environment - A case study in Spain	84
Designing optimal supply chains for anaerobic bio-digestion/energy generation complexes with distributed small farm feedstock sourcing	83
Spatial decision support system to evaluate crop residue energy potential by anaerobic digestion	82
Robust facility location problem for bio-waste transportation	81
Site selection for landfill gas extraction plant by fuzzy analytic hierarchy process and fuzzy analytic network process in the city of Najafabad, Iran	81
CyberGIS-BioScope: A cyberinfrastructure-based spatial decision-making environment for biomass-to-biofuel supply chain optimization	71
A systemic approach for dimensioning and designing anaerobic bio-digestion/energy generation biomass supply networks	63
Optimal location of centralized biodigesters for small dairy farms: A case study from the United States	63
Optimal renewable energy systems for industries in rural regions	58
Parametric Optimization of Linear and Non-Linear Models via Parallel Computing to Enhance Web-Spatial DSS Interactivity	31

3. Results and discussion

Deterministic techniques are optimization models based on mathematical programming while multicriteria techniques evaluate conflicting criteria in decision making. In this study, emphasis will be given to the multicriteria techniques AHP, fuzzy and FDEMATEL as they were found in the analysed articles.

After reading the articles, the techniques used to locate biodigester were summarized in mathematical programming, multicriteria, GIS and others (scenario analysis, barycenter, NPV and P-Graph). Graph

1 was elaborated, which contains the optimization methods and their respective frequencies found in the literature.

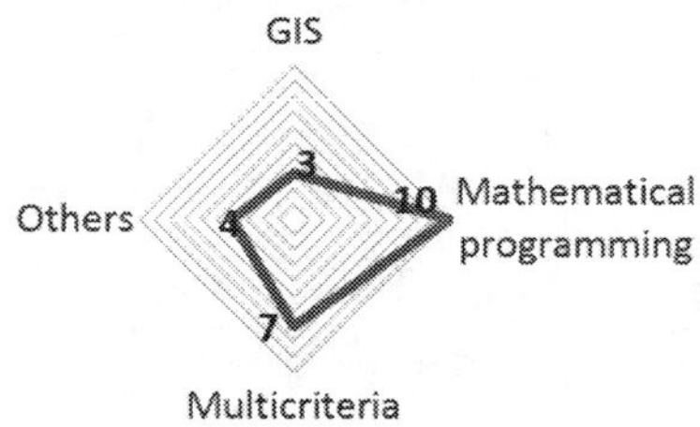

Graph 1. Optimization methods.

It was obtained as a result that the multicriteria and mathematical programming methods were the most used contained in 10 and 7 articles respectively. The use of only GIS techniques were found in 3 articles. The other methods (scenario analysis, barycenter, NPV and P-Graph) were observed once in the remaining articles.
After applying the MO it was observed that the countries with more publications are Spain and the United States with 4 articles, followed by Colombia, Czech Republic, Greece, Iran and Italy with 2 articles and with only one article are Brazil, Ireland, Mexico, Portugal and Turkey, as shown in graph 2.

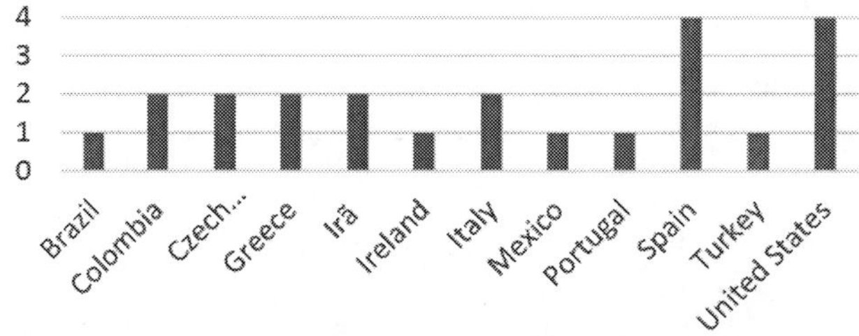

Graph 2. Countries with more publications.

Regarding the years of publications, it was observed that the theme of biodigester localization increased until 2017, with a small drop in 2018, a period that had 2 publications less than the previous year. It is interesting to note that the subject of the study has few papers in the literature, presenting in 2017 the largest number of publications on the subject, as shown in graph 3.

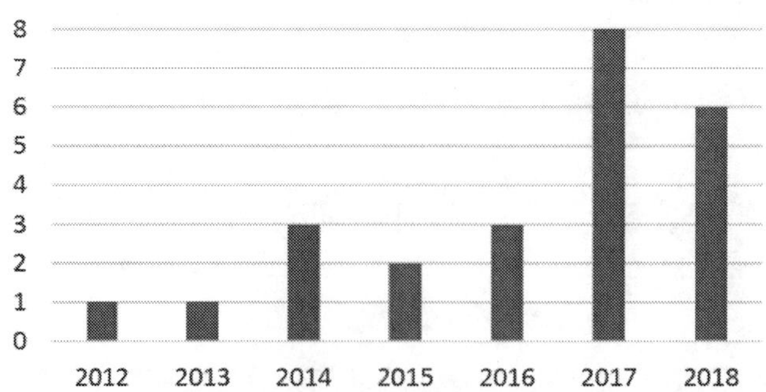

Graph 3. Publication year.

The journal that published more articles addressing the location of the biodigester was Renewable Energy with 4 publications, followed by Applied Energy with 3 publications, Journal of Cleaner Production with 2 publications and other journals with only 1 publication, according to graph 4.

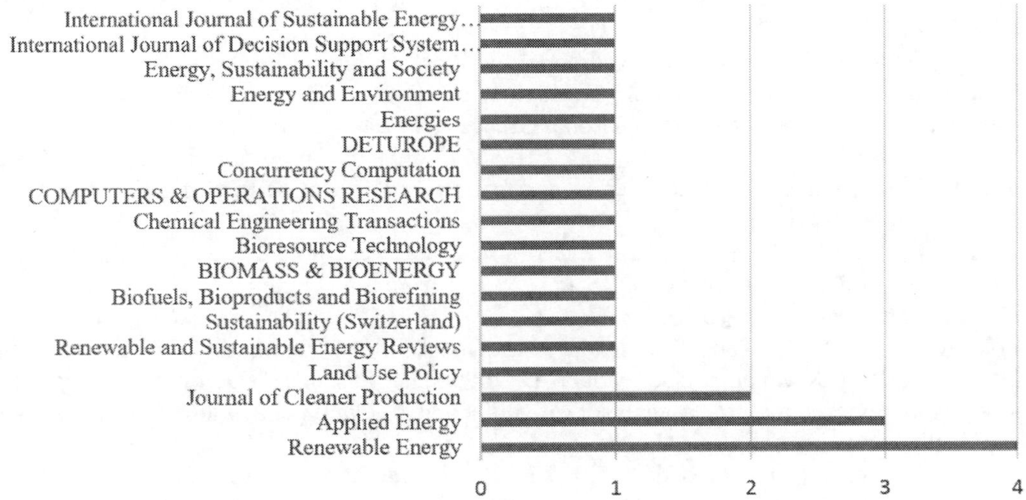

Graph 4. Papers per journal.

4. Conclusion and suggestions

Renewable energy is an alternative to combat environmental degradation, and it is a solution for environmental issues because it reduces CO2 emissions. Thus, bioenergy appears as a choice of renewable energy and has gained relevance. The location of the anaerobic digester is important to guarantee its construction and operational viability. Therefore, studies are being developed to assist in this stage of the planning project, in this sense, aiming at consolidating the optimization techniques used by the authors and covering the literature gap with the consolidation of these, this article was developed, obtaining as the main result that the mathematical programming (41,16%) and multicriteria (29,16%) techniques are the most used by the literature to develop models for optimizing the location of the biodigester. For future studies, it is recommended to consolidate the environmental, social and economic parameters used in the models proposed by the authors to locate anaerobic digester.

References

[1] Khademalhoseiny M S, Ahmadi Nadoushan M and Radnezhad H 2017 Site selection for landfill gas extraction plant by fuzzy analytic hierarchy process and fuzzy analytic network process in the city of Najafabad, Iran *Energy Environ.* **28** 763–74

[2] Iakovou E, Karagiannidis A, Vlachos D, Toka A and Malamakis A 2010 Waste biomass-to-energy supply chain management: A critical synthesis *Waste Manag.* **30** 1860–70

[3] Balaman T Y and Selim H 2014 A network design model for biomass to energy supply chains with anaerobic digestion systems *Appl. Energy* **130** 289–304

[4] Ramachandra T V, Krishna S V and Shruthi B V 2005 Decision support system to assess regional biomass energy potential *Int. J. Green Energy* **1** 407–28

[5] Mayerle S F and de Figueiredo J 2016 Designing optimal supply chains for anaerobic bio-digestion/energy generation complexes with distributed small farm feedstock sourcing *Renew. Energy* **90** 46–54

[6] López-Díaz D C, Lira-Barragán L F, Rubio-Castro E, Ponce-Ortega J M and El-Halwagi M M 2017 Optimal location of biorefineries considering sustainable integration with the environment *Renew. Energy* **100** 65–77

[7] Taifouris M R and Martín M 2018 Multiscale scheme for the optimal use of residues for the production of biogas across Castile and Leon *J. Clean. Prod.* **185** 239–51

[8] Jeong J S and Ramírez-Gómez Á 2018 Optimizing the location of a biomass plant with a fuzzy-DEcision-MAking Trial and Evaluation Laboratory (F-DEMATEL) and multi-criteria spatial

IOP Conf. Series: Materials Science and Engineering 1196 (2021) 012013 doi:10.1088/1757-899X/1196/1/012013

decision assessment for renewable energy management and long-term sustainability *J. Clean. Prod.* **182** 509–20

[9] Ioannou K, Georgios T, Arabatzis G, Andreopoulou Z and Zafeiriou E 2018 A Spatial Decision Support System Framework for the Evaluation of Biomass Energy Production Locations: Case Study in the Regional Unit of Drama, Greece *Sustainability* **10** 531

[10] Silva S, Alçada-Almeida L and Dias L C 2017 Multiobjective programming for sizing and locating biogas plants: A model and an application in a region of Portugal *Comput. Oper. Res.* **83** 189–98

[11] Mukherjee D, Cromley R, Shah F and Bravo-Ureta B 2015 Optimal location of centralized biodigesters for small dairy farms: A case study from the United States *Int. J. Sustain. Energy Plan. Manag.* **8** 3–16

[12] Liberati A, Altman D G, Tetzlaff J, Mulrow C, Gøtzsche P C, Ioannidis J P A, Clarke M, Devereaux P J, Kleijnen J and Moher D 2009 The PRISMA statement for reporting systematic reviews and meta-analyses of studies that evaluate healthcare interventions: explanation and elaboration *BMJ* **339**

[13] Pagani R, Kovaleski J and Resende L 2015 Methodi Ordinatio: a proposed methodology to select and rank relevant scientific papers encompassing the impact factor, number of citation, and year of publication. *Scientometrics* **104** 1–27

Selective Bioflocculation of Ultrafine Hematite Particles using the Yeast Candida stellata

M C Camarate[1], A G Merma[1], C R Simões[1] and M L Torem[1,*]

[1]Department of Chemical and Materials Engineering, Pontifical Catholic University of Rio de Janeiro, Rua Marquês de São Vicente 225, Gávea, Rio de Janeiro, RJ 22453-900, Brazil.

Corresponding author email: torem@puc-rio.br

Abstract. The loss of ultrafine particles has been one of the biggest problems in mineral recovering during concentration process. One alternative to solve this problem is the use of biological raw material in processes like flotation, coagulation and flocculation, however, there are few studies related to ore treatment involving the bioflocculation process. This research aims at a study of the processing of ultrafine hematite particles using the biosurfactant extracted from the yeast Candida stellata. The bioflocculation experiments will be evaluated by jar-test assay, evaluating the influence of parameters like pH, solid concentration and biosurfactant concentration. The interaction of the bioreagent onto the mineral surface will be evaluated by FTIR (Fourier Transform Infra-red) analyzing the functional groups absorbed and surface tension measurement. This research aims as a sustainable route for the recovery of ultrafine hematite particles using a biodegradable material with high efficiency meeting with the requirements demanded for this purpose.

Keywords: Hematite, Quartz, Bioflocculation, Biosurfactant, Iron Ore.

1. Introduction

In mineral processing, the size of the particle is one of the biggest problems about mineral recovering. Depending on the nature of the mineral, the concentration process will not work effectively. For iron ores, froth flotation is not worthy in size particles less than 10 μm [1-4]. In these cases, an operation like coagulation, flocculation, magnetic separation or leaching are more useful for hematite recovery [5-6].

The flocculation can be realized by chemicals or by microorganisms and their byproducts, which absorb onto the mineral surface and changes their characteristics (electrostatic charge and hydrophobic feature), act like surface modifiers [7-10]. The bioflocculation has some advantages for being a eco-friendly, low cost and flexible operation and has a high efficiency to separate the valuable mineral from the gangue [5]. Yeasts are eukaryotic microorganisms and they are delimited by a plasma membrane containing phospholipids, which have proteins in their composition. [11] The biosurfactant extracted from the yeasts are commonly used in the food and pharmaceutical industry, but in mineral processing, there are few studies about it. The Candida stellata can produce a biosurfactant called sophorolipids, a glycolipid formed by a hydrophilic part (sepharose) and a hydrophobic part (a long chain of fatty acid) [12-15].

In this work, the biosurfactant extracted from the yeast Candida stellate is used for the selective flocculation of hematite particles. In the flocculation assays, we evaluated the best conditions of pH, solids concentration and biosurfactant concentration for the hematite recovery. The interaction of the biosurfactant onto the mineral surface is evaluated by FTIR and surface tension.

2. Materials and Methods

Mineral samples
The quartz samples were obtained from a local supplier in Belo Horizonte, Minas Gerais State - Brazil, and the mineral was comminuted in a jaw crush and dry-ground in a porcelain mortar to achieve the required particle size. The hematite sample was provided by Sigma-Aldrich in a powder form (>96%, 5 µm).

Biosurfactant Extraction and yeast culture
The Candida stellata strain was provided by the collection of Reference Bacteria on Health Surveillance (CBVRS) – Oswaldo Cruz foundation. The solid medium and the broth composition was by YMG type, that consists of 20.0 g/L glucose, 5.0 g/L peptone, 3.0 g/L malt extract, 3.0 g/L yeast extract and 20.0 g/L agar (only for solid medium). A six days incubation in a rotatory shaker at 151 rpm, 28° C is imposed on the broth. After that time, a centrifugation operation (3200g, 8 min) is needed to separate the culture medium from the cells, requiring a re-suspending with deionized water. According to Moreau et al. [16], the extraction of the biosurfactant begins with a centrifugation and a re-suspending of the culture medium in ethanol for one day and then passed to an autoclave (20min, 1 bar, 121 °C) for sterilization. After that, another centrifugation was realized to separate the soluble part from the cells and dried for 24 hours at 45 °C and then the material was dissolved in deionized water and a filtration (8 µm) was done to separate de insoluble part, finally getting the final solution. The biosurfactanct concentration was obtained by dry-weight measurement.

Flocculation Tests
The flocculation tests were conducted by jar-test and the minerals were studied individually and analyzed the influence of pH, solids concentration and biosurfactant concentration. The experiments were made from 5 to 120 minutes time, in a 500 mL graduated cylinder, a 5 minutes time for the interaction of the biosurfactant and the mineral and 20 minutes of vigorous mixing. After that, the flocculation was started so the time and clarified height were recorded.

For the pH experiments, the solids and the biosurfactant concentrations were fixed. The range of pH was from 3 to 11. The solids concentration influence, the best condition of pH and the biosurfactant concentration were constant. The values of solids concentration vary from 0,1% (0.50 g) to 2.0% (10.0 g). The last tests were made in order to evaluate the influence of biosurfactant concentration, fixing the best conditions of the last tests. The values vary from 50.0 mg/L to 200.0 mg/L. A solution of NaOH and HCl was used for the adjustment of pH.

Surface Tension Measurement
DCA-200S tensiometer (Surface Electro optics) was used for the surface tension measurement. For the first experiment, the concentration of the biosurfactant was analyzed in a range from 0 to 250 mg/L at pH 3. The critical micelle concentration (CMC) was obtained. Aliquots of NaOH and HCl were used to adjust the pH values.

Fourier Transform Infrared – FTIR
Nicolet FTIR 2000 spectrophotometer (Attenuated total reflection) was used for the FTIR analysis. The pre-treatment of the samples was a drying at temperature of 50 °C. The analyses were made before and after the interaction of the biosurfactant with the minerals.

3. Results and Discussions

Bioflocculation of hematite and quartz

The agglomeration of particles depends on the balance of forces between the particles (attractive and repulsive), the electrostatic repulsion keeps the minerals in suspension before the interaction, because they had a surface charge. According to the literature, the higher flocculation of hematite is expected near of it isoelectric point (IEP), around 7, and the quartz flocculation will be higher near it isoelectric point, around 2, but the experiments were conducted in higher pH values [17-18].

In the experiments with pH 3 before the interaction, the quartz recovery (~63%) was higher than the hematite recovery (~54%) due to the proximity of the quartz IEP. After interacting with the biosurfactant the flocculation results suffered a modification. The recovery of hematite increased from ~54% to ~91%. At pH 7, the effect of the biosurfactant is very unnoticeable, related to a non-adsorption of the biosurfactant onto the mineral surface.

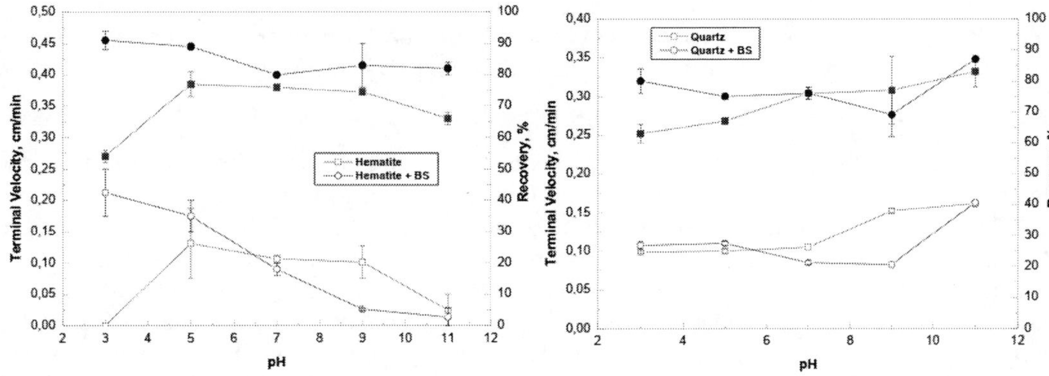

Figure 1. Recoveries and terminal velocities of hematite and quartz particles before and after interacting with the biosurfactant as a function of pH. Solid concentration: 0.10% and Biosurfactant concentration: 50 mg/L.

The quartz bioflocculation had a similar profile as the hematite experiments, having a higher recovery at pH 3 than pH 7, this could be related to a higher adsorption at pH 3 than at pH 7. But compared to the hematite, the increment of the quartz recovery was lower (around 15%) while the hematite recovery grows around 40%.

The best response for this stage were the experiments at pH 3, but it doesn't show a selective separation of the hematite from quartz, according to the recovery. However, the terminal velocities could help to explain the separation. In acidic conditions, the terminal velocities in both minerals had grown after interacting with the biosurfactant, being higher for the hematite at pH 3. At pH 3, the terminal velocities of hematite and quartz were around 0.22 cm/min and 0.12 cm/min, respectively (obtained in 20 and 40 minutes, respectively) hence that the separation of hematite from quartz could be achieved during the beginning of the flocculation.

Figure 1 shows the recovery and the terminal velocities of hematite and quartz before and after the interaction with the biosurfactant, respectively.

Figure 2 shows the recovery and terminal velocities of the hematite and quartz mineral as a function of the solids concentration. The higher hematite recoveries were at 0.25% (1.25 g) and 0.50% (2.50 g) >95% and for quartz, the maximum recovery was at 2.00% (10.0 g), ~90%. Previous studies indicated that solids concentration has influence on the collisions in the initial moments of the flocculation due to high sediment concentration and low turbulence. However, high solids concentration promotes an increment of the flocs breakage and influence the settling rate. Therefore, a medium solids concentration is ideal for the operation [19].

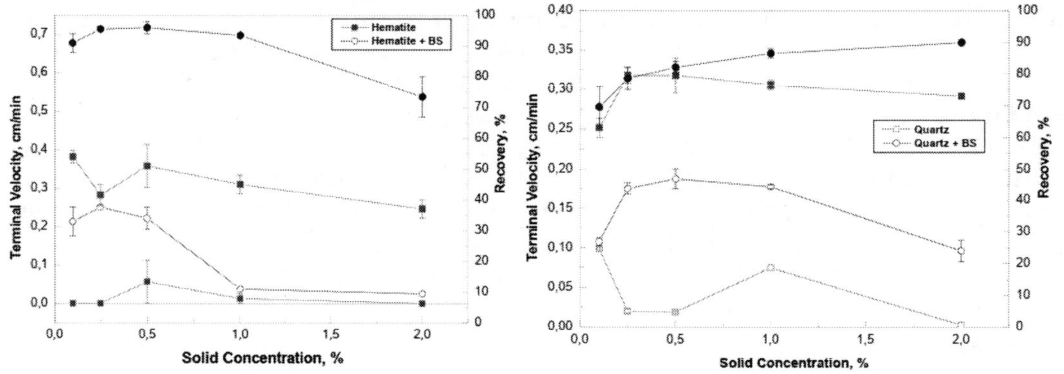

Figure 2. Recoveries and terminal velocities of hematite and quartz particles before and after interacting with the biosurfactant as a function of solids concentration. pH: 3 and biosurfactant concentration: 50 mg/L.

Figure 3 shows the flocculation tests of hematite and quartz as a function of biosurfactant concentration, at pH 3 and solids concentration of 0.25% (1.25 g), before and after interacting with the biosurfactant. For the hematite recovery, the best responses were at 75 mg/L biosurfactant concentration (~99%) and 100 mg/L (~95%) and for the quartz, the best conditions were ate 150 mg/L and 200 mg/L biosurfactant concentration (~85%). These results suggest that the biosurfactant had an affinity with the hematite and adsorbs into the hematite surface, making it hydrophobic. For a selective flocculation of the hematite particles, a 75 to 100 mg/L biosurfactant concentration is enough to achieve the maximum recovery. Using a higher concentration indicates a lower recovery of hematite, which can be explained by a restricted adsorption into the hematite surface, reducing the hydrophobicity [20].

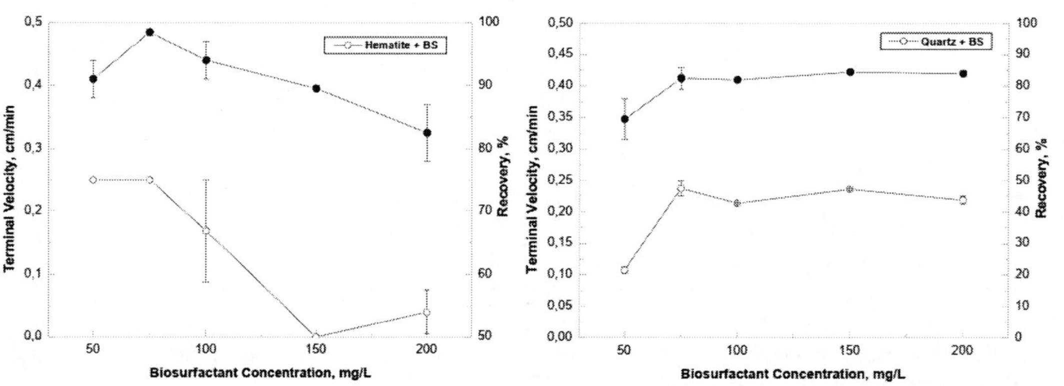

Figure 3. Recoveries and terminal velocities of hematite and quartz particles before and after interacting with the biosurfactant as a function of biosurfactant concentration. pH: 3 and solid concentration: 0.10%.

Surface Tension

For the analysis of the biosurfactant surface tension as a function of concentration, after mathematical regression, the critical micellar concentration around 150 mg/L was founded. The surface tension was decreased to 30 mN/m. These results are accorded to previous works [17-18]. Figure 4 shows the surface tension as a function of the concentration.

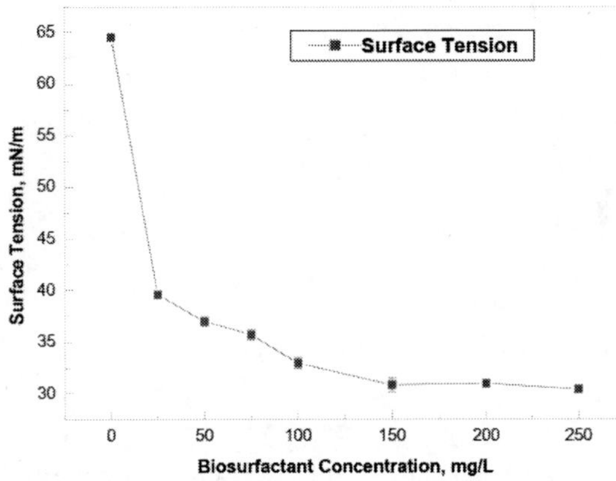

Figure 4. Surface tension of the biosurfactant as a function of biosurfactant concentration.

FTIR Analysis
The infrared analyses were studied between 3500 and 100 cm-1 because the characteristic bands for both minerals are found below 1000 cm-1 and the biosurfactant extracted from microorganisms have complex compositions with several functional groups, which can be seen in the region of infrared spectrum between 3500 and 1000 cm-1 [21-23].

Figure 5 and 6 show the FTIR spectra of hematite and quartz particles before and after the interaction at pH 3, respectively. The FTIR spectra of the hematite, Figure 5, evidence strong adsorption onto the mineral surface. According to literature, this result suggests a chemisorption driven force, in order to remove weak attached substances. After the interaction, several peaks were included, the 2924 and 2853 peaks correspond to alkyl C-H stretch, the peak between 1780 and 1630 correspond to carbonyl stretching absorption and the peak at 1090 corresponds to phosphate group in teichoic acid [17; 24-25].

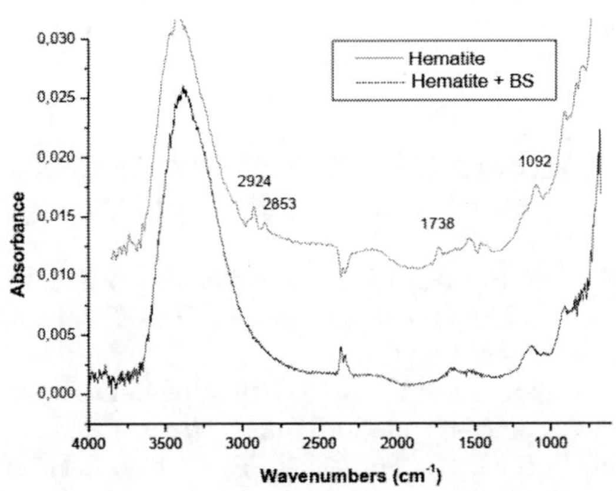

Figure 5. FTIR spectra of hematite before and after interacting with the biosurfactant pH 3.

The FTIR spectra of quartz was observed, no changes of the quartz after the interaction, Figure 6, evidencing a weak adsorption of the biosurfactant, concluding a higher selectivity of the biosurfactant onto the hematite surface.

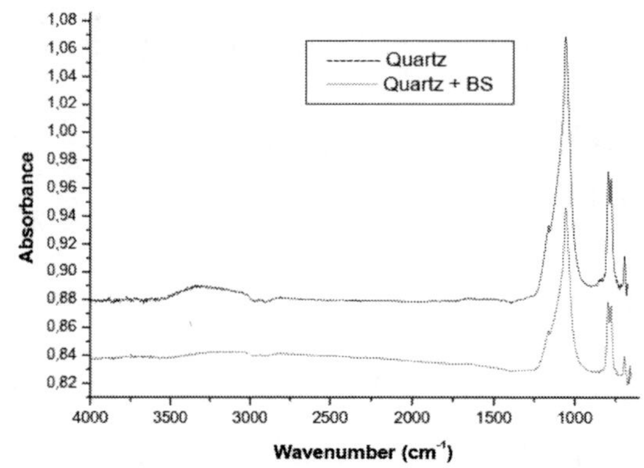

Figure 6. FTIR spectra of quartz before and after interacting with the biosurfactant pH 3.

4. Conclusions

The biosurfactant extracted from the yeast Candida stellata behaves as a good flocculant for hematite particles compared to the quartz. The best conditions for the jar test flocculation assays were: pH 3 (~91%), a 0.50% (1.25 g) solids concentration (~96%) and a 75 mg/L biosurfactant concentration (~99%), besides having the best terminal velocities. The interaction of the biosurfactant onto the hematite surface was confirmed in the FTIR analysis, confirming the non-adsorption onto the quartz surface. The surface tension measurement confirms the surfactant properties of the biosurfactant compounds, owning a CMC of 150 mg/L at pH 3. From this research, further studies can be carried out in order to obtain more information about hematite biofloculation technique using a biosurfactant extracted from a yeast. Other parameters that could be analyzed for its influence are the time of process conditioning, volume of the test tube and the mechanical stirring time provided before the flocculation process.

5. References

[1] Tadesse B, Albijanic B, Makuei F, Browner R. 2018. Recovery of fine and ultrafine mineral particles by electroflotation- A review. *Mineral Processing and Extractive Metallurgy Review*, 1-15, Taylor and Francis.

[2] Hacha R R, Torem M L, Merma A G, Coelho V F. 2018. Electroflotation of fine hematite particles with R. opacus as a biocollector in a modified Partridge-Smith cell. *Minerals engineering*, **126**, 105 – 115. Elsevier.

[3] Rubio J, Capponi F, Matiolo E, Rosa J J. 2004. Advances in flotation of copper and molybdenum sulphide ore fines. *In: XXI National Meeting of Ores Treatment and Extractive Metallurgy*, vol. **2**, **pp**. 69–78. Florianópolis-SC.

[4] Peng Y, Liang L, Tan J, Sha J, Xie G. 2015. Effect of flotation reagent adsorption by different ultra-fine coal particles on coal flotation. *Int. J. Miner. Process*. **142**, 17–21. Elsevier.

[5] Hosseini M R, Sadeghieh S M, Azizinia M R, Tabatabaei S H. 2019. Biological separation of quartz from kaolinite using Bacillus licheniformis. *Separation Size and Technology*. Elsevier.

[6] Zhang J, Sun W, Gao Z, Niu F, Wang L, Zhao Y, Gao Y. 2018. Selective flocculation separation of fine hematite from quartz using a novel grafted colopymer flocculant. *Minerals*, **8**, 227, 1 – 11. MDPI.

[7] Harajima T, Aiba Y, Farahat M, Okibe N, Sasaki K, Tsuruta T, Doi K. 2012. Effect of microorganisms on flocculation of quartz. *International Journal of Mineral Processing*, **102-103**, 107 – 111. Elsevier.

[8] Vilinska A, Rao H. 2010. Leptosririllum ferrooxidans-sulfide mineral interactions with reference to bioflotation and bioflocculation. *Trans. Nonferrous Met. Soc.* China, **18**, 1403 – 1409. Science Press.

[9] Selim K A, Rostom M. 2018. Bioflocculation of (iron oxide-silica) system using Bacillus cereus bacteria isolated from Egyptian iron ore surface. *Egyptian Journal of Petroleum*, **27**, 235 – 240. Elsevier.

[10] Abdel-Khalek N A, Selim K A, Amir A, Faharat M M, El-Hendaway H H, Elbaz R M. 2015. Enhancement of Selective Bioflocculation of (Quartz- Pyrolusite_Hematite) System Using Paenibacillus polymyxa Bacteria. *Bio Technology*, **89**, 36712-36717. Elixir.

[11] Fukuda E K, Vasconcelos A F D, Matias A C, Barbosa A, de M, Dekker R F H, Silva M, de L C da. 2009. Fungal cell wall polysaccharides: purification and characterization. *Semina: Ciências Agrárias*, **30** (1), 117-134. CAB Direct.

[12] Garcia M, Esteve-Zarzoso B, Cabellos J M, Arroyo T. 2018. Advances in the study of Candida stellata. *Fermentation*, 4, **74**, 1-22. MPDI.

[13] Ciani M, Ferraro L. 1998. Combined use of immobilized Candida stellata cells and Saccharomyces cerevisiae to improve the quality of wines. J. *Appl. Microbiol.* **85**(2), 247–54.

[14] Kurtzman C P, Price N P J, Ray K J, Kuo T M. 2010. Production of sophorolipid biosurfactants by multiple species of the Starmerella (Candida) bombicola yeast clade. *FEMS Microbial Lett*, **311**(2), 140-6.

[15] Daverey A, Pakshirajan K. 2009. Production of sophorolipids by the yeast Candida bombicola using simple and low cost fermentative media. *Food Res Int.*, **42**(4), 499-504.

[16] Moreau R, Powell M J, Singh V. 2003. Pressurized Liquid Extraction of Polar and Nonpolar Lipids in Corn and Oats with Hexane, Methylene Chloride, Isopropanol, and Ethanol. *Journal of the American Oil Chemists' Society*, **80**, 1063–1067.

[17] Puelles J G S. 2016. *Hematite flotation using a crude biosurfactant extracted from R. opacus. Master Dissertation*. Chemical Department, Pontifical Catholic University of Rio de Janeiro.

[18] Olivera C A C. 2018. *Hematite-quartz system flotation using the soluble biosurfactant extracted from R. erythropolis. Doctoral Thesis*. Chemical Department, Pontifical Catholic University of Rio de Janeiro.

[19] Lai H, Fang H, Huang L, He G, Reible D. 2018. A review on sediment bioflocculation: Dynamics, influencing factors and modelling. *Science of the Total Environment*, **642**, 1184-1200. Elsevier.

[20] Shrimali K, Atluri V, Wang Y, Baccuwar S,Wang X, Miller J D. 2018. The nature of hematite depression with corn starch in the reverse flotation of iron ore. *Journal of Colloid and Interface Science*. Elsevier.

[21] Wang Y, Maramatsu A, Sugimoto T. 1998. FTIR analysis of well-defined α-Fe2O3 particles. *Colloids and Surfaces. A: Physicochemical and Engineering Aspects,* **134**, 281-297. Elsevier.

[22] Maity D, Agrawal D C. 2007. Synthesis of iron oxide nanoparticles under oxidizing environment and their stabilization in aqueous and non-aqueous media. *J. Magn. Magn. Mater.* **308** (1), 46–55.

[23] Liu W, Liu W, Wei D, Li M, Zhao Q, Xu S. 2017. Synthesis of N,N-Bis (2-hydroxypropyl) laurylamine and its flotation on quartz. *Chem. Eng. J.*; **309**:63–9.

[24] Garip S, Gozen A C, Severcan F. 2009. Use of Fourier transform infrared spectroscopy for rapid comparative analysis of Bacillus and Micrococcus isolates. *Food Chem.* **113** (4), 1301–07. Elsevier.

[25] Hacha R R, Torem M L, Merma A G, Coelho V F. 2018. Electroflotation of fine hematite particles with Rhodococcus opacus as a biocollector in a modified Partridge-smith cell. *Mineral Engineering*, **126**, 105 – 115. Elsevier.

[26] *Corresponding author: Pontifical Catholic University of Rio de Janeiro/ Department of Chemical and Materials Engineering. Head – Mineral Processing Research Group, Rua Marquês de São Vicente, 225 - Gávea - 22451-900, Rio de Janeiro - RJ – Brazil Tel. + 55 (21) 3527-1723. Email: torem@puc-rio.br

Electroflotation of iron ore fines using biosurfactant

Carolina Simões[1], Ronald Rojas[1], Marcelo Camarate[1], Maurício Torem[1,*]

[1]Department of Chemical and Materials Engineering, Pontifical Catholic University of Rio de Janeiro, Rua Marquês de São Vicente 225, Gávea, Rio de Janeiro, RJ 22453-900, Brazil

***Email:** torem@puc-rio.br

Abstract. Conventional flotation shows low recovery of the fine particles due to the low probability of bubble-particle adhesion and collision, what can be solved with the electroflotation process. Electroflotation uses oxygen and hydrogen microbubbles (<100μm) generated from water electrolysis. In addition, this process can become a biotechnology using a biosurfactant. The present work aims to evaluate the recovery of hematite fine particles from an iron ore using the electroflotation process with a biosurfactant obtained from *Rhodococcus opacus*. The tests were conducted with an iron ore (-38+20μm) in a Partridge-Smith modified electroflotation binary cell. The parameters used in these tests were current density (16 mA/cm^2), agitation (300rpm), electrolyte concentration (0.2mol/L), conditioning time (5min) and flotation time (10 min). The pH range between 3 and 11, biosurfactant concentration (50 to 800mg/L) were evaluated and up to now, the results show that the electroflotation process was able to concentrate the hematite of the iron ore.

Keywords: Eletroflotation, biosurfactant, iron ore and hematite.

1. Introduction

Mineral processing has a large amount of waste which is in tailings dams nowadays. These dams occupy big areas, deforesting the local vegetation and may contaminate the groundwater and even if there is a risk of rupture bringing irreparable damages to environment and population (Fontes, et al. 2016). Tailings and slimes of the iron ore concentration (fine and ultrafine particles) are stored in these dams. The recovery of fine (F) and ultrafine (UF) particles continues to generate challenges for the mineral sector, since conventional flotation processes present low recovery. This behaviour can be attributed to low probability of bubble-particle collision and adhesion, difficulty in overcoming the energy barrier, mechanical drag, high adsorption of reagents, low selectivity, surface oxidation, and mineralogical alteration [9]. Another important variable for the flotation process of the F and UF particles is the bubble size. The smaller the bubble size the better the recovery efficiency of F and UF particles due to a large surface area of the bubbles[20]. Some technologies have been studied for recovery of the F and UF particles based on the concept of increasing the efficiency of bubble-particle adhesion and collision. One of these techniques is the electroflotation [19].

The electroflotation process generates bubbles of oxygen and hydrogen smaller than 100 μm from the electrolytic decomposition of the water. This process allows the control of several factors in production and size of bubbles such as: current density, material and electrode type, electrolyte concentration and system pH [19]. This process can become a biotechnology when the synthetic surfactants are replaced by biosurfactants (BS) with similar characteristics. Biosurfactants are amphipathic substances equal to synthetic surfactants, however, had different origin, are obtained from the microorganisms such as bacteria, yeasts, and fungi in different carbon sources [1]. An important characteristic of BS is the high degradability when compared to conventional reagents used in flotation [14]. There are different applications already widely disseminated to the uses for microorganisms in industry in leaching and oxidation processes, however, few studies focused on mineral flotation. Therefore, this study aims to evaluate the electroflotation of fine particles of iron ore using a BS extracted from *Rhodococcus opacus*.

2. Materials and methods

2.1. Iron ore sample and electrodes
The iron ore sample went through the jaw crusher and ball mill to reduce the particle size. Different types of analyses such granulometric (wet-sieving), mineralogical (X-ray diffraction) and chemical (volumetry using potassium dichromate and X-ray fluorescence) were performed. Platinum electrodes were used for the anode and the cathode. The total area of the electrode is 25 cm^2, with the wire spacing of 0.16 mm and the wire diameter of 0.08 mm.

2.2. Preparing and production of biosurfactant
The biosurfactant uses to the electroflotation tests was obtained from *Rhodococcus opacus*. First, the bacterium was cultured in solid medium (Yeast Malt Glucose – YMG) in Petri plates. After that, they were used to create a liquid medium (YMG) in a 250 ml Erlenmeyer, staying in the rotating shaker for 7 days. The bacterial solution was centrifuged at 4500 rpm for 10 min (CIENTEC CT-5000). The biomass was washed with deionized water twice and stored in 98% ethyl alcohol for 24 hours. And then, the biomass-alcohol solution was placed in an autoclave (CS PRISMATEC) for 20 min at 1 atm for extraction of BS. The BS solution stayed at the greenhouse for approximately 48 hours. Finally, the BS was solubilized in deionized water and filtrated. The not-soluble BS was discarded and the soluble BS was used in the electroflotation process.

2.3. Electroflotation tests
Electroflotation of iron ore was performed on a modified Partridge-Smith binary cell with a volume of 380 ml as shown in Figure 1.

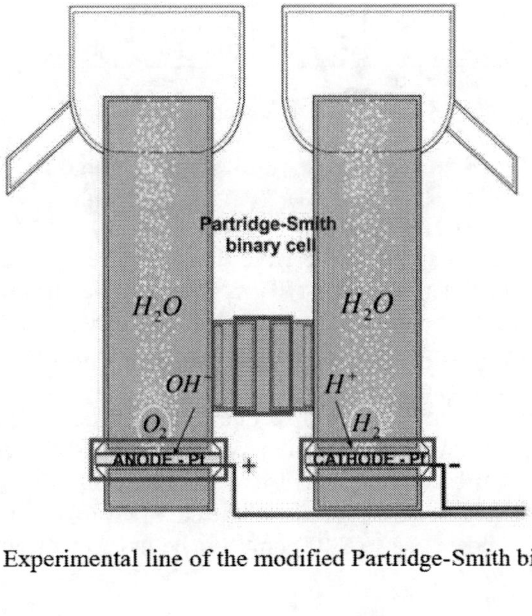

Figure 1. Experimental line of the modified Partridge-Smith binary cell.

At first, 190 ml of the indifferent electrolyte sodium sulfate (Na_2SO_4) was added in cell. A tension was applied to the electrodes initiating electrolysis, generating hydrogen gas bubbles at the cathode and oxygen gas bubbles at the anode. After that, was necessary stabilize the system with the operating conditions used in this study. At the same time, the iron ore was conditioned with the BS in a glass Beaker for 5 min in agitation. After the conditioning, the solution composed by the BS and the iron ore were added to the cathode cell and the total volume was filled with the electrolyte. The electroflotation lasted for 10 minutes. White tests (without BS), were also carried out under the same operating conditions to determinate the mechanical drag. The operating conditions used in the electroflotation test are : pH: 3, 5, 7, 9 and 11, electrolyte concentration of 0.2 mol / L, biosurfactant concentration of 50, 100, 300 and 500mg / L, 1 g of iron ore mass, current density 16mA / cm^2, shaking 300 rpm, hydrogen gas bubble (H_2), conditioning time 5 min and electroflation time 10 min.

3. Results and discussion

3.1. Sample of iron ore

The sample of iron ore was submitted to the granulometric analysis by CILLAS 1190. The results of the particle size fraction used (-38 +20 μm). The average particle size was of 30 μm. The results of the chemical analysis by X-ray fluorescence (FRX) indicated that the sample was consisted by 82.5% hematite and 16.5% quartz. The mineralogical analysis made by the X-ray diffraction (Figure 2) showed that the sample was composed of hematite and quartz.

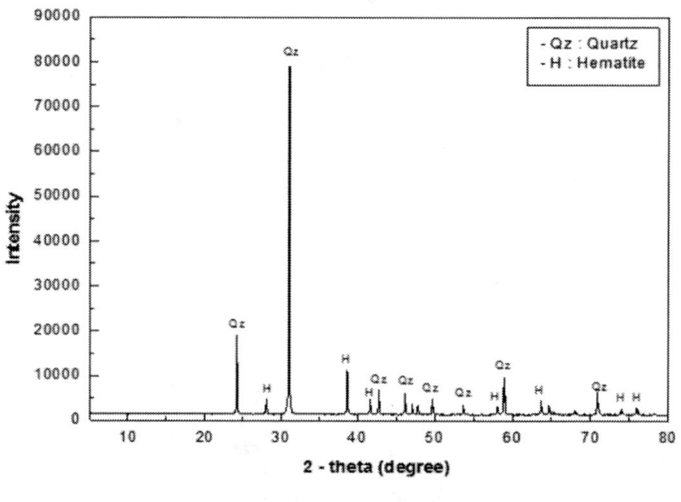

Figura 2. X – ray difracttion of iron ore.

3.2. Fourier transform infrared (FTIR) spectroscopy

FTIR analysis of hematite, quartz and iron ore were done (Figure 3). The mineral hematite spectrum shows the adsorption bands between 450 and 470 cm^{-1} that can be attributed to the vibration mode of the Fe-O group, characteristic of the hematite phase. The adsorption band in 540 cm-1, and the bands between 670 and 680 cm^{-1} are associated with the vibration mode of the Fe-O group [10]. The mineral quartz spectrum shows bands ranging from 400 to 1200 cm^{-1} that can be attributed to Si-O bonds [8]. It was observed that the 1085.74 cm^{-1} vibration is due to the asymmetrical stretching of the Si-O, the adsorption band of 778.49 cm^{-1} corresponds to the symmetrical Si-O stretch vibration, the adsorption band of 690.49 cm^{-1} can be attributed to symmetrical Si-O flexion vibration, and the adsorption band of 460.18 cm^{-1}can be attributed to asymmetric vibration of Si-O bond [8]. The iron ore spectrum shows that the absorbance bands close to 3400 cm^{-1}can be associated with water molecules [6]. The adsorption bands between 1083.84 and 797.96 cm^{-1}can be related to the Si-O bonds, and the bands of 544.13 and 463.23cm^{-1}can be attributed to Fe-O groups.

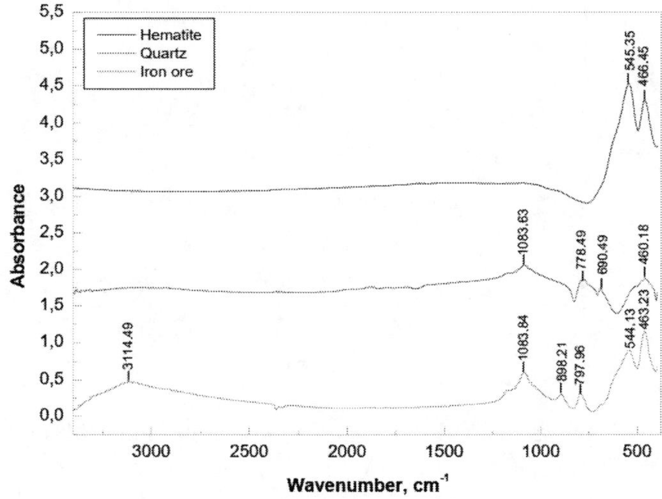

Figure 3. Infrared Spectrogram.

3.3. Electroflotation Test

White test (without BS) were performed under the same operating conditions (Table 1) to determine the entrainment. Under these conditions, entrainment was less than 5% of total mass.

3.3.1.Effect of pH. The pH effect in the systemis very important for flotation processes, mainly for electroflotation. This variable influences the mineral surface because the surface charge is determined by the solution pH [16]. The system pH can also influence the microorganism surface properties. Some functional groups present in the biosurfactant can go through changes depending on the pH system, these changes can influence the mineral/BS interaction [6]. The tests were performed at a pH range of 3 to 11 with BS concentration of 300 mg/L and current density of 18 mA/cm^2 (Figure 4). Under these conditions, recovery around 78% and Fe grade close to 59.4%at pH 3 were observed. The recovery decreased as the pH increased up to pH value of 7. This behavior can be attributed to the electrostatic interactions between the hematite and the BS in this pH range [3]. This can be also explained with the help of hematite zeta potential. The hematite PIE is generally found at pH values from 5 to 8. The pH values below of the hematite PIE exhibit positive surface charge and the pH values above of the hematite PIE exhibit negative surface charge [11]. The BS presents anionic characteristics allowing a greater interaction in acidic medium (below the PIE) with hematite favoring the recovery. The not-soluble surfactant species are found in higher concentrations. At pH 11the recovery was around 51.33% with an iron grade of 51.37%.

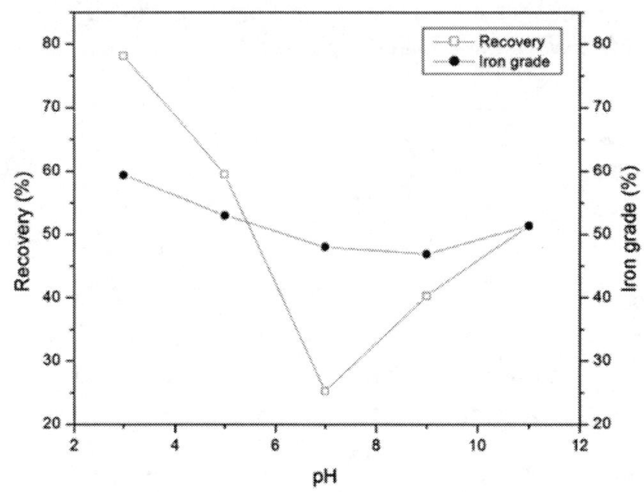

Figure 4. Effect of pH on the grade and recovery. BS concentration: 300 mg/L; current density: 16 mA/cm2.

3.3.2.Effect of biosurfactant concentration. This evaluation was performed with different BS concentrations from 50 mg/L to 800 mg/L at pH 3 and current density 18 mA/cm2. The recovery depends directly of the BS concentration (Figure 5), the increase of the BS concentration favors the increase of the recovery up to 300 mg/L. In this concentration value was achieved a recovery of 78% with iron grade of 59.4%. Above this concentration, the recovery decreases, this reduction can be attributed to the critical micellar concentration (CMC) of the BS. The biosurfactant CMC was around 100 mg/L [11]. CMC is the surfactant concentration, where organized molecular assemblies known as micelles are formed, decreasing the interactions of the biosurfactant with the mineral surface and consequently negatively affecting the flotation recovery [13].

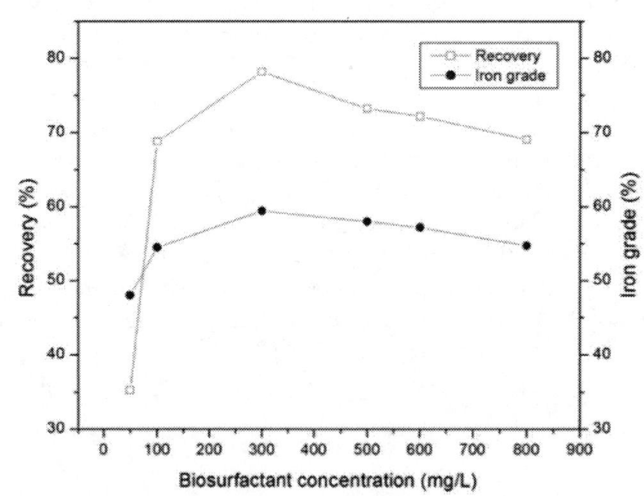

Figure 5. Effect of BS concentration on the iron grade and recovery. pH 3; current density: 16 mA/cm^2.

4. Conclusion

The results obtained in the electroflotation study of fine particles of iron ore with the use of the BS extracted from *Rhodococcus opacus* allow concluding that the recovery was satisfactory. Satisfactory conditions of electroflotation were found at pH 3 and with a BS concentration of 300 mg/L, under these conditions a recovery of 78% and an iron grade of 59% were achieved. Further studies are needed to understand the behaviour of the electroflotation process of fine particles of iron ore using the BS extracted from *Rhodococcus opacus*.

5. References

[1] Bezerra M. (2012). Estudo da produção de biossurfactantes sintetizados por Pseudomonas aeruginosa AP029-GVIIA utilizando manipueira como fonte de carbono. *Doctoral thesis, Federal University of Rio Grande Do Norte*. Natal/RN, 125.

[2] Botero A. (2007). Bioflotation of Magnesite calcite and barite using Rhodococcusopacus as a bioreagent, *Doctoral thesis, Pontifical Catholic University, Rio de Janeiro*.

[3] Bueno B Y M, Torem M L, Molina F, de Mesquita L M S. (2008). Biosorption of lead(II), chromium(III) and copper(II) by R. opacus: Equilibrium and kinetic studies. *Minerals Engineering*, **21**(1), 65–75. https://doi.org/10.1016/j.mineng.2007.08.013Correa Bicca, F., Colombo Fleck, L., & Záchia Ayub, M. A. (1999). Production of Biosurfactant By Hydrocarbon Degrading. Measurement, 231–236. Retrieved from http://mmbr.asm.org/cgi/doi/10.1128/MMBR.67.4.503-549.2003

[4] Ekmekyapar F. (2006). Biosorption of copper (II) by nonliving lichen biomass of Cladoniarangiformishoffm, *Journal of Hazardous Materials*, v. **137**, n. 1, **p**. 293-298.

[5] Maity D, Agrawal D C. (2007). Synthesis of iron oxide nanoparticles under oxidizing environment and their stabilization in aqueous and non-aqueous media, *Journal of Magnetism and Magnetic Materials*, v.**308**, **pp**.46-55.

[6] Merma A G, Torem M L, Morán J J V, Monte M B M. (2013). On the fundamental aspects of apatite and quartz flotation using a Gram positive strain as a bioreagent. *Minerals Engineering*, **48**, 61–67. https://doi.org/10.1016/j.mineng.2012.10.018.

[7] Mesquita L M S, Lins F A F, Torem M L.(2002) Mineral biobeneficiation: potentialities of microorganisms as flotation reagents, *Mineral technology series*, ISSN 0103-7382. CETEM.

[8] Oliveira C A C. (2018) Flotation of the hematite-quartz system using the soluble biosurfactant produced by Rhodococcuserythropolis, *Doctoral Thesis - Chemical and Materials Engineering, Pontifical Catholic University, Rio de Janeiro*.

[9] Peng Y, Liang L, Tan J, Sha J, Xie G. (2015). Effect of flotation reagent adsorption by different ultra-fine coal particles on coal flotation. *International Journal of Mineral Processing*, **142**, 17–21. https://doi.org/10.1016/j.minpro.2014.12.005

[10] Prakash S, Das B, Venugopal R. (1999). Surface properties of Indian hematite and bauxite and their coating mechanism with colloidal magnetite. *Journal of Scientific and Industrial Research*, **58**(6), 436–442.

[11] Puelles J G S. (2016). Flotation of hematite using biosurfactant extracted from Rhodococcusopacus, Master's Dissertation - Materials Engineering, *Chemical and Metallurgical Processes*, Pontifical Catholic University, Rio de Janeiro.

[12] Rao S R. (2004). Surface Chemistry of Froth Flotation, Kluwer Academic/Plenum Publishers, 2°nd edn, 744 **p**.

[13] Santos D K F, Rufino R D, Luna J M, Santos V A, Sarubbo L A. (2016). Biosurfactants: Multifunctional biomolecules of the 21st century. *International Journal of Molecular Sciences*, **17**(3), 1–31. https://doi.org/10.3390/ijms17030401.

[14] Sanwani E, Chaerun S, Mirahati R, Wahyuningsih T. (2016). Bioflotation: Bacteria-Mineral Interaction for Eco-friendly and Sustainable Mineral Processing. *Procedia Chemistry*, **19**, 666–672. https://doi.org/10.1016/j.proche.2016.03.068.

[15] Stratton H M. (2002). Cell surface hydrophobicity and mycolic acid composition of Rhodococcus strains isolated from activated sludge foam, *Journal of Industrial Microbiology and Biotechnology*, v. **28**, n. 5, **pp**. 264-267.

[16] Stumm W. (1992). *Chemistry of the solid-water interface: processes at the mineralwater and particle-water interface in natural systems,* John Wiley & Son Inc.

[17] Sun W. (2011). Hydrogen bubble flotation of fine minerals containing calcium, *Mining Science and Technology*, China, v. **21**, n. 4, **pp**. 591-597.

[18] Szymanska A, Sadowski Z. (2010). Effects of biosurfactants on surface properties of hematite,Adsorption, v.16, pp. 233–239.

[19] Wei S, Liang M, Yuehua H, Dong Y, Zhang G. (2011). Hydrogen bubble flotation of fine minerals containing calcium. *Mining Science and Technology*, **21**(4), 591–597. https://doi.org/10.1016/j.mstc.2011.01.002.

[20] Yao J, Yin W, Gong E. (2016). Depressing effect of fine hydrophilic particles on magnesite reverse flotation. *International Journal of Mineral Processing*, **149**, 84–93. https://doi.org/10.1016/j.minpro.2016.02.013

6. Acknowledgements

The authors acknowledge PUC-Rio (Pontifical Catholic University of Rio de Janeiro); CNPq (National Council for Scientific and Technological Development); CAPES (Coordination for the Improvement of Higher Level Personnel); and FAPERJ (Rio de Janeiro State Research Foundation) for financial and technological support.

Best response dynamics in waste-to-energy plants' price setting problem

I Eryganov[1,*], R Šomplák[2], J Hrdina[1] and V Nevrlý[2]

[1] Institute of Mathematics, Faculty of Mechanical Engineering, Brno University of Technology, Technická 2896/2, Brno, 616 69, Czech Republic.
[2] Institute of Process Engineering, Faculty of Mechanical Engineering, Brno University of Technology, Technická 2896/2, Brno, 616 69, Czech Republic.
* **E-mail:** xperyga00@vutbr.cz

Abstract. This paper deals with the task of waste energy recovery and effective price setting of waste-to-energy plants. Its main contribution is the invention of a new approach to pricing of waste disposal in a particular waste management network. This novel method, which is based on a problem of finding the Nash equilibrium, takes into account the amounts of waste production of cities, capacities of waste-to-energy plants and their locations. The best response dynamics algorithm enables the computation of Nash equilibrium for a game of waste-to-energy plants in a normal form with numerous players and multiple strategies in comparison with other standard algorithms. The algorithm is able to find the Nash equilibrium for a sufficiently big step size between possible gate fees. The results of this work are applicable in the forecasting of prices of waste-to-energy plants' services for real waste management networks.

Keywords: waste management, waste-to-energy plant, gate fee, game theory, Nash equilibrium, best response.

Introduction

Waste management is an important and relevant topic that nowadays dynamically develops due to a significant increase in the world's attention to modern environmental problems. This accretion can be easily illustrated by an increasing number of various articles about waste management in recent years: according to the Web of Science search of the keyword "waste management" there are 1,955 articles on this topic in 2011 and 4,950 articles in 2018 [1]. The main issues of waste management are monitoring and regulation of collection, transportation, treatment and disposal of waste [2]. Waste management is closely connected to the concept of circular economy because one of their common goals is an effective treatment of solid waste.

The Circular Economy Package, which is a circular economy initiative adopted by the European Commission, sets up a series of goals which dwell in a decrease of the amount of solid waste that is being landfilled and an increase of its material and energy recovery. However, actual capacities of already built waste treatment facilities in some EU countries can be insufficient for that and new facilities should be built. Before this is done, an analysis of the feasibility of possible investments, waste transportation optimization and waste market modelling are needed.

Thus, various tools to support waste management planning have to be considered. They are usually built on optimization methods that are part of operations research techniques and used in design of sustainable supply chains. An article about application of optimization in designing of cost-effective waste management network in Thailand [3] and a study of optimization of routing in waste management network [4] can serve as examples. The review of the articles about applications of the above-mentioned means and assessment of the current state of the waste management studies are presented in [5]. In this article, we will focus on waste that is not suitable for material use; in this case, a highly effective waste-to-energy (WtE) plant technology should be considered.

At the Institute of Process Engineering of Brno University of Technology, a computational tool NERUDA [6] was developed supporting decision making in waste management. From the government's point of view, the results of its implementation are optimal and there are no complications. However, in case of solving a particular project applied to a certain locality, some of the parameters such as gate fees, which, for example, can correspond to a required return on investment, have to be fixed in order to maximize the revenues of WtE plants. That fact does not correspond to a real conflict of interests on a waste market and can be interpreted only as an analysis of a particular scenario.

Thus, the next logical step is an application of game theory that can by its nature provide a more realistic insight into market modelling and is a rational extension of existing approaches. By now, there is a certain lack of methods that use this area of mathematics to solve waste management problems. The majority of existing articles mainly use the basics of game theory. A brief literature review of the application of the game theory in waste management is presented at the end of the introduction, after the considered problem is described.

This article focuses on an identification of the price resulting state in an already built network (locations, productions and capacities are already given) that can be used in strategical planning (which plants should be built), forecasting of cash flow (how much will cities pay for waste disposal) and overall analysis of the waste management network (what capacities of the plants are sufficient and what locations are preferable).

In the game theory exist two basic approaches: cooperative and non-cooperative. The first one assumes that players can somehow agree on their decisions in order to increase their payoffs. However, in a price setting problem cooperation would mean the existence of the illegal collusion about price level that could damage the prosperity of consumers. Therefore, the non-cooperative approach, which describes a situation when each player pursues only his or her welfare without paying attention to profits of other players, will be used.

An approach to price setting problem used in the article mostly corresponds to [7]. WtE plants with specified capacities are located in an area and individually set their gate fees in order to maximize their income. Waste producers, who are represented by cities, are cooperatively minimizing their total costs with respect to the gate fees and transportation costs. The game occurs at the moment, when waste producers are trying to choose prices that will change demand for their prosperity. The primary purpose of the research is to find a stable set of gate fees such that none of WtE plants would like to change their gate fee. Particularly in this article, the well-known Nash equilibrium is considered to be such a stable point. The outcome can be used in solving further problems such as the interaction between cities or their usage of routes in waste management network to deliver waste to certain WtE plants.

In [8], the Stackelberg game of WtE incentive policy with a government in the leading role and dairy farms in the following role is discussed and solved. However, in the price setting problem of WtE plants, there is more than one player in the leading role, and a compromise between them is studied. The problem of finding the Nash equilibrium arises in [9] for effective funding of the recovery and recycling system and in [10] for invention of coordination strategies between participants of the market of the waste electrical and electronic equipment disposal in China. Nevertheless, in the above-mentioned articles, the considered payoff functions are continuous and differentiable; that assumption enables to analytically find the Nash equilibrium using first derivatives of the payoff

IOP Conf. Series: Materials Science and Engineering 1196 (2021) 012016 doi:10.1088/1757-899X/1196/1/012016

functions. In the model considered in this article, the payoff functions fail to meet the assumption of the continuity and differentiability. The evolutionary game theory approach performed in [11] cannot be used because the considered model is deterministic.

The contribution of this article is that a previous approach (based on an extensive search on small strategy spaces) to non-cooperative behavior of WtE plants in price setting problem [7] is improved by increasing cardinality of the sets of possible gate fees for which equilibrium can be found in adequate time. A novelty that enables such an improvement lies in using an algorithm based on the best response dynamics which will be useful even in case of many WtE plants in an arbitrary area. The sequential best response dynamics algorithm is based on the natural idea of non-cooperativity: every player independently sequentially maximizes his or her payoff function and this process, iteratively repeated, leads to an equilibrium state.

Materials and methods

All the concepts used in this section can be found in the standard book on game theory courses [12]. Considering the price setting a normal form game, where strategies of players are the gate fees, the main interest revolves around an analogy of stability when no player would like to change his or her decision. It will lead to the concept of a strategy profile of stable prices which corresponds to the Nash equilibrium and can be used in a further study of waste management problems. Thus, the main concern will be the computation of the Nash equilibrium.

Model description

Let $N = \{1, \dots, n\}$ be a set of WtE plants; $w_1^{\mathrm{c}}, \dots, w_n^{\mathrm{c}}$ denotes their capacities and $C_1^{\mathrm{g}}, \dots, C_n^{\mathrm{g}}$ denotes their strategy sets (sets of possible gate fees) with an element $c_j^{\mathrm{g}} \in C_j^{\mathrm{g}}, j \in N$. The set of producers is $M = \{1, \dots, m\}$. Their waste productions are $w_1^{\mathrm{p}}, \dots, w_m^{\mathrm{p}}$. Transportation costs are given by the matrix $[c_{i,j}^{\mathrm{t}}]$, where $c_{i,j}^{\mathrm{t}}$ represents the cost of waste transportation from the producer $i \in M$ to the plant $j \in N$. In the following expressions $x_{i,j}$ denotes the amount of waste sent by the producer $i \in M$ to the WtE plant $j \in N$ in tonnes. For each producer $j \in N$, the payoff function π_j is defined as $\pi_j(c_1^{\mathrm{g}}, \dots, c_n^{\mathrm{g}}) = \sum_{i \in M} c_j^{\mathrm{g}} x_{i,j}^*$, where $x_{i,j}^* \in \{x_{i,j}^* : i \in M, j \in N\}$, such that

$$\{x_{i,j}^* : i \in M, j \in N\} = \arg \min_{x_{i,j}: i \in M, j \in N} \sum_{j \in N} \sum_{i \in M} \left(c_{i,j}^{\mathrm{t}} + c_j^{\mathrm{g}}\right) x_{i,j} \qquad (1)$$

$$s.t. \sum_{i \in M} x_{i,j} \le w_j^{\mathrm{c}}, \ \forall j \in \mathrm{N}, \qquad (2)$$

$$\sum_{j \in N} x_{i,j} = w_i^{\mathrm{p}}, \ \forall i \in M,$$
(3)

$$x_{i,j} \ge 0, \ \forall i \in M, \ \forall j \in N.$$
(4)

The previous equations describe cooperative minimization of total costs by cities and the fact that they have to dispose of all waste they produce and cannot exceed the capacities of WtE plants. A linear cost minimization problem, the solution of which is partially included in the players' payoff function, has to be solved in order to compute their payoff for a given strategy profile. However, the solution $\{x_{i,j}^* : i \in M, j \in N\}$ is not necessarily unique, thus a choosing rule, which will work equivalently for all of the players, should be established. There are two most common ways of choosing a particular solution: optimistic and pessimistic. The optimistic approach is the confidence of a player that if total costs of using his or her services equals the total costs of using services of another player for one of cities, then this city will always choose to work with him or her. Analogically, the pessimistic one describes the situation when a player expects payoff corresponding to the worst possible scenario. In this article the latter approach will be used.

By now two of three necessary milestones to define the normal form game of WtE plants have been established: the set of players $N = \{1, \dots, n\}$ and their payoff functions $\pi_j(c_1^g, \dots, c_n^g), j \in N$ have been defined. Only one question remains: what are the strategy sets of WtE plants?

The previously defined payoff functions are not differentiable or continuous. As a result, their derivatives cannot be described in order to analytically find the Nash equilibrium, and there is no special necessity to consider continuous strategy sets. This fact leads to an algorithmic way of finding the Nash equilibria. Moreover, a choice of discrete strategy spaces will help to avoid problems with the computation of maxima of payoff functions which occur in the case of continuous strategy sets. Therefore, in this article, the strategy sets of players will be described as discrete sets $C_j^g = \{kq, k \in N_0, k < K\}, j \in N$, where $q > 0$ is a step and K is a sufficiently large positive integer. In the next section, possible algorithms for finding the Nash Equilibrium will be discussed.

Algorithms

The first algorithm is a direct search of the Nash equilibrium. By definition of the Nash Equilibrium a strategy profile $c^{*,g} \in C_N^g = \Pi_{j \in N} C_j^g$, such that $\pi_j\left(c_j^{*,g}, c_{-j}^{*,g}\right) \geq \pi_j\left(c_j^g, c_{-j}^{*,g}\right), \forall c_j^g \in C_j^g, \forall j \in N, c_{-j}^{*,g} = \left(c_i^{*,g}\right)_{i \in N \setminus j}$, has to be found. In each step of the algorithm, an arbitrary strategy profile is chosen and checked whether the previous condition is satisfied for all possible strategies of each player. The problem arises when there are too many possible combinations: if 40 players with two strategies are considered, in the worst-case scenario it means checking 2^{40} possible combinations.

The idea of the second algorithm, which can be found in [13], dwells in the removal of dominated strategies. A strategy $c_j^{1,g} \in C_j^g$ is dominated by a strategy $c_j^{2,g} \in C_j^g$ if $\pi_j\left(c_j^{2,g}, c_{-j}^g\right) \geq \pi_j\left(c_j^{1,g}, c_{-j}^g\right), \forall c_{-j}^g \in C_{-j}^g$, and $\pi_j\left(c_j^{2,g}, c_{-j}^g\right) > \pi_j\left(c_j^{1,g}, c_{-j}^g\right)$ for some $c_{-j}^g \in C_{-j}^g$. Therefore, in the first step, all the dominated strategies of the first player are removed, and his or her strategy set is reduced to a set of undominated strategies. In the second step, the same approach is applied for the second player using the new reduced strategy set of the first player and so forth. If such an algorithm leads to a singleton, then this strategy profile is the Nash equilibrium. However, the issue remains: when there are 40 players and every player has at least two strategies, 2^{40} combinations have to be computed and checked in the first step of the algorithm.

The third algorithm is called sequential best response dynamics. Its detailed description can be found in the already mentioned text [13] or more specifically in [14]. The main idea of the algorithm is natural: a process starts at a given point and at each iteration every player chooses a strategy from his or her best response correspondence $B_j\left(c_{-j}^g\right) = \{c_j^g \in C_j^g : \pi_j\left(c_j^g, c_{-j}^g\right) \geq \pi_j\left(c_j^{\prime,g}, c_{-j}^g\right), \forall c_j^{\prime,g} \in C_j^g\}$. The new starting strategy profile for the next player is obtained from a chosen strategy of the previous player. If the algorithm converges to some strategy profile, then this profile is the Nash equilibrium. The advantage of the algorithm is apparent: at each step, only the maximum of a given player's payoff function has to be computed. In order to compute the maximum of the player's payoff function, all $K + 1$ points in his or her strategy set will be tried out with fixed prices of rivals. Thus, in one full iteration (the moment when all players would make their decision) $|N|(K + 1)$ linear programs have to be solved to find out a maximal payoff of each player. However, the main disadvantage of this algorithm is the fact that it can get stuck in a cycle. Figure 1 explains the main principles of the algorithm.

Comparing all the previous algorithms, it was decided to choose the sequential best response dynamics in order to solve the realistic exemplary problem due to its satisfactory computational complexity.

Case study

In this section results of applying sequential best response dynamics to the exemplary problem will be presented. The data that will be used in a case study does not describe any real region. However, the values of model parameters are realistic and were chosen in such way that they correspond to average figures valid for developed countries of the EU. The primary purpose of the case study is to show the functionality of the algorithm and non-triviality of the resulting equilibrium due to different operating conditions. The results of the computation of the Nash equilibria will be demonstrated starting with a large step between gate fees in strategy sets, and then it will be gradually reduced.

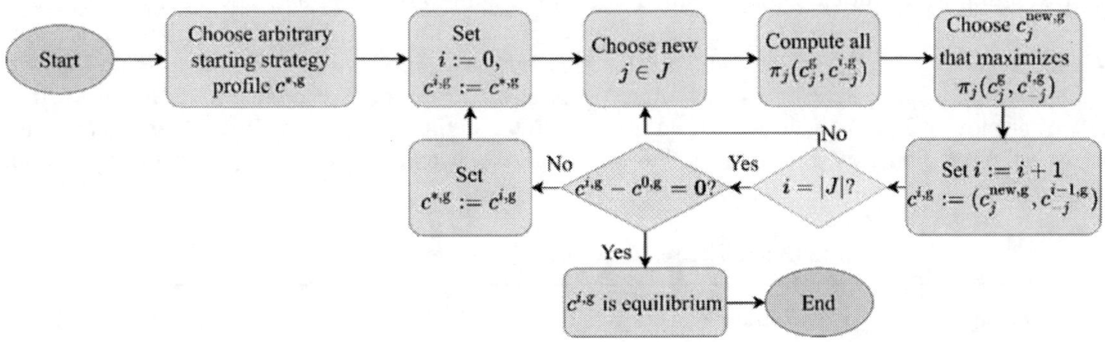

Figure 1. Flowchart of the algorithm based on the sequential best response dynamics.

Problem description

The exemplary problem is given by the data in table 1. In order to create an artificial upper bound in the strategy sets of players, the existence of a WtE plant with a fixed gate fee and capacity which will meet the total production of waste in the network will be considered (in real life that would mean already built foreign WtE plants, with price that cannot be impacted). Particularly, plant K with the capacity of 1,550 kt will have a fixed gate fee of 125 €. Productions and capacities are given for a period of one year.

Table 1. The network data of the exemplary problem.

Producer\WtE plant	Transportation costs (€/t)											Waste production (kt)
	A	B	C	D	E	F	G	H	I	J	K	
1	10	35	40	25	30	25	15	5	5	10	40	170
2	15	35	35	15	20	25	20	15	5	5	50	130
3	20	35	25	5	15	20	30	25	10	5	60	110
4	15	45	40	30	20	15	10	5	5	10	50	200
5	15	40	35	10	10	10	10	15	5	5	60	150
6	30	45	30	5	10	15	20	35	10	5	60	160
7	20	45	45	25	15	5	5	5	5	10	70	180
8	30	55	50	15	10	5	10	15	5	5	80	250
9	35	50	40	5	5	5	15	20	10	5	90	200
WtE plant capacity (kt)	50	75	275	150	125	250	100	175	200	200	1,550	

As best response dynamics could not answer the question about the uniqueness of equilibria, three starting strategy profiles (lower bound, middle of strategy sets and upper bound) will be tried out for each chosen step in order to find out if the algorithm converges to the same result. Obviously, it is not sufficient to make a general conclusion about the uniqueness of the found equilibria. Analogically, if the algorithm gets stuck in a cycle for all of the considered starting strategy profiles, the existence of equilibria for the given step cannot be disproved because a more complex and detailed analysis of the situation is needed.

Results and discussion

The results of the algorithm execution are presented in table 2. The step size of 22.5 €/t is the smallest for which the algorithm converges and after reducing it, a cycle begins to occur for each starting point. Thus, the equilibrium for that step (the penultimate row in table 2) will be considered as the most precise solution of the price setting problem. From the stable gate fees and information about used capacities an overall analysis of the situation can be made. The rationality of WtE plants guarantees that plant K with a fixed gate fee will not be used by waste producers. WtE plants B and C with inconvenient locations (represented by high transportation costs) are still competitive; however, their gate fees are lower than the gate fees of their rivals and C is the only plant with 18 % of capacity left unused. All WtE plants except B and C have the same gate fee, which points out similar operating conditions.

Table 2. The results of the algorithm execution.

Step size [€]	Starting point [€]	The highest possible price [€]	Time of execution [s]	Gate fee [€]									
				A	B	C	D	E	F	G	H	I	J
100	0	400	1.72	100	100	100	100	100	100	100	100	100	100
100	200	400	2.36	100	100	100	100	100	100	100	100	100	100
100	400	400	2.99	100	100	100	100	100	100	100	100	100	100
50	0	300	1.41	100	100	100	150	100	150	150	150	150	150
50	150	300	2.78	100	100	100	150	100	150	150	150	150	150
50	300	300	4.57	100	100	100	150	100	150	150	150	150	150
25	0	300	2.6	125	100	125	125	125	150	150	150	150	150
25	150	300	5.24	125	100	125	125	125	150	150	150	150	150
25	300	300	8.58	125	100	125	125	125	150	150	150	150	150
22.5	0	315	2.89	135	90	112.5	135	135	135	135	135	135	135
22.5	157.5	315	6.98	135	90	112.5	135	135	135	135	135	135	135
22.5	315	315	12.1	135	90	112.5	135	135	135	135	135	135	135
Capacity used for step size 22.5 €/t [%]				100	100	82	100	100	100	100	100	100	100

In the considered exemplary problem the difference between the total waste production and the total capacity of the all WtE plants with the non-fixed price is rather small, this is the reason why percentages of the used capacities are so high and the operation of the whole network will be sustainable. However, during strategic planning in the big size waste management network, problems with a small use of capacity could occur. This fact will lead to some requirements: for example, considered configuration of WtE plants will be built only if more than 60 % of each plant's capacity will be used. At this point a further complication arises: elimination of the WtE plant with the smallest use of capacity is not always optimal (from the customers' point of view) and could lead to an increase in total waste producers' expenses. Therefore, all possible eliminations should be compared and the optimal one should be selected. If the problem occurs again, this process should be repeated iteratively. Such an approach could enable a stable compromise between the prosperity of producers and WtE plants.

Observing the results, some general statements about best response dynamics application and the found Nash equilibria of the considered price setting problem can be made: for sufficiently large steps the algorithm converges to the Nash equilibrium, which corresponds to the lowest possible gate fees (except zero); with reducing the step size, for some WtE plants it becomes more profitable to deviate to some greater gate fees, but the algorithm is still able to find the Nash equilibrium; from a particular step size, the algorithm gets stuck in a cycle with at least two WtE plants constantly changing their gate fees and there exists a possibility, that a change of gate fee is always profitable for a small step size and no stable point exists.

In this article, only the basic sequential best response dynamics algorithm with exhaustive search method optimization on discrete sets has been used. Therefore, there are ways of improvement of the considered approach. Instead of exhaustive search method, more rational ways of finding the maxima of payoff functions using bilevel optimization techniques, which can gradually reduce the time of computation, can be used. Moreover, such techniques could enable us to find optimality conditions of players from which the equilibrium can be analytically found.

Ways of choosing the best solution should be discussed in further studies of this issue. Analogically to approach used in this article, the solution found for the smallest step size, for which the algorithm still converges, could be considered the most precise. Thus, the use of heuristics in best response dynamics to reduce the step for which algorithm is able to find the equilibrium should also be considered: players can admit a decrease in their payoffs in order to avoid cycle or can be satisfied with points in a neighborhood of the actual Nash equilibrium.

The further development of the research in this area will focus on the application of an invented method on real data and detailed analysis of results based on actual operating conditions in a region.

Conclusion

In this article, the game theory has been applied to the process of setting gate fees by WtE plants in order to show possible applications of this area of mathematics in the analysis of the sustainable waste management networks. The mathematical model of conflict between WtE plants has been described and the corresponding normal form game has been defined. Moreover, the aim to find the Nash equilibrium of this normal form game has been set in order to get an outline of the possible stable outcome of the mentioned process. After that, algorithms for finding the Nash equilibria were discussed and compared. The sequential best response dynamics has been chosen as an appropriate algorithm due to its computational complexity: it is able to find the Nash equilibrium of the normal form game with many players and large strategy sets in adequate time. Results of applying best response dynamics to the problem with realistic waste management data have been presented together with the brief analysis of the exemplary waste management network based on the found equilibria. Moreover, observation of the results has enabled the establishment of ideas about some consistent pattern of the Nash Equilibria values for specific step sizes in strategy sets. This idea has been described in three statements which are part of subsection 3.2. Future research on that topic might be devoted to mathematical proofs of those statements, implementing rational bounds in the strategy sets, searching for conditions of existence and uniqueness of the Nash equilibria for the WtE plants' games in a normal form and improving the algorithm by using heuristics or bilevel programming methods. The main contribution of the invented approach is that due to its speed it enables us to effectively set gate fee in an arbitrary waste management network and can be used in the iterative process of design of sustainable waste management networks described in subsection 3.2. This could lead to the network design in which expenses of waste producers are minimal, and all plants use most of their capacities with a sustainable gate fee.

References

[1] Web of Science n.d. Retrieved from https://apps.webofknowledge.com/
[2] Anca N Iuga 2016 IOP Conf. Ser.: Mater. Sci. Eng. 161 012086
[3] Olapiriyakul S, Pannakkong W, Kachapanya W and Starita S 2019 Multiobjective optimization for sustainable waste management network design J. *Adv. Transp.* 2019 **pp** 1–15
[4] Sahoo S, Kim S, Kim B-I, Kraas B, and Popov A 2005 Routing optimization for waste management Interfaces **35**(1) **pp** 24–36

[5] Barbosa-Póvoa A P, da Silva C and Carvalho A 2018 Opportunities and challenges in sustainable supply chain: an operations research perspective Eur. J. *Oper. Res.* **268**(2) pp 399–431
[6] Šomplák R, Pavlas M, Kropáč J, Putna O and Procházka V 2014 Logistic model-based tool for policy-making towards sustainable waste management Clean Technol. *Environ. Policy* **16**(7) **pp** 1275–86
[7] Osicka O 2016 Game theory in waste management MSc thesis Brno university of technology
[8] Zhao N and You F 2019 Dairy waste-to-energy incentive policy design using Stackelberg-game-based modeling and optimization *Appl. Energy* **254**
[9] Zhang D, Cao Y, Wang Y and Ding G 2020 Operational effectiveness of funding for waste electrical and electronic equipment disposal in China: an analysis based on game theory Resour. *Conserv. Recycl.* **152**
[10] Guo L, Qu Y, Tseng M-L, Wu C and Wang X 2018 Two-echelon reverse supply chain in collecting waste electrical and electronic equipment: A game theory model Computers & Industrial Engineering 126 **pp** 187–95
[11] Chen J, Hua C and Liu C 2019 Considerations for better construction and demolition waste management: identifying the decision behaviors of contractors and government departments through a game theory decision-making model J. *Cleaner Prod. 212* **pp** 190–9
[12] Owen G 2013 Game theory (Bingley: Emerald)
[13] Moulin H 1984 Dominance solvability and Cournot stability Mathematical Social Sciences **7**(1) **pp** 83–102
[14] Matsui A 1992 Best response dynamics and socially stable strategies *Journal Of Economic Theory* **57**(2) **pp** 343–62

Acknowledgments
The authors gratefully acknowledge the financial support provided by ERDF within the research project No. CZ.02.1.01/0.0/0.0/16_026/0008413 "Strategic Partnership for Environmental Technologies and Energy Production" and FME BUT within the research project No. FSI-S-20-6187.

Evaluating the Ecological Sustainable Transportation System with An Application of Radial Basis Function Neural Network

Yan Zhuang[1], Chunjiao Dong[1,*], Jianpei Qian[1], Shengyou Wang[1], Song Xue[1]

[1] Key Laboratory of Transport Industry of Big Data Application Technologies for Comprehensive Transport, Beijing Jiaotong University, Beijing 100044, China

***Email:** cjdong@bjtu.edu.cn

Abstract. The ecological and sustainable development of urban transportation system is critical to improving the overall quality of life and achieving the goals of climate protection and sustainable development. The study focused on factors from level of social and economic development, the quality of transport service and urban ecological environment to develop a comprehensive assessment indicator system. A novel method for evaluating the ecological sustainable transportation system based on the radial basis function neural network（RBFNN）was proposed. Results of comparative experiments between RBFNN and the conventional BP neural network (BPNN) showed that the accuracy of RBFNN was 8.3% higher than that of BPNN while the training time was 27.2s lesser and the Root Mean Square Error (RMSE) was 0.15 smaller. In the case of Shenzhen, China, the proposed model gave a reasonable evaluation, which implied that RBFNN brought forward a new perspective for further research.

Keywords: Ecological sustainable transportation system; Radial basis function; Neural network; Assessment indicator system; Evaluation model

1. Introduction

Sustainable development is of great strategic significance for the development of human society. As a crucial part of the national economy, the transportation industry is an energy-intensive industry and needs to go further for ecological and sustainable development. There has been a trend in exploring different methods to evaluate the ecological and sustainable urban transportation development system. The main popular evaluation methods used are as follows: [1] applied the data envelopment analysis (DEA) to assess the aggregated efficiency of resource use and environment improvement. Simulation-based optimization (SBO) is incorporated to seek the evaluation and optimization of the sustainable transportation system [8]. Xu and Song [9] proposed a comprehensive evaluation method of sustainable transportation with entropy method. Fuzzy logic is adopted to evaluate the performance of transport sustainability [6]. Egbue [2] found a multimetric index to access the adoption of sustainable transportation by combining demographic determinants as well as behavioral and attitudinal measures. Jia and Zhang [5] studied the evaluation of the urban low-carbon and ecological transportation system by using the gray correlation model. The traditional methods could arise a lot of disadvantages because of the susceptibility of human factor. It is mainly manifested in the unintentional exaggeration or reduction of the effect of certain indicators, so the evaluation result cannot truly reflect the actual situation of the object being evaluated. The methods mentioned were proved to be either subjective and fuzzy or slow in convergence speed and easy to fall into the local optimal solution with complex coding. Multilevel neural network for the evaluation of the sustainable development of urban transportation can reduce error in the assessment and thus is considered to be an important direction of the development of the evaluation method. It trains neural networks by collecting a mass of samples to assess the ecological and sustainable development of urban transportation system as a rule [7]. However, since all training samples constitute the hidden layer neurons directly, the larger the number of training samples, the more complex the neural network structure will be [4]. The BP neural network has a long running time because each adjustment parameter has the same influence on the output result and needs to be propagated back according to the error [3]. In order to shorten the running time and simplify the problem of the complex structure of the neural network due to the large training samples, this paper applied the radial basis neural network (RBFNN) to the evaluation model.

2. Assessment indicators of ecological sustainable transportation System

Assessment of ecological sustainable transportation system is not merely putting the ecological and sustainable indicators into the traditional evaluation system of urban transportation, but a comprehensive assessment aimed to reflect the connotation of the ecological sustainable transportation development, such as accessibility, social justice and so on. When determining the assessment indicator system and evaluation method, principles of systematization, scientificity, operability and practicability ought to be followed. In this context, this paper divides the assessment indicator system of ecological sustainable transportation system into three layers (Figure 1). The first layer is the overall goal of the ecological sustainable transportation system assessment, and the second layer includes three subsystems consisting of corresponding sub-indicators from the level of social and economic development, the urban ecological environment and the quality of transportation service as the third layer. The level of social and economic development mainly reflects the supporting role of transportation plays in social economy; The quality of transportation service mainly reflects the serviceability, fairness and universality of transportation to the whole society. The urban ecological environment mainly reflects the impact of transportation industry on the natural ecological environment.

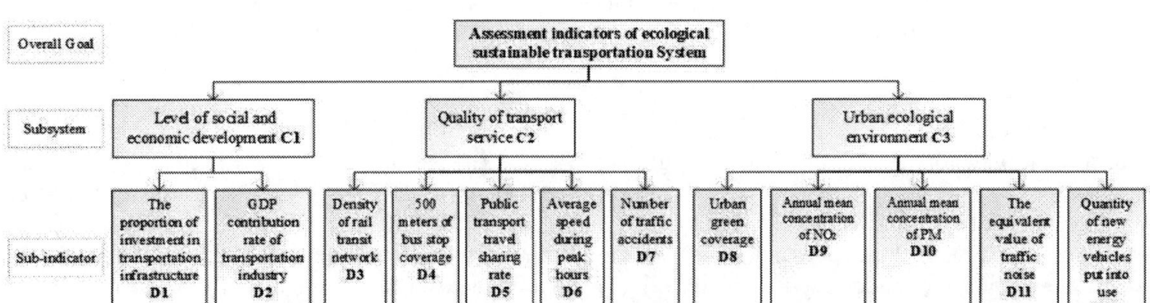

Figure 1. Assessment indicators of ecological sustainable transportation system

3. Evaluation methods and model

3.1. Evaluation methods

For the individual sub-indicators interacting with each other in each subsystem, a radial basis function (RBF) neural network model is adopted to obtain their weights through sample training. As to the independent subsystems in the assessment system, neural network models are not suitable to adopt due to the small sample size. In order to minimize the effects of inadequate training data samples, the weights of different subsystems are obtained by the analytic hierarchy process (AHP) instead. They were achieved by constructing comparative judgment matrixes between subsystems and sub-indicators integrating sophisticated suggestions of experts and literature. In this way, the updated weight of each indicator in the assessment system can be determined by weighted multiplication. Finally, the composite index reflecting the development level of ecological and sustainable transportation can be obtained.

3.2. Evaluation model

Radial basis function is a non-negative real value function whose radial symmetric value only depends on the distance from the center point. Supposed μ_i is the gaussian center of the ith node in the hidden layer and σ_i is the extension constant for the ith node, then:

$$\phi\left(\|x_i - \mu_i\|\right) = \exp\left(-\frac{\|x_i - \mu_i\|^2}{2\sigma_i^2}\right) \tag{1}$$

The basic idea of the radial basis function neural network is that by changing the transformation function in the hidden layer into the radial basis function, the hidden layer can transform the input vector and transform the input data from low-dimensional space to high-dimensional linear separable. The smaller the distance between the input vector and the center point $\|x_i - \mu_i\|$, the larger the output of the network, and the final output is:

$$y_i = \sum_{i=1}^{S^1} \omega_{ij}\phi\left(\|x_i - \mu_i\|^2\right), j = 1,2,\text{K},R \tag{2}$$

In this paper, the generalized RBF neural network modeling mainly involves three kinds of parameters: the basis function center and extension constant, and the weight of the output node. Suppose the input space with R dimensions have X_n $(n = 1,2,3,\cdots N)$ input samples. The specific algorithm steps are as follows:

(1) Initialization. Select S^1 different vectors as the initial clustering center $q_i(0)$ $(i = 1,2,3,\cdots S^1)$, that is, S^1 is the number of hidden nodes.

(2) Calculate the Euclidean distance between each sample point in the input space and the center point of the cluster $\|X_n - \mu_i(s)\|$ $(n = 1,2,3,\cdots N)$, where $\mu_i(s)$ represents the center of the sth iteration.

(3) Similarity matching. Let i^* represent the subscript of the winning hidden node, and determine the classification $i^*(X_n)$ of each input sample X_n according to its minimum Euclidean distance from the clustering center, that is when $i^*(X_n) = \min_i \|X_n - \mu_i(s)\|$, X_n is classified as class i^*, consequently divide the whole sample into S subsets $U_1(n), U_2(n), \cdots U_{S^1}(n)$. Each subset constitutes a clustering domain typically represented by the clustering center.

(4) Apply competitive learning rules to adjust and update all clustering centers.

$$\mu_i(s+1)=\begin{cases}\mu_i(s)+\eta\left[X_n-\mu_i(s)\right] & i=i^* \\ \mu_i(s), & \text{otherwise}\end{cases} \tag{3}$$

Add 1 to the value of s and turn to step 2. Repeat the process until $\left|\mu_i(s+1)-\mu_i(s)\right|\langle\varepsilon$.

(5) Determine the expansion coefficient. After determination of the center μ_i of each cluster, the expansion constant n of the corresponding radial basis function $\sigma_i=\frac{d_{max}}{\sqrt{2I}}$ can be determined according to the distance between the centers, where, d_{max} is the maximum distance between the selected cluster centers.

(6) Output weights. The pseudo-inverse matrix method is used to calculate the weights between the hidden layer and the output layer. Let's define the matrix $B=\left(b_{rn}\right)_{R\times N}$, where, $b_{rn}=\exp\left(-\frac{\left\|X_n-q_i\right\|^2}{2\sigma_i^2}\right)$ $\left(r=1,2,3,\cdots R;\ n=1,2,3,\cdots N\right)$. The weight matrix $\omega=\left(\omega_{ni}\right)_{N\times I}$ of each node in the output layer is calculated as $\omega=B^+T$, where B^+ is the pseudo-inverse matrix of matrix B, and T is the output expectation matrix in the training data set $T=\left(t_{ri}\right)_{R\times I}$.

4. Case Study

4.1. Data and samples

In this study, data during 2011 and 2018 is collected from Municipal Bureau of Ecology and Environment, Shenzhen Government Online, Municipal Bureau of Urban Management and Comprehensive Law Enforcement and other official websites of Shenzhen. The evaluation results are divided into five levels (V={I, II, III, IV, V}) and the corresponding theoretical outputs are 1,2,3,4,5. The level of ecological sustainable transportation development is determined by the level of each indicator. When all sub-indicators are at the same level, the level of ecological sustainable transportation system belongs to the level as well. 100 samples were generated randomly of different levels by changing the seed of random function irregularly.80 groups of sample data were used as the training set of radial basis function neural network, while 20 groups of data were used as the test set for comparison and verification.

4.2. Establishment, training and verification of RBFNN model

RBF neural network design function newrb (), simulation function sim () and inverse function pinv () in MATLAB toolbox were utilized to establish a radial basis function neural network model. In order to demonstrate the applicability of RBF neural network to the evaluation of ecological sustainable transportation system, a comparative evaluation experiment was conducted between RBFNN model and traditional BPNN model. Taking the second level of development as an example, 20 test samples were input into the established RBFNN model and BPNN model respectively to obtain the images of the predicted value and the expected output value, as shown in Figure 2.

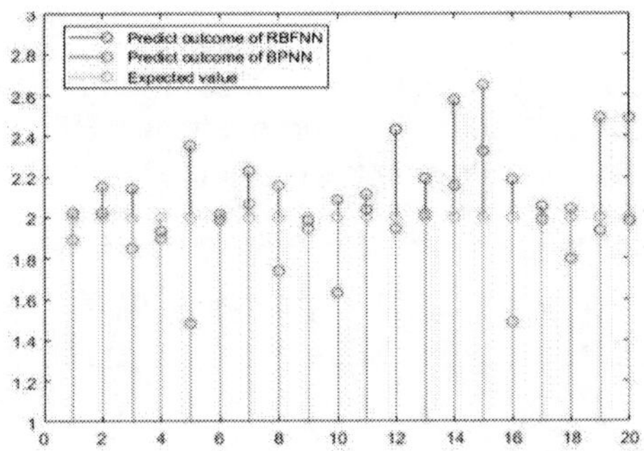

Figure 2. Comparison between the predicted output and the expected output of RBFNN

(Accuracy of RBFNN=95%,$RMSE_{RBF}$=0.28 ; Accuracy of BPNN=85%,$RMSE_{RBF}$=0.43)

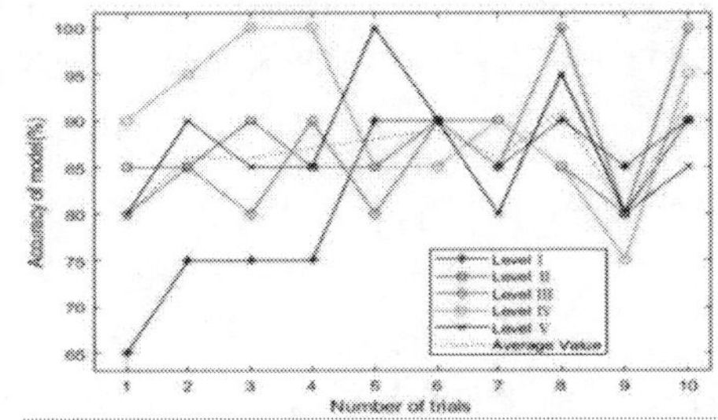

Figure 3. The predicted results of 10 repetitions at different levels in RBFNN (Average accuracy=87%)

Due to the contingency of a single experiment, this paper repeated 10 training experiments on test samples of different levels to obtain 10 result models, and recorded the accuracy of each model. Results are shown in Figure 3.

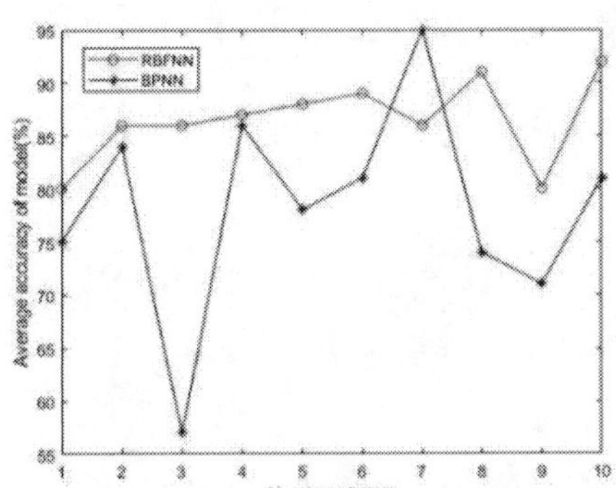

Figure 4. Comparison in prediction accuracy

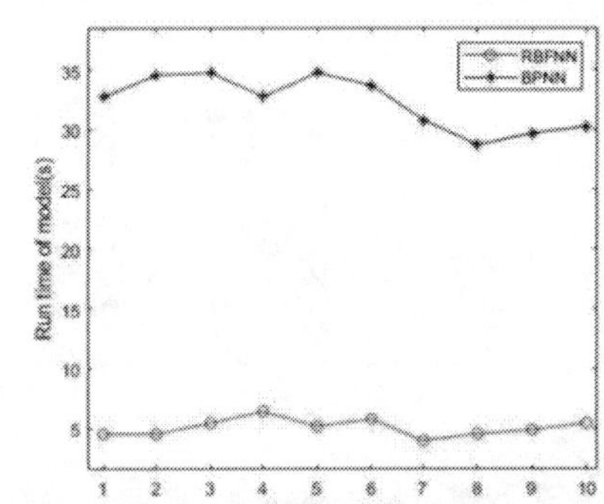

Figure 5. Comparison in run time

Table 1. The prediction accuracy of repeated experiments

	RBFNN	BPNN	Difference
Accuracy (%)	86.5	78.2	8.3
Run time (s)	5.1	32.3	27.2

It can be concluded from Figure 4, Figure 5 and Table 1 that RBF neural network model has its unique advantages in accuracy, running time and stability. BPNN is much slower in operation speed since it has to be trained according to the error backpropagation algorithm. Therefore, it is demonstrated that the RBF neural network model constructed in this paper has high reliability and fitting ability for sample data.

4.3. Results and discussion

Collected data was entered into the RBF neural network model to obtain the weight of each sub-indicator at first. Weights of three subsystems were calculated to be 0.21,0.38 and 0.41 respectively by AHP. Then the updated weights of indicators can be calculated as shown in the following table (Table 2).

Table 2. The weight of indicator of ecological sustainable transportation system

Sub-Indicator	D1	D2	D3	D4	D5	D6	D7	D8	D9	D10	D11	D12
Weight in RBFNN	0.41	0.59	0.14	0.16	0.21	0.26	0.23	0.19	0.17	0.15	0.23	0.26
Updated weight	0.08 6	0.12 4	0.053	0.06 1	0.08 0	0.099	0.08 7	0.07 8	0.07 0	0.06 2	0.09 4	0.10 7

Therefore, the development level of ecological and sustainable transportation in Shenzhen over the years can be calculated as shown in the following table:

Table 3. Development of ecological sustainable transportation system in Shenzhen

Year	2011	2012	2013	2014	2015	2016	2017	2018
Composite Index	2.29	2.99	3.03	3.57	3.87	3.95	4.06	4.17
Level	II	II	III	III	III	III	IV	IV

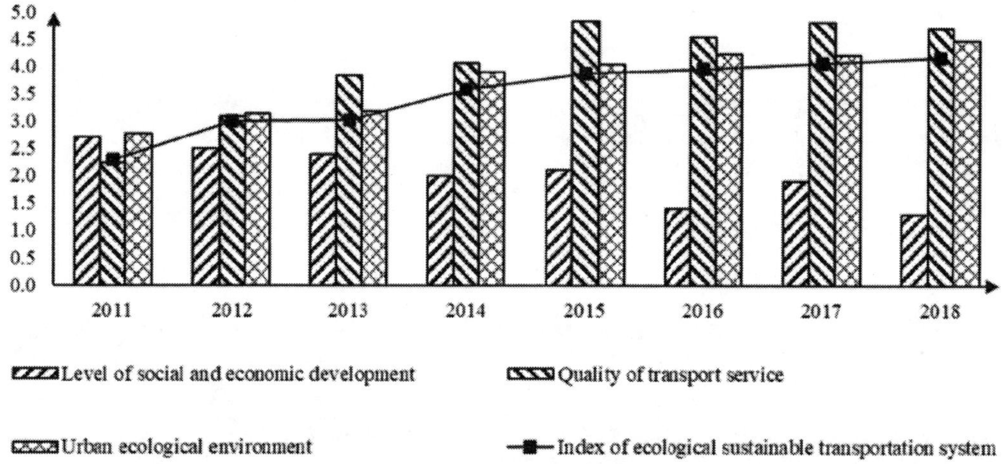

Figure 6. Development of ecological sustainable transportation system in Shenzhen.

As can be seen from Table 3 and Figure 6, the index of ecological sustainable transportation system in Shenzhen maintained a stable upward trend from 2011 to 2018 and a fast growth in 2012 and 2014. Among the three subsystems, the urban social and economic development subsystem shows a declining trend, while the urban ecological environment subsystem and the traffic service quality subsystem show a rising trend year by year. It indicates that in recent years, the development of ecological sustainable transportation of Shenzhen runs well, which further implies the social and economic development continues to play a supporting role in the economy, the quality of transportation service is constantly improving, and the urban ecological environment is better improved.

4.4. Suggestions

Since the 26th Summer Universiade in 2011, Shenzhen has seized the opportunity to accelerate the demonstration and promotion of new energy vehicles, paying attention to the refined design of traffic facilities, and comprehensively improving the traffic environment. However, we still need to note that as good as the evaluation of ecological sustainable transportation system in recent years looks, it actually just meets the standard of level Ⅱ and still lags behind the level Ⅰ. Moreover, urban expansion, population suburbanization and the previous unreasonable planning of road traffic system and land use all aggravate the phenomena of 'separation of workplace and residence', making commuting time longer and traffic jam worse during rush hours.

Given the existing problems of Shenzhen's ecological sustainable transportation system, this paper puts forward the following suggestions: 1) Keep intensifying investments in public transport to guarantee its priority status; Optimize the bus operation network and strengthen the traffic connection between east and west area to highlight social equity; 2) In the renewal of old towns and the movement of new towns, attention should be paid to the reasonable planning of urban land and balanced spatial allocation of basic public goods such as educational and medical facilities. 3)Upgrade and reform the existing road traffic system, and enhance cooperation between different traffic modes to solve the "last kilometer problem " of connection with public transport. 4) Draw on advanced traffic design and management experience at home and abroad like bringing in 'the smart luminous zebra crossing' to improve road traffic safety.

5. Conclusions

The paper chose corresponding indicators from social economy, service quality of transportation and the ecological environment at three different levels to build up a comprehensive indicator system for ecological sustainable assessment of urban transportation system. Furthermore, a radial basis function neural network (RBFNN) was established as foundation of the evaluation model. Empirical study indicated that the novel approach developed in this study outperformed the traditional BP neural network in abilities of approximation, classification and learning speed. Finally, the development of ecological sustainable transportation system in Shenzhen over the years is evaluated, and corresponding suggestions for improvement are put forward. Properties of RBF neutral network might perform better with a more rigorous and objective approach in the future. The robust RBFNN did pave a new way to study the ecological and sustainable development of transportation system.

6. References

[1] Bian Y, and Yang F (2010). Resource and environment efficiency analysis of provinces in China: A DEA approach based on Shannon's entropy. *Energy Policy*, **38**(4), 1909-1917.

[2] Egbue O, Long S, and Samaranayake V A (2017). Mass deployment of sustainable transportation: evaluation of factors that influence electric vehicle adoption. *Clean Technologies and Environmental Policy*, **19**(7), 1927-1939.

[3] He L, Zhu H, and Gao Z (2018). Performance Evaluation of Asphalt Pavement Based on BP Neural Network. *NeuroQuantology*, **16**(6).

[4] Huang J, Wang Y, Liu Z, Guan B, Long D, and Du X (2016). On modeling microscopic vehicle fuel consumption using radial basis function neural network. *Soft Computing*, **20**(7), 2771-2779.

[5] Jia J M, and Zhang R H (2013). Evaluation on urban low-carbon and ecological transportation system. *In Applied Mechanics and Materials* (Vol. 361, pp. 1967-1972). Trans Tech Publications Ltd.

[6] Rajak S, Parthiban P, and Dhanalakshmi R. (2016). Sustainable transportation systems performance evaluation using fuzzy logic. *Ecological Indicators*, **71**, 503-513.

[7] Shu Y, and Xu G H (2018). Multi-level dynamic fuzzy evaluation and BP neural network method

for performance evaluation of Chinese private enterprises. *Wireless Personal Communications*, **102**(4), 2715-2726.

[8] Song M, Yin M, Chen X M, Zhang L, and Li M (2013). A simulation-based approach for sustainable transportation systems evaluation and optimization: theory, systematic framework and applications. *Procedia-Social and Behavioral Sciences*, **96**, 2274-2286.

[9] Xu Z J, and Song Y K (2014). Index System and Method of Evaluation to Sustainable Technology with Fuzzy Comprehensive Evaluation. *In Applied Mechanics and Materials* (Vol. 448, pp. 4429-4433). Trans Tech Publications.

Is it possible to achieve waste free construction sites in Norway?

Selamawit Mamo Fufa[1,*], [1]Christoffer Venås, [1]Marianne Kjendseth Wiik

[1]SINTEF Community, Børrestuveien 3, NO-037 Oslo, Norway

*Email: Selamawit.fufa@sintef.no

Abstract. The construction industry is one of the world's largest consumers of energy (ca. 40%) and natural resources (ca. 40%). Norwegian statistical data shows that waste from construction, rehabilitation and demolition is steadily increasing. Norway alone generated approximately 1.9 million tonnes of waste in 2017, around 1% more construction waste compared to 2016. Of this, only 34% of the waste was recycled, which is 8% less than in 2016, and 34% below the EU's requirements (70% reuse and recovery of all non-hazardous construction and demolition waste). There are on-going initiatives in Norway to address this and realise 'waste free' construction sites, with an ambition to contribute towards global, national and regional environmental impact reduction targets. The aim of this paper is to discuss challenges and opportunities for waste free construction sites in Norway and lessons learnt from Norwegian fossil and emission free construction sites. This paper concludes that Norwegian waste-free construction is ambitious but possible.

Keywords: Construction site, Construction waste, Emission free, Waste free construction, Zero waste construction.

1.Introduction

Construction sites represent between 5-10% of total emissions in cities [6]. In Oslo, the emissions relating to construction sites are approximately 7% of the city's total emissions1. The construction phase emissions occur during the early stages of a building's life cycle, and given the magnitude, one can question whether new construction can contribute to reaching environmental goals, no matter how energy efficient buildings are during operation [1]. This shows the importance of minimising embodied construction emissions. Today most Norwegian construction sites generates 40-60 kg/m2 waste [15]. Several projects have set a goal of waste generation below 25 kg/m2. However, the construction industry has identified several challenges to reduce today's levels of waste to a significantly lower level. One major challenge is related to complex projects that involve many stakeholders with different needs and practices. Waste reduction requires extensive planning throughout the project period and close collaboration between different actors, which is both time and resource intensive. Human factors, such as perceptions, attitudes, behaviours and expectations of the involved multiple stakeholders is another key barrier. Most actors in the construction industry are unsure of the profitability and environmental

[1] https://www.smartcitiesworld.net/news/news/cities-commit-to-clean-construction-to-cut-building-emissions-by-up-to-44-per-cent--4665

IOP Conf. Series: Materials Science and Engineering 1196 (2021) 012018 doi:10.1088/1757-899X/1196/1/012018

benefits of waste reduction measures. It is cheaper to buy new materials and easier to send leftovers to waste treatment facilities. Waste reduction and reuse of products are key principles in the waste prevention and management hierarchy to reduce waste, emissions (from production, transport and waste treatment of additional new products) and the potential adverse impacts of waste on environment and health. A recent Nordic study shows a significant positive environmental impact by reducing the resource usage by 20%, which is equivalent to 900,000 tonnes of GHG emissions, and a potential social and financial benefits for private companies equivalent to 1.7% annual growth [13]. Reducing waste from construction sites will play a major role in fulfilling local, regional, national and global environmental goals, while addressing several of the UN sustainable development goals (such as SDG 11 Sustainable cities and communities; SDG12 Responsible production and consumption and SDG13 Climate action and waste and pollution reduction impacts).

The aim of this study is to evaluate and discuss the challenges and opportunities for waste free construction sites in Norway; as well as review the lessons learnt from Norwegian fossil and emission free construction sites and discuss how emission reduction measures can enable waste free construction. This paper starts by outlining the background and definitions. It then presents four Norwegian case studies in addition to the planned and actual waste generation results. This is followed by a discussion on challenges, opportunities and lessons learnt before presenting concluding remarks.

2.Background

The Norwegian construction sector is a front runner in taking responsibility for setting environmental goals which enable the industry to develop low emission solutions and the utilisation of transparent data and communication schemes to follow building performance (such as BREEAM [3] and EPDs [2]). There is a growing interest in Norway on fossil and emission free construction site activities [1][5][7][10][17]. Oslo municipality is the main driver behind this initiative and requires all public construction projects to be fossil (as a minimum requirement) or emission free since 2017. Oslo municipality use public procurement as a tool, by setting achievable ambitious requirements in competition tenders. This enables the rapid development and implementation of new market-oriented fossil and emission free solutions.

Currently, there are no legal requirements for waste reduction from construction sites. Existing requirements made by authorities and builders are passive and focus on waste sorting, and not on waste prevention or reduction. The construction industry has now taken the lead to address waste challenges through a joint initiative called "waste free construction site"[3]. Eleven key public builders have gathered in this initiative to identify challenges and opportunities, motivate the development of new solutions and facilitate public support schemes for innovative projects with targets for waste reduction. The initiative also plans to initiate a new requirement for waste free construction sites by 2022.

The term waste free and zero waste are often used interchangeably when discussing waste from building sites. A waste free construction site is a construction site that does not generate any waste in any of its construction site activities. However, some actors in Norway have used the concept of waste-free construction sites for achieving less than 40 kg/m2 of waste. In order to harmonise definitions, the Norwegian waste-free construction site initiative define waste free construction sites as construction sites that do not have any waste from construction site activities (including the production of the materials). The scope of the definition is limited to waste from construction of new buildings and focuses on five construction waste fractions, namely wood, gypsum, plastic, packaging and pipes. [12].

However, achieving zero waste from construction sites is challenging. Bærum municipality has taken the first steps by focusing on one of the waste fractions and is working towards smart packaging solutions to achieve no packaging waste [4]. Similarly, Stavanger and Sandnes municipalities are also underway to acquire two pilot projects where waste reduction is a priority [16].

---

[2] https://www.epd-norge.no/?lang=en_GB

[3] https://innovativeanskaffelser.no/avfallsfrie-byggeplasser/

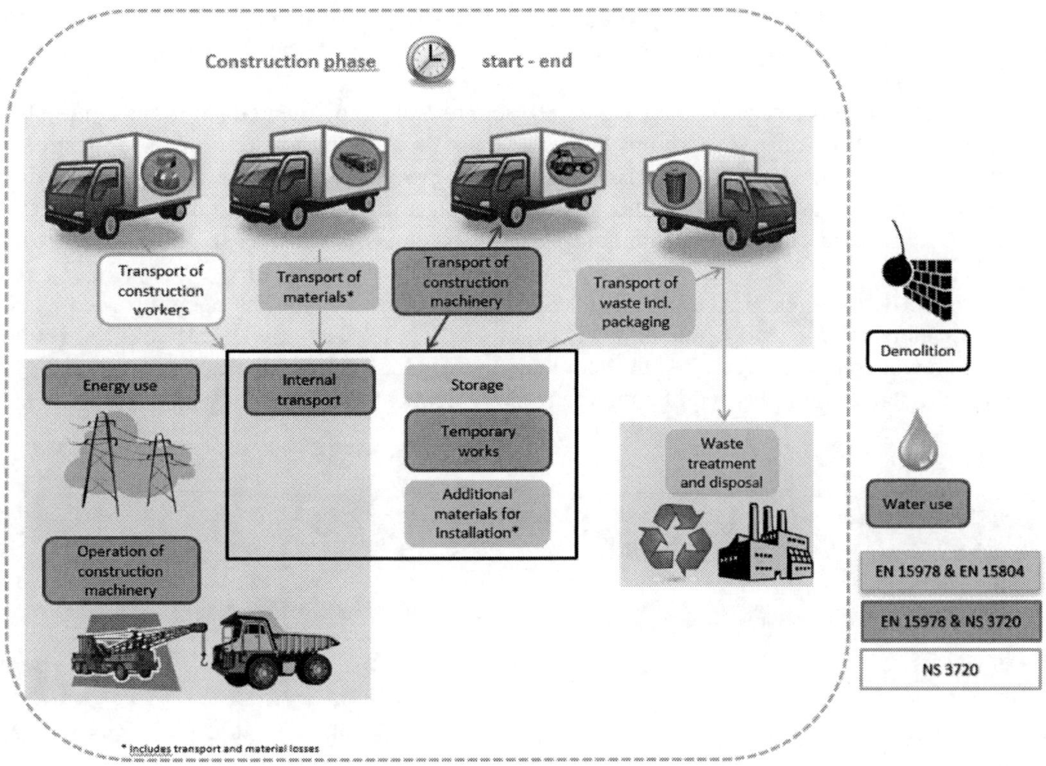

Figure 1: System boundary for emission free construction sites (Fufa, Wiik, Mellegård, & Andresen, 2019).

Waste is one component under construction site activities, and waste free construction sites support the development of emission free construction sites. The joint initiative sees that similar, in-depth work is needed to realise waste free construction sites. In spring 2019, contractors and suppliers have shown an interest in developing new collaborative waste reduction solutions through a series of dialogue meetings under the joint initiative for waste free building sites network [12]. However, they also expect transparent and achievable requirements in the procurement process and incentives to support new cost-intensive innovative working methods.

3.Materials and methods

This study follows the construction phase of four Norwegian building case studies (Campus Evenstad, in Hedmark, Lia nursery school in Oslo, as well as a small and a large construction site in Bærum). The methodology involves collecting data on waste plans, waste generation and waste sorting, visiting the construction sites, and performing qualitative interviews with key stakeholders. The two Bærum construction sites are on-going projects, whilst Campus Evenstad was completed in 2016 and Lia nursery school in 2017.

The two Bærum construction sites are traditional construction sites which are selected by Bærum municipality for a qualitative study to investigate the current practice of the ongoing "smart packaging construction sites" project. The small construction project is for a new social service building with a total floor area of 1500m^2 over three floors. The building consists of a traditional construction with a steel load-bearing frame and hollow deck concrete slabs. The large construction project is also for a new social service building with a total floor area of 17000m^2 consisting of several smaller buildings over two to three floors. The buildings consist of solid wood construction with either a brick facade or treated wood panels. The aim is to understand the importance of packaging and packaging waste in construction

IOP Conf. Series: Materials Science and Engineering 1196 (2021) 012018 doi:10.1088/1757-899X/1196/1/012018

and to contribute to the parallel development of waste-free construction sites. Interviews have been carried out with the construction manager and supervisor. In addition, a group interview with the suppliers has been conducted.

In contrast, Campus Evenstad and Lia nursery school have implemented solutions aiming to achieve low emission construction sites. Campus Evenstad consists of an administration and educational school building with a total floor area of 1141m^2. The building consists of a solid wood construction, wood fibre insulation, and an untreated timber cladding. It was the most ambitious pilot project from the Norwegian ZEB research centre, which compensates the emissions related to materials "M", operation "O" and construction "C" through on-site renewable energy generation (ZEB-COM) [9][17]. Whilst Lia nursery school is the first fossil free construction site in Norway and has been awarded as Norway's most environmentally friendly nursery school in 2018 [11]. The total floor area is 1600m^2. The construction consists of prefabricated timber elements with timber interior and exterior cladding, hollow core concrete slabs, and light weight concrete roof elements.

Figure 2: Visualisation of Campus Evenstad (left, by Ola Roald Arkitekter) and photograph of Lia nursery school (right, by SINTEF).

Both Campus Evenstad and Lia nursery school evaluate emissions from the following construction site activities: transport of building materials; transport and operation of construction machineries; waste management including transport of waste from construction site activities to the treatment facilities, waste processing (recycling or incineration) and waste disposal; energy use; temporary works and person transport (see Figure 1). All demolition works and waste belong to the previous life cycle and are not included in the analysis. The waste data was collected from the waste plan filled out by contractor. Further detailed information about the two case studies can be found in [7][11][17].

4.Results

Table 1 presents the type and amount of construction waste planned for, generated and sorted on-site from the four case studies. The total waste generated per m^2 total floor area results are 27.1 kg/m^2 for Campus Evenstad, 32.1 kg/m^2 for Lia nursery school, 10.3 kg/m^2 for Bærum small and 24.3 kg/m^2 for Bærum large. More than 80% of total waste generated in Campus Evenstad is sorted and consists of untreated wood (47%), mixed waste (20.1%), concrete (16.2%) metals (11.8%) and gypsum (3.2%), whilst in Lia nursey school more than 82.5% of total waste is sorted and consists of untreated wood (47%), mixed waste (17.5%), gypsum (14.9%), metals (6.7%), paper (6.4%) and plastic (5.8%). In Bærum small, 64% of total waste is sorted and consists of untreated wood (52%), gypsum (27%), plastic (8%), mixed waste (8%) and mineral wool insulation (3%). In Bærum large, 86% of total waste is sorted and consist of gypsum (34%), untreated wood (29%), mixed waste (14%), concrete (11%) and plastic (4%).

Table 1. Total planned and actual construction waste for four Norwegian construction sites.

Type of building	Evenstad*	Lia nursery	Bærum small**	Bærum large**

Total floor area (m^2)	1141	1600		1500		17000	
Construction Waste (kg)	Planned	Planned	Actual	Planned	Actual	Planned	Actual
- Untreated wood	14522	29030	28140	17000	8060	147650	120700
- Paper, cardboard and carton	365	1500	3840	2000	-	14770	6240
- Glass	130	0	0	0	-	0	-
- Iron and other metals	3640	7061	3980	7000	340	29530	15170
- Gypsum based materials	1000	10199	8900	8000	4200	73830	141010
- Plastic	30	1850	3440	4000	1160	7380	16310
- Concrete, brick, Leca and other heavy building materials	5000	15000	0	15000	-	147650	45440
- Electric and electronic waste	10	0	0	500	-	1480	1148.5
- Mineral wool insulation	0	0	0	900	500	1480	6900
- Mixed construction waste	6200	6136	9920	6000	1200	36910	58530
- Hazardous or special waste	21	0	1040	600		1480	1057
Total construction waste	**30918**	**71476**	**59260**	**61000**	**15460**	**462160**	**412505.5**
Total construction waste - sorted	24697	65340	48300	55000	14260	425250	353975.5
Total construction waste generated per total floor area (kg/m^2)	**27.1**	**38.7**	**32.1**	**40.7**	**10.3**	**27.2**	**24.3**
Sorting fraction (%)	80	91	82.5	90	92	92	86

NB: Packaging waste is included under 'wood', 'plastic' and 'paper, cardboard and carton' fractions.
* The actual waste data for Evenstad was not available at the time of writing this paper.
**At the time of writing this paper, the Bærum construction sites were under the final stages of construction and will have some additional wastes from internal finishes (e.g. gypsum, cardboard, plastic) not accounted for here.

The interviews with the construction site personnel in Bærum projects provided insight to the practical implications of waste requirements and potential packaging waste measures. Both projects in Bærum focus solely on sorting on-site, and not on waste reduction. The informants did not have strong beliefs in the ability of achieving packaging-free construction or waste free construction and questioned the profitability and environmental benefits of leaving today's practice. Substantial changes in current working methods, giving less flexibility for on-site operations and better planning of the construction logistic activities are described as important measures for potential waste reduction solutions. The need for public procurement and stricter requirements from public authorities was also highlighted as the most important drivers for change in the future.

5.Discussion

From the results, it is possible to see that all case studies have managed to reduce the total construction waste generated per total floor area, excluding Campus Evenstad since data on actual waste generated was not available. Lia nursery school and Bærum large have not been able to meet their planned sorting fractions, and it may be considered that this is a consequence of reducing 'sortable' waste. On the other hand, Bærum small has been able to improve its sorting fraction. Through interviews, it was found that since Bærum small is a smaller project with a less complex structure and has a site layout with available dry and safe temporary storage for materials. This enables easy implementation of good waste management practices, where sub-contractors and suppliers followed-up to ensure good sorting practices.

The Norwegian waste free construction site initiative's definition limits the scope to waste from construction of new buildings and focuses initially on five construction waste fractions: wood, gypsum, plastic, packaging and pipes [12]. It is interesting to note that all case studies have high fractions of waste from untreated wood, gypsum and plastic, as well as mixed construction waste and concrete. Packaging wastes are integrated under untreated wood, paper, cardboard and carton and plastic, whilst pipe wastes are integrated under iron and other metals and plastic. Developing a stepwise ambition level definition, such as in the Norwegian ZEB research centre for zero emission buildings [9] or in the emission free construction site project (Figure 1), can be a useful approach to realising waste free construction site ambitions. Recommendations include establishing a waste free construction site ambition definition system using benchmarks for total construction waste generated per total floor area, sorting fractions and the waste hierarchy system. This will highlight the amount of waste generated from the construction process and the treatment method used for waste generated. It will also enable stakeholders to identify measures for reducing waste and increase the fraction of waste material recovered. This approach will help harmonise the definition of waste free construction and facilitate for transparent communication of results between construction projects. However, in order to establish benchmark values, either national statistics or a larger body of case studies is required.

More time and resources were used in the early project phases for the planning of both Campus Evenstad and Lia nursery school, in order to create a common understanding of the project ambitions and discuss potential measures and challenges to achieve low emission construction sites. There was a close cooperation between actors, whereby the ambition for low emission construction, concepts, lessons from previous experiences, accessibility of machineries and equipment in the market, and potential challenges and opportunities were discussed at an early stage. This enabled both projects to achieve their ambitions. As waste is one part of construction activities, both case studies consider ambitious waste minimisation targets (for example using prefabricated elements in Lia and use of locally available low carbon materials in Campus Evenstad). The two Bærum construction projects focus on delivering projects on time and budget without spending more time in the early planning. For example, the marginal higher cost of using pre-cut gypsum boards in Bærum small, and the risk of longer delivery times in Bærum large was described as the reasons why traditional methods were chosen. Furthermore, the informant believed that the most important success criterion for waste reduction would be to ensure an optimal quantity in stock, continuously produce properly and with good quality and avoiding temporary storage of materials. It is recommended that all construction projects should deliver a waste plan, containing the total amount and type of onsite construction waste that is going to be generated during the construction period to local authorities before construction starts. The waste plan can then be used as a tool during the construction phase to actively reduce and sort waste fractions. Through interviews it was confirmed that waste plans are not actively used to plan waste operations or used to follow up waste generation during construction. The need for digitized tools to plan and follow-up waste during construction has been discussed.

Green public procurement is common practice in the Norwegian building sector, whereby the public sector uses its position in the market to set strong demands on developers, contractors and suppliers. Oslo municipality sets a good example in setting requirements for fossil and emission free building sites to reduce ca. 30% of the city's total construction site emissions. Lia nursery school was the first construction site to implement the fossil free construction site ambition. The lessons learnt from this project were then used in other fossil free and emission free construction sites around Norway. This can also be used as a reference when setting requirements for waste free construction sites. Setting clear, measurable and realistic environmental requirements or targets and incentives in the procurement process encourages different stakeholders to consider new and innovative solutions. Lack of knowledge and experience amongst stakeholders, high costs and lack of access to waste-free solutions are considered as important barriers across the case studies. There is a need to replace traditional solutions based on linear thinking and consider new technologies and methods aimed at optimising operations.

Early planning with consideration of external factors, collaboration and good dialogue between all stakeholders involved (to share knowledge, expertise and other resources) and choices of construction methods (such as prefabricated elements and modules) are the main factors which enable all projects to achieve their ambitions. Purchase of pre-cut and prefabricated materials/elements in the desired dimension, effective utilisation of areas, efficient construction site lay out, on-time and efficient delivery (to reduce unnecessary movement and storage of materials), proper material handling with dry and clean storage areas if temporary material storage is needed (to avoid material damage) are some measures that should be considered. An agreement with suppliers to return packaging, more on-time delivery of materials without packaging, bulk deliveries to avoid unnecessary use of small packaging are some examples of packaging waste reduction measures which can be implemented in current practice. The economic and environmental consequences of each waste reduction measures should be considered to make informed choices.

There is a need for developing circular supply value chain systems, with new working methods, allowing close collaboration between different stakeholders for the successful realisation of waste free construction sites. There is also a need to develop practical solutions, new ways of working profitably, develop competence, collect best practices through pilots, and create a platform for the dissemination of knowledge and information between different actors. Combining theoretical and practical knowledge through pilot testing, documentation of practical advancements and best practices from different projects will help to fill the current knowledge gap. There is a need to develop and apply research-based knowledge and methodologies which enable the evaluation of potential cost savings, environmental and social benefits when choosing different waste reduction measures. Using a life cycle thinking approach is central to evaluate the sustainability of these solutions. Performing cradle-to-grave life cycle assessment will enable to avoid problem shifting from one life cycle phase to another or from one environmental indicator to another. Governmental involvement is central to strengthen incentives, further develop regulations and promote public and private procurements. Upcoming new requirements should be realistic, transparent, measurable, comparable and incentivised so that ambitious actors can develop and implement innovative solutions. It is also important to provide recommendations based on lessons learnt for procurement procedures, regulations and incentives for the realisation of waste free construction sites. Lessons learnt from previous and on-going projects on fossil or emission free construction sites can be used as an example.

6.Conclusions

This study has presented the findings from four Norwegian construction sites. The results show that achieving absolute zero waste may be challenging to achieve soon but is possible. It requires new digitalised waste reduction solutions, good early planning, new working methods and close collaboration between actors to create a closed circular supply chain. There is also a need for harmonised ambition levels, definitions and system boundaries which enable stakeholders to evaluate, document and follow up waste reduction measures. The study also highlights considerable scope of work for further analysis.

References

[1] Antti S, Jukka H, and Seppo J (2012). A scenario analysis of the life cycle greenhouse gas emissions of a new residential area. *Environmental Research Letters,* **7**(3), 034037.

[2] Aragones M P, and Serafimova T (2018). *Zero emission construction sites: The possibilities and barriers of electric construction machinery. Bellona Europe.*

[3] BREEAM-NOR. (2016). *BREEAM-NOR New Construction 2016. Technical Manual SD5075NOR, Norwegian Green Building Council.*

[4] Bærum kommune. (2019). Avfallsfrie byggeplasser – Emballasjesmart byggeplass. Presentasjon av forprosjektet Emballasjesmart byggeplass. Retrived from https://innovativeanskaffelser.no/avfallsfrie-byggeplasser-emballasjesmart-byggeplass/.

[5] DNV GL (2018). Emission-reduction potential of fossil- and emission-free building and

construction sites. Report no. 2018-0367, Rev. 1-ENG. 2018a.

[6] DNV GL (2019). Perspectives on Zero Emission Construction. Climate Agency, City of Oslo. Report no. 2019-0535, Rev. 1.

[7] Fufa S M (2018). GHG emission calculation from construction phase of Lia barnehage. SINTEF Notes nr. 29. ISBN:978-82-536-1586-8.

[8] Fufa, S M, Mellegård S, Wiik M K, Flyen C, Hasle G, Bach L and Idsøe, F. (2018). Utslippsfrie byggeplasser - State of the art. Veileder for innovative anskaffelsesprosesser. SINTEF Fag rapport nr. 49. ISBN:978-82-536-1589-9.

[9] Fufa S M, Schlanbusch R D, Sørnes K, Inman M and Andresen I (2016). A Norwegian ZEB Definition Guideline. The Research Centre on Zero Emission Buildings. ZEB Project report no 29.

[10] Fufa S M, Wiik M K and Andresen I (2019). Estimated and actual construction inventory data in embodied GHG emission calculations for a Norwegian zero emission building (ZEB) construction site. In P Kaparaju, R J Howlett, J Littlewood, C Ekanyake and L Vlacic (Eds.), *Smart Innovation, Systems and Technologies* Switzerland: Springer Nature Switzerland.

[11] Fufa S M, Wiik M K, Mellegård S and Andresen I. (2019). Lessons learnt from the design and construction strategies of two Norwegian low emission construction sites. *Eart and Environmental Science,* 352.

[12] Halogen. (2019). Avfallsfrie byggeplasser. Bærekraftige byggeplasser gjennom digitalisering og industrialisering av byggebransjen.

[13] Høibye L and Sand H (2018). Circular economy in the Nordic construction sector. Identification and assessment of potential policy instruments that can accelerate a transition toward a circular economy. TemaNord 2018:517.

[14] Multiconsult (2018). Review of implementation of fossil free building sites. Document no. 10206471-TVF-RAP-001. 2018.

[15] Nordby A S and Wærner E R (2017). Hvordan planlegge for mindre avfall. En veileder for å redusere avfallsgenerering i byggprosjekter. Norwegian Green Building Council.

[16] Sandnes Eiedomsselskap KF (2019). Første steg mot avfallsfrie byggeplasser. Retrived from https://innovativeanskaffelser.no/wp-content/uploads/2019/04/forste-steg-mot-avfallsfrie-byggeplasser-sekf.pdf.

[17] Wiik M K, Sørensen Å L, Selvig E, Cervenka Z, Fufa S M and Andresen I (2017). ZEB Pilot Campus Evenstad. Administration and educational building. As-built report. The Research Centre on Zero Emission Buildings. ZEB Project report no 36.

Using Tanflos SG to Remove Fine and Ultra-Fine Iron Mineral Particles from Water

Ian Werner, Elivone Lopes da Silva and Mauricio Leonardo Torem*

Pontifícia Universidade Católica do Rio de Janeiro, Departamento de Engenharia Química e de Materiais, Brazil
*Email: torem@puc-rio.br

Abstract. One of the biggest challenges in the mineral scenario is concentrating and removing fine particles, as consequence of that, there are a high concentration of mineral inside the Ganga and this reality needs to be changed. Seeking to contribute significantly, this research used the process of DAF (dissolved air flotation), because in this process the bubbles are less than 100 μ, what is fundamental because the fines and ultrafine particles. For the pretreatment was used the biosurfactant Tanfloc SG as much as flocculant and coagulant. Other techniques such as FTIR (Fourier-Transform infrared), Zeta potential and contact angle was used to show the mineral and the biosurfactant characteristics. As a conclusion, the IEP (isoelectric point) was close to pH 8,5 for Tanfloc, but for hematite, with Tanfloc the pH went down to 5,3 - 6,5 and the DAF depends the biosurfactant concentration, the best results were between 140 and 175 mg/L.

Keywords: Fine, ultra-fines, biosurfactant, DAF, Tanfloc SG.

1. Introduction

Dissolved air flotation (DAF) can be an efficient alternative for treatment of fine iron particles, because it is possible to generate bubbles less than 100μ. Reports are confirming, collision efficiency and adhesion tend to increase with decreasing bubble size [5]. Moreover, this process only works in the presence of chemical agents capable of altering the surface characteristics of the particles.

Tanfloc SG biosurfactant was used because it is a cationic organic polymer of vegetable origin. This biosurfactant has low molar mass and positive electric charge, acting on colloids by neutralizing charges and forming bridges between these particles an is efficiency covers pH range between 4.5 and 8.0 [4], without changing it, precisely because it does not consume the alkalinity of the medium [9].

The objective of this work is to separate hematita from quartz of iron ore fines by dissolved air flotation using the biosurfactant Tanfloc SG. Several parameters and variables, that influence hematite selectivity and recovery, such as pH and bioreagent and depressor concentration, were evaluated throughout this work to determine the optimal conditions where the highest selectivity and flotability occurs.

2. Materials and methods

Iron ore, hematite and quartz were purchased from Mineração Zé da Estrada Ltda. Chemical analyses showed 54% of Fe in the iron ore and 99% of purity in the hematite and quartz samples. Biosurfactant Tanfloc SG was obtained from TANAC. Sodium hydroxide was obtained from Merck, hidrocloric acid was purchased from Vetec and toluene and potassium chloride was obtained from Próquimios. For flotation, an Aquaflot Dissolved Air Flotator was used on a laboratory scale. It has 3 L of water capacity, using a pressure in the range of 400 to 600 kPa. A needle valve was used to adjust the outlet of the supersaturated water portion.

2.1. FTIR (Fourier-Transform infrared)
Infrared spectra were performed on the Scientific Nicolet 6700 FT-IR FTIR spectrophotometer and followed the KBr pellet method. The spectra were performed to identify the functional groups present in the polymer, pure mineral and mineral after interaction with the polymer. In order to obtain the pellets, particle or bioreagent size reduction was first performed using mortar and pestle in order to obtain a homogeneous mixture between the sample and the KBr. The proportion of sample and KBr was 1/100 (w / w). Then, the homogeneous mixture pellet was obtained by compression. Finally, the obtained tablet was analyzed in the FTIR apparatus. Spectra were obtained at a resolution of 4 cm^{-1} using 120 scans.

2.2. Zeta Potential
Zeta potential measurements were made for polymer and mineral sample. These studies were performed on a Malvern Zetasizer Nano-type microelectrophoresis device at PUC-RIO mineral and environmental technology laboratory. The measurements aimed to study the possible mechanisms of interaction between the polymer and the mineral surface. Zeta potential measurements were made for polymer, mineral and mineral after interaction with the polymer. All Zeta potential measurements were performed within a pH range of 2 to 10.

2.3. Contact Angle
Contact angle measurements were performed for mineral samples before and after interaction with the biosurfactant. These studies were done on the Surface Electro Optics DCA 200 SERIES device at PUC-Rio mineral and environmental technology laboratory. Washburn method was used to obtain contact angle values in pH range from 7,0 to 8,0 in duplicate with the powder samples. The solvent used to determine Washburn constants was Toluene.

2.4. Flotation
Firstly, coagulation and flocculation tests were performed to verify flake formation. The working range of pH was between 5.0 and 11.0 and the biosurfactant concentration was between 35 and 200 mg L-1, maintaining a fixed iron ore mass of 4.0 g. At the end of each test, with the highest biosurfactant concentration, the mixture was flotated to preliminarily assess the flotability of each pH. This test is illustrated in figure 5.

Based on these preliminary coagulation and flocculation results, flotation tests were performed to evaluate ore recovery at each fixed pH with the Tanfloc SG concentration variation.

First, water and iron ore are added, which are placed under agitation. Then the biosurfactant was added and vigorously stirred for 5 minutes. After this interval, stirring was reduced and maintained for 10 minutes to keep the solid material suspended for flake formation. At the end of these 10 minutes the flotation was performed and the flotated material was scraped and filtered.

3. Results

3.1. FTIR (Fourier-Transform infrared)
One of the important aspects of infrared spectroscopy analysis is the possibility of identifying the functional groups of the collector and their structure. With this, one can understand more about how the polymer interacts with the mineral.

These characteristics can be observed by the FTIR spectra. Table 1 shows the main characteristic peaks of Tanfloc. The absorption band is preserved around 3369 cm^{-1} which corresponds to the stretch vibration of the O-H and N-H groups present in the polymer. Other important factors are strong carbonyl bands near 1713 cm^{-1}.

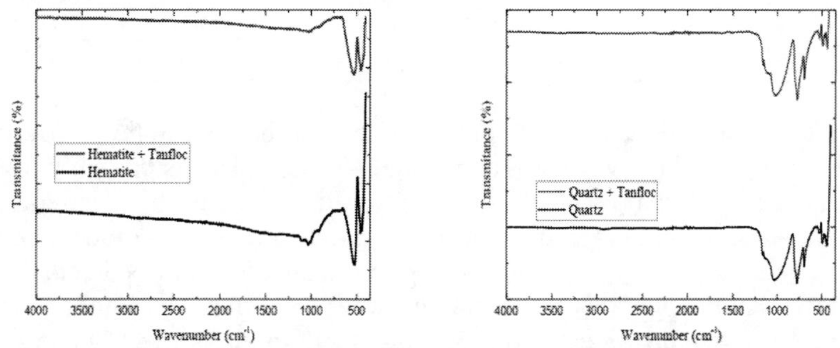

Figure 1. FTIR hematita and Quartz with and without Tanfloc SG

This band is in the range of carbonyls attached to alkyl groups or carboxylic acids. Due to the interval, it is difficult to confirm with infrared spectroscopy alone. The bands in 1609, 1233, 1021 and 830 cm^{-1} refer to aromatic groups. At 1609 cm^{-1} there are stretches of conjugated alkenes or the aromatic ring itself. At 1233 cm^{-1} the strong bands may refer to the ν (C-N) or ν (Ph-O) stretches of alkyl amines or aromatic ethers. The bands at 1021 and 830 cm^{-1} refer to the δ (C-H) folds both inside and outside the plane of the aromatic ring. Finally, the 1111 cm^{-1} band refers to the stretching of aliphatic cyclic ethers directly linked to the aromatic ring and at 1300 cm^{-1} there is evidence of the presence of alcohols, with phenol grouping being the most likely. Thus, one can assume which are the major functional groups present in the tanfloc structure, being similar to the tannin structure as expected.

Table 1. Characteristic peaks of Tanfloc.

Tanfloc SG	Attribution
3369	ν (O-H) + ν (N-H)
1713	ν (C=O)
1609	ν (C=C)
1463	δ ($C-H_2$)
1354	ν (C-N)
1300	ν (C-O)
1233	ν (C-N) + ν (Ph-O)
1111	ν (C-O-C)
1021	δ (C-H)
830	δ (C-H)

Table 2 shows that the pure hematite mineral and after conditioning with the biosurfactant. The bands below 1000 cm^{-1} correspond to the iron oxide stretching modes, with the bands around 673, 539, 458 and 439 cm^{-1} corresponding to the Fe-O bond stretching modes [1][7].

Table 2. Pure hematite mineral and with Tanfloc SG

Hematite	Hematite + Tanfloc SG	Attribution
439	-	ν (Fe-O)
447	-	ν (Fe-O)
458	453	ν (Fe-O)
519	-	ν (Fe-O)
522	-	ν (Fe-O)

530	532	ν (Fe-O)
539	-	ν (Fe-O)

After the interaction of hematite with the biosurfactant, it is possible to observe the overlapping of the bands below 1000 cm^{-1} by two bands at 532 and 453 cm^{-1}. This is probably due to the interaction of Tanfloc with hematite, so that there is no significant difference in the different vibrational stretching modes of Fe O bonds, integrating with the different existing rotational modes. Thus, there is formation of the two predominant bands. Not enough resolution to observe Tanfloc bands.

Table 3 shows that the pure quartz mineral and after conditioning with the biosurfactant. The bands around 465 and 692 cm^{-1} correspond to the asymmetrical folding of Si-O-Si and symmetrical folding of Si-O bond. The bands around 792 and 1090 cm^{-1} correspond to the symmetrical Si-O and asymmetrical stretching of the Si-O bond [3].

It is not possible to observe the same phenomenon of band overlap as in the hematite spectrum after conditioning, as well as the Tanfloc bands.

Table 3. Pure quartz mineral and with Tanfloc SG

Quartz	Quartz + Tanfloc SG	Attribution
465	468	δ (Si-O-Si)
692	691	δ (Si-O)
792	792	ν (Si-O)
1089	1092	ν (Si-O)

3.2. Zeta Potential

Figure 2 shows the results obtained from the Tanfloc SG zeta potential measurements in the presence of potassium chloride (KCl) as electrolyte at concentrations of 10^{-3} mol/L and 10^{-4} mol/L at different pHs.

The biosurfactant presented an isoelectric point (IEP) at pH around 8,5, observing that for pH values below 8,5 the zeta potential is positive and for values higher than this the potential measured is negative. At both concentrations, the same IEP was identified without major differences in the magnitude of the Tanfloc SG potential associated with the compression effect of the electric double layer.

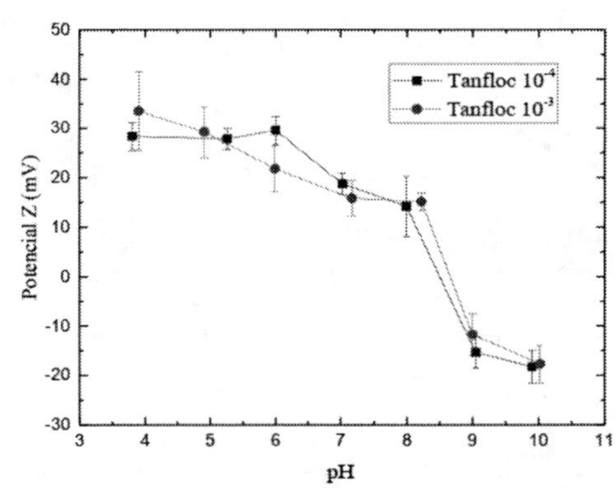

Figure 2. Tanfloc SG zeta potential

Figure 3 shows the hematite potential zeta profiles before and after interaction with the biosurfactant. Prior to interaction, hematite zeta potential values are close to 10 mV in the acid range and around -20 mV. These values show how hematite can interact with an acidic anionic surfactant and basic cationic surfactant.

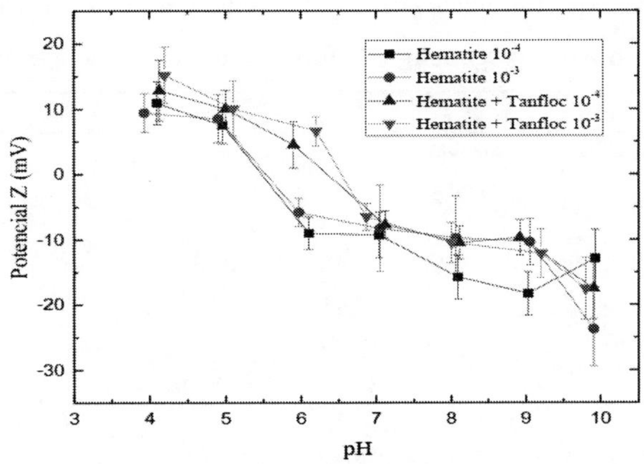

Figure 3. Hematite Pura e Hematita com Tanfloc SG – zeta potential

Following interaction with the biosurfactant, a change in the surface properties of the mineral is observed. There is a shift in the pH value corresponding to the IEP pH around 5,3 to 6,5 after interaction. This result may be due to the adsorption on the surface of the mineral caused by electrostatic forces, since hematite has a less positive charge than Tanfloc.

Figure 4 shows the zeta potential profiles of quartz before and after interaction with the biosurfactant. Prior to interaction, the quartz zeta potential values are close to the EIP in the acid range and around -20 and -30 mV for 10^{-4} and 10^{-3} mol/L electrolyte concentration, respectively.

These values show that quartz particles are negatively charged and how quartz can interact with cationic surfactant throughout the pH range. In the literature, there are reports in which IEP is around pH 2,0 [2]. After interaction with the biosurfactant, the pH corresponding to the PIE pH around 4,2 is observed.

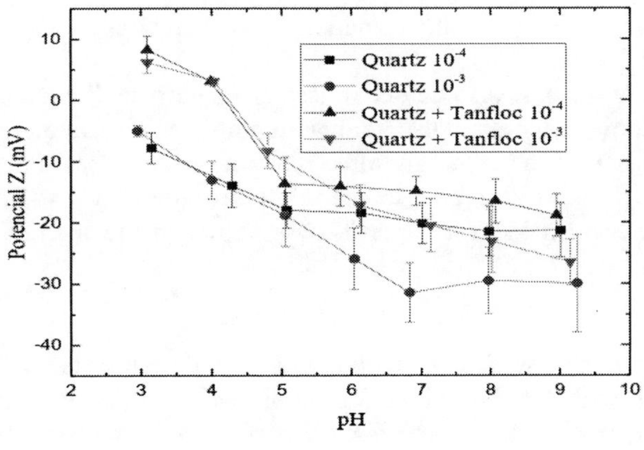

Figure 4. Quartz potential zeta

3.3. *Contact angle*

Contact angle measurements were performed to prove the hydrophobicity of minerals before and after interaction with Tanfloc SG biosurfactant. Both pure mineral and biosurfactant interacting mineral (140 mg/L) measurements were performed at pH 7,0, 7,5 and 8,0 and are shown in Table 4.

Table 4. Contact angle before and after interaction with Tanfloc SG.

pH		**7,0**			**7,5**			**8,0**	
Sample	**C ($x10^{-3}$)**	**θ (°)**	**Standard deviation**	**C ($x10^{-3}$)**	**θ (°)**	**Standard deviation**	**C ($x10^{-3}$)**	**θ (°)**	**Standard deviation**
Hematite	3,43E-03	72,5	±4,5	3,43E-03	72,0	±4,0	3,43E-03	81,0	±1,0
Hematite + Tanfloc	3,80E-03	85,0	±1,0	3,12E-03	88,0	±3,8	4,00E-03	62,0	±1,2
Quartz	3,13E-03	65,0	±7,0	3,13E-03	66,5	±2,5	3,13E-03	63,0	±1,0
Quartz + Tanfloc	2,67E-03	70,0	±2,0	3,92E-03	81,0	±1,5	3,74E-03	89,0	±1,0

Prior to the interaction with the biosurfactant, hematite and quartz surfaces presented lower values than interacted minerals. Quartz presented a value below hematite, around 65º and 72º, respectively. It is noted that the contact angle at pH 8,0 is much higher for hematite.

Regarding interacted minerals, their contact angle values are generally higher than pure minerals. Hematite samples show values above 80° for pH 7,0 and 7,5, decreasing at pH 8,0.

Quartz samples, on the other hand, have increasing values of pH 7,0 to 8,0, ranging from 70º to 89º. Thus, the biosurfactant would have attributed a hydrophobic character to the minerals after interaction, with values varying in relation to the pH of the measurement. These contact angle values for pure minerals are high compared to literature reports [2].

One factor that may have interfered with these higher values is the particle size of the mineral samples (<20μm). There are reports of decreased contact angle due to increasing particle diameter, happening exactly the opposite in this case. It is possible that smaller particles may generate smaller capillaries compared to larger sample sizes, leading to an increase in water mass in the samples during measurement [10].

The larger contact angle may also be due to less efficient packaging and possible higher porosity of the system, as the preparation method used only contemplates pipe impacts on the table.

The values obtained for the minerals after interaction are also high, and it is possible that the combination of low particle size and increased hydrophobicity by Tanfloc may be responsible. Nevertheless, the increase in contact angle is observed with the interaction of the mineral with the biosurfactant.

3.4. *Flotation Tests*

Prior to the flotation tests, some coagulation and flocculation tests were performed to verify the pH range that flakes form with Tanfloc. Flake formation occurred that the pH range between 6,5 and 9,0 (Figure 5). From these results, flotation tests were performed at the best pHs.

Figure 5. Coagulation and flocculation

The effect of Tanfloc concentration on the flotability of fine iron ore particles is shown in Figure 6. The study was performed at a concentration of 30 to 200 mg/L, a particle size range of -20 μm and at a pH of 6,5 up to 9,0.

First, preliminary results of flotability were sought in order to observe the best conditions to perform the tests. According to the results obtained, the maximum buoyancy values for a concentration of 175 mg / L at pH 7.0 with also high results at pH 8.0. A considerable result at pH 9.0 is also observed with 140 mg / L Tanfloc.

It was not possible to verify the decrease in flotability at high biosurfactant concentrations, which may be due to the formation of aggregates during the biosurfactant adsorption on the mineral surface, resulting in the decrease of the effective adsorption area.

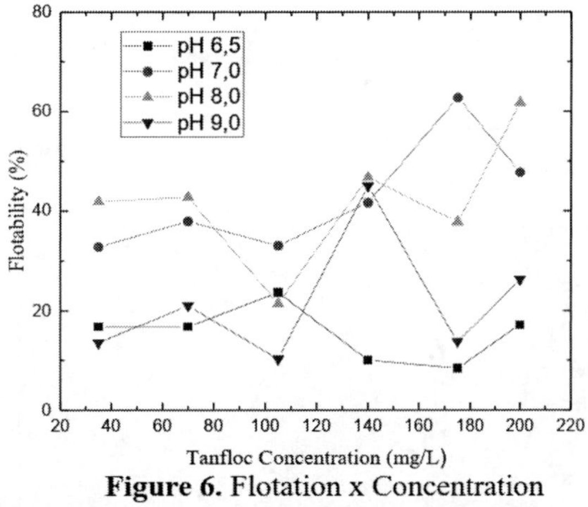

Figure 6. Flotation x Concentration

From the preliminary results, flotability tests were performed at pH 7,0, 7,5 and 8,0 with a concentration of 175 mg/L of Tanfloc SG, and this range presented the best preliminary results. These results are presented in Figure 7.

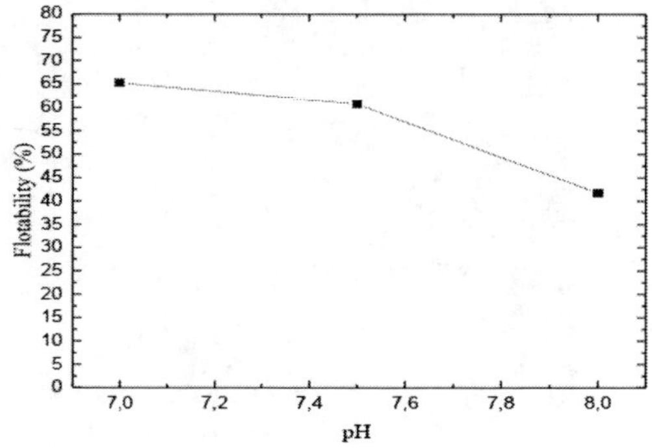

Figure 7. Flotation x pH

In order to obtain selectivity in iron ore flotation, sodium metasilicate (Na_2SiO_3) was used in an attempt to depress the quartz present in the iron ore sample. The effect of depressant concentration on the flotability of fine iron ore particles is shown in Figure 8

The study was performed at a concentration of 0,04 to 200 mg/L at pH 7,0 and biosurfactant concentration of 140 mg/L. These conditions were chosen based on the previous results because it is one of the conditions with the highest flotability.

It can be seen that the flotability decreases considerably with increasing depressor concentration, converging to between 40 and 30% flotability at higher concentrations. A decrease in flotability is expected as the quartz ore particles present are not expected to float under these conditions. Depreciation with the increase is also justified by the possible depreciation of the hematite particles as there is a large excess of depressor in the flotation cell.

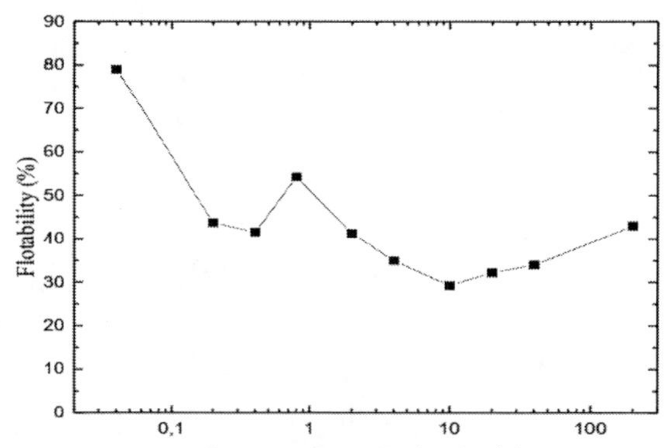

Figure 8. Flotation x depressor concentratio

The iron content was analyzed in some selected results to observe if there is concentration of the mineral of interest with the use of this depressant. The results of the analyzes are shown in table 5. It was possible to concentrate iron ore from 54 to 60% with a single flotation step, and concentrations above 4,00 mg/L depressor showed no significant differences in iron recovery.

Table 5. Concentration of depressor in the mineral.

Depressor Concentration (mg/L)	Iron % (% p/p)
0,00	53,76

0,04	54,98
4,00	59,85
200,00	60,51

4. Conclusion

Possible functional groups present in Tanfloc SG and minerals were identified by infrared spectroscopy, identifying the Fe-O clusters of hematite and Si-O of quartz. In addition to O-H, N-H, C-N, Ph-O groups, conjugated alkenes, aromatic groups and aliphatic cyclic ethers present in tannin-based structures.

Zeta potential profile measurements for biosurfactant, mineral and mineral / biosurfactant interaction showed the following isoelectric points (IEP): close to pH 8,5 for Tanfloc, while the hematite EIP was close to pH 5,3 and quartz below pH 3,0.

The hematite / Tanfloc interaction shifted the EIP from the hematite from pH 5,3 to pH 6,5. The quartz / Tanfloc interaction also showed a change in the quartz IEP to pH around 4,2, which indicated interactions of an electrostatic nature.

Contact angle measurements identified that the hematite surface has a more hydrophobic character compared to quartz before interaction with biosurfactant, ranging from 72 to 80° for hematite and around 63° for quartz. After interaction with Tanfloc SG hematite presented higher contact angle values at pH 7,0 and 7,5 (85 and 88°) compared to quartz (70 and 81°), being the value obtained at pH 8, The largest for quartz, 89° for quartz and 62° for hematite.

The dissolved air flotation of iron ore using Tanfloc SG depends on the biosurfactant concentration and the pH of the mixture, with the best flotability results at pH 7.0 with Tanfloc concentration between 140 and 175 mg/L for 4 g. ore with flotation values above 50% for a single flotation step. Besides concentrating the ore from 54 to 60% with sodium metasilicate as a quartz depressant.

Acknowledgment

The authors thank the CNPq (National Council for Scientific and Technological Development) and the prestigious PUC-Rio (Pontifícia Universidade Católica de Rio de Janeiro) for their financial support.

References

[1] Chernyshova I V (2007). Size-dependent structural transformations of hematite nanoparticles. 1. Phase transition. . *Physical Chemistry Chemical Physics*, **9**(14), 1736-1750.

[2] De Mesquita L M (2003). Interaction of a hydrophobic bacterium strain in a hematite–quartz flotation system. *International Journal of Mineral Processing*, **71**(1-4), 31-44.

[3] Li X Z (2017). Flotation separation of quartz from collophane using an amine collector and its adsorption mechanisms. . *Powder technology*, **318**, 224-229.

[4] Mangrich A S (2013). Química verde no tratamento de águas: uso de coagulante derivado de tanino de Acacia mearnsii. *Revista Virtual de Química*, **6**(1), 2-15.

[5] Miettinen T R (2010). The limits of fine particle flotation. *Minerals Engineering*, **23**(5), 420-437.

[6] Quast K (2017). Literature review on the use of natural products in the flotation of iron oxide ores. . *Minerals Engineering*, **108**, 12-24.

[7] Rendon J L (1981). IR spectra of powder hematite: effects of particle size and shape. *Clay Minerals*, **16**(4), 375-382.

[8] Tadesse B A (2019). Recovery of fine and ultrafine mineral particles by electroflotation–A review. *Mineral Processing and Extractive Metallurgy Review*, **40**(2), 108-122.

[9] TANAC (2013). *TANAC.* Retrieved from TANAC: http://www.tanac.com.br/pt-br/produtos/aguas

[10] Valencia G A (2018). Applicability of the modified Washburn method to contact angle measurements of calcium carbonate. *Cerâmica*, **64**(370), 197-206.

Evaluating the Microstructure of Photoluminescent Concrete Pavement Containing Strontium-Aluminate, Acrylic and Recycled Waste Glass

Ahmed Seyab Maqsud[1], Shohel Amin[1*] and Rahat Iqbal[2]

[1]Energy, Construction Environment, Coventry University, Priory Street, Coventry CV15FB, UK
[2] Interactive Coventry Ltd., Coventry University Technology Park, Puma Way, Coventry, CV1 2TT, UK
*** E-mail:** Shohel.Amin@coventry.ac.uk

Abstract. This paper constructs the photoluminescent concrete pavement (PhotoCP) mixing the photoluminescent material (Strontium-Aluminate) with recycled waste glass and transparent acrylic to visible the neighbourhood streets without streetlights. The non-destructive analyses of photoluminescent materials were conducted using the X-Ray Fluorescence, X-Ray diffraction, and Scanning Electron Microscopy instruments to understand the behaviour of atoms in photoluminescent materials when they interact with radiation. The compressive strength test examined the load bearing capacity of PhotoCP. A 30cm x 230cm test bed was constructed at a neighbourhood street in Peshawar, Pakistan to assess the impact of photoluminescent materials on lighting the neighbourhood street. The non-destructive analyses and compressive strength test show that PCP specimens have good interlocking capability, structural strength and durability. The testbed experiment observed the illuminance of PhotoCP for a period of 6 to 8 hours with highest lumen intensity of 1-3 lux from sunset to 8:30 pm.

Keywords: Strontium Aluminate, RWG, XRF, XRD, SEM image, compressive strength.

1. Introduction

The electricity costs for street lighting are a significant expenditure for municipalities. There is a substantial increase in the electricity cost around the world particularly through the rapid increase in the cost of oil and gas [1]. For promoting energy efficiency streetlighting, studies focused on the consumption, production patterns and program costs of streetlights minimising the electricity consumption [2-3]. To achieve the short-term energy efficiency of urban streetlighting, municipalities sometimes adopt a 'part-night' lighting system particularly in the residential areas where lights of relatively lower importance (in respect of traffic accident and crime rate) are dimmed or switched off at times during the night when they are least likely to be needed [1,4]. The energy efficient streetlights and 'part-night' lighting system reduce the energy consumption and carbon emission; however, the municipalities are still struggling to light up the urban streets due to budget deficit leaving many urban streets in dark in developing countries. The luminescent road marks can ensure safety and security at urban access roads where traffic volume at night is low.

The photoluminescence phenomenon of pavement materials can be a source of road illumination at night by constructing the road surface with phosphorescent materials. Strontium-Aluminate is a photoluminescent material that is white, nontoxic and non-radioactive powder commonly used in artistic or decorative purposes. The slower time-scale emission of light from the Strontium-Aluminate ($SrAl_2O_4$), transition from 4f to 5d of the Europium (Eu^{2+}), make it a good candidate for the phosphorescent road material [5]. Several studies discuss on the emission intensity and luminescence capacity of Strontium-Aluminate [5-10]. Bone [6] partially substituted the Calcium (Ca^{2+}) and Strontium (Sr^{2+}) sites in the matrix for enhancing the emission intensity and luminescence of $SrAl_2O_4$ with Eu^{2+} and Dysprosium (Dy^{3+}). Similarly, Katsumata et al. [5] investigated the phosphorescence characteristics of $Sr_3Al_2O_6$, $SrAl_2O_4$, $SrAl_4O_7$, and $SrAl_{12}O_{19}$ polycrystalline ceramic doped with Eu^{2+} and Dy^{3+}. Luitel et al. [8] examined the Eu^{2+} and Dy^{3+} doped phosphors and illustrated the causes of higher brightness and slow-scale light emission from the Dy^{3+} doped phosphors comparing to Eu^{2+} doped phosphors. Praticò et al. [10] experimented the photoluminescent phenomenon of dense-graded and open-graded friction courses and the surface modifications (friction and texture) due to the coating of photoluminescent material.

This paper constructs the photoluminescent concrete pavement mixing the photoluminescent material (Strontium-Aluminate) with recycled waste glass (RWG) and transparent Acrylic to visible the neighbourhood streets without the streetlights. The RWG is widely used in glass–ceramics, tiles for flooring or lining, foam glass and insulating or reinforcing fibre [11]. There are some evidences of using the RWG in the concrete pavements [12-14], however this study, for the first time, applies it in the photoluminescent concrete pavement for refracting the light rays to the Strontium-Aluminate particles. The Acrylic was used as the transparent binder of Strontium-Aluminate and coarse aggregates in the concrete pavement. The glass particles in the concrete refract the daylight as well as the light emission from the Strontium-Aluminate particles. The United Nations estimated that the annual volume of solid waste was 200 million tonnes of which 7% were made up of glass [15]. A test bed was constructed at the neighbourhood street in Peshawar, Pakistan where 113 tons of glass waste is generated daily from municipal waste [16].

2. Materials and Experiments

2.1. Photoluminescent Materials

The strontium aluminate, acrylic and glass powder were mixed with cement and aggregates to manufacture the photoluminescent concrete pavement (Figure 1a and 1b). Acrylic is a transparent thermoplastic (methyl methacrylate) and acts as a transparent binder for the Strontium-Aluminate and coarse aggregates of concrete pavement (Figure 1c and 1d). Acrylic binder resin enhances the stability, photoluminescence and surface absorption of photoluminescent concrete pavement [17]. The glass powder was used in the concrete mixture instead of sand because glass particles refract the ray of light to the Strontium-Aluminate particles enhancing the invisible ultraviolet energy from the natural light source (Figure 1e and 1f). The recycled glass wastes were collected from the local recycle stores and grounded into glass powder. The glass powder was passed through no. 20 sieve.

Three groups of photoluminescent concrete cubes were manufactured mixing cement, glass powder, Strontium-Aluminate and coarse aggregates to understand the significance of Strontium-Aluminate and glass powder on the photoluminescence phenomenon of photoluminescent concrete pavement (Figure 2). Group 1 specimens were designed with 2 inches top layer of cement, glass powder, Strontium-Aluminate and coarse aggregates mixture. The four inches bottom layer was made of cement, glass powder and coarse aggregates (Figure 3). Sand replaced glass powder at the bottom 4 inches layer of Group 2 specimens to understand the impact of glass particles on refracting the light rays (Figure 3). Group 3 specimens are the standard (base) concrete cubes with 1:2:4 ratios of cement, sand and coarse aggregates, respectively (Figure 3).

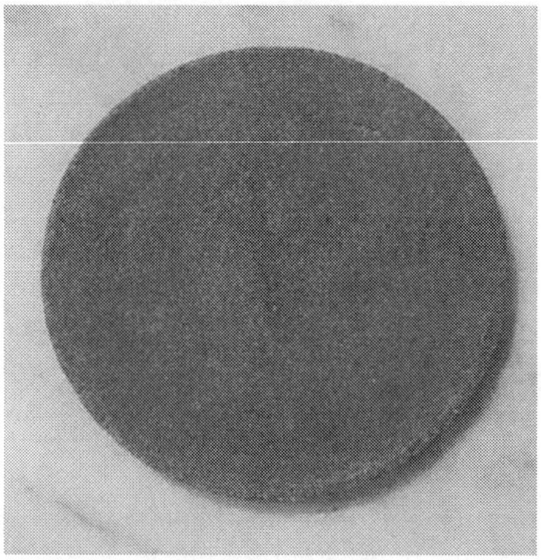

(a) Cement, Strontium Aluminate, and glass powder (in light)

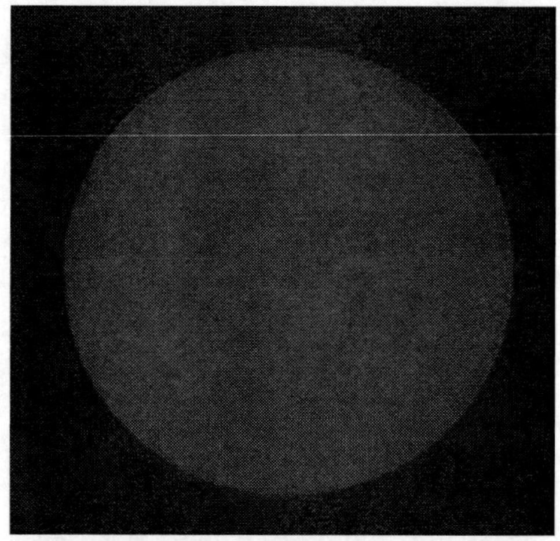

(b) Cement, Strontium Aluminate, and glass powder (in dark)

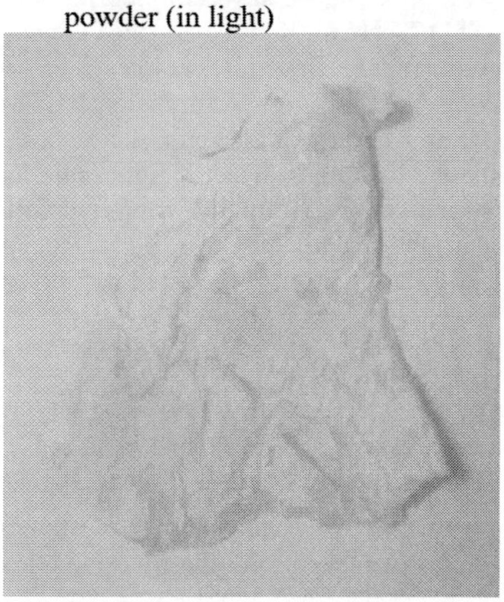

(c) Acrylic mixed with Strontium-Aluminate (in light)

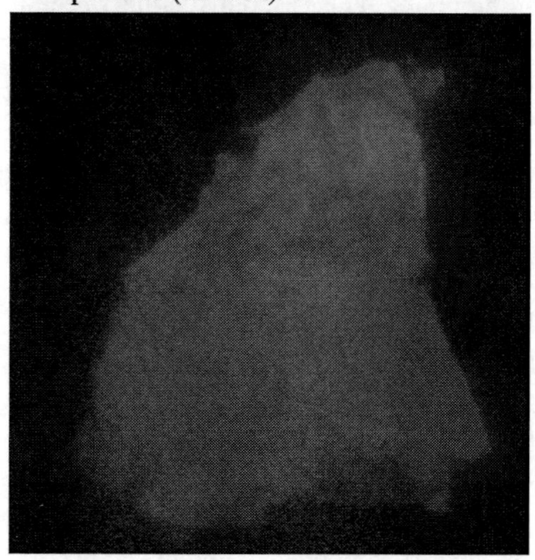

(d) Acrylic mixed with Strontium-Aluminate (in dark)

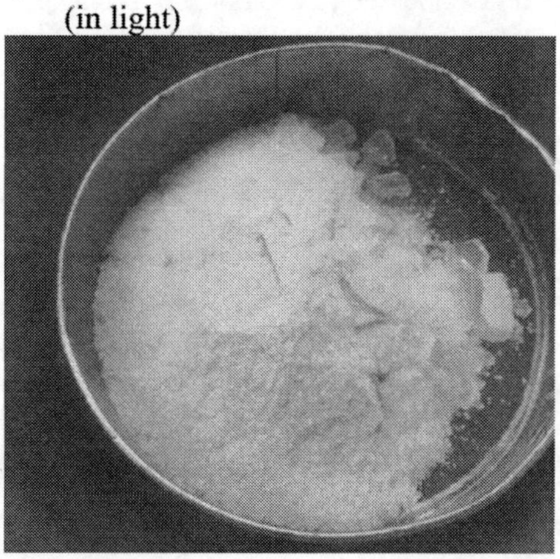

(e) Recycled glass waste powder

(f) Recycled glass powder passed through 20 no. sieve

Figure 1. Photoluminescent materials

Figure 2. Concrete cube samples (6x6x6 in) of three groups

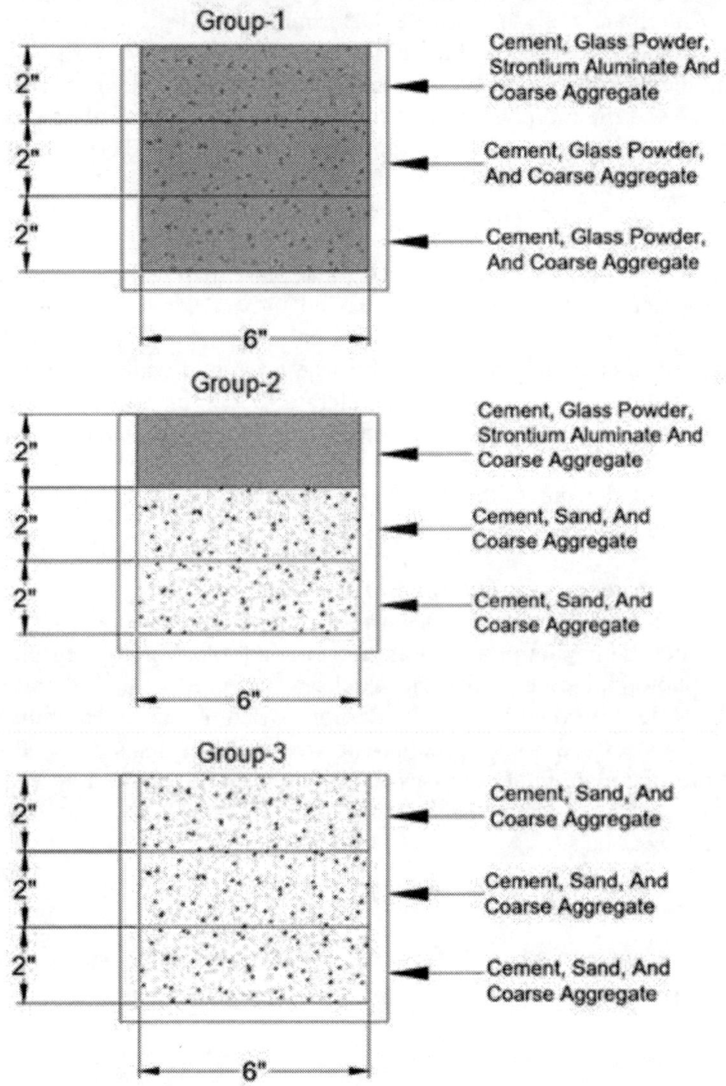

Figure 3. Schematic diagrams of concrete specimens of three groups

IOP Conf. Series: Materials Science and Engineering 1196 (2021) 012020 doi:10.1088/1757-899X/1196/1/012020

2.2. Laboratory experiments

2.2.1 Chemical experiments

The non-destructive analyses of photoluminescent materials were conducted using the X-Ray Fluorescence (XRF), X-Ray diffraction (XRD), and Scanning Electron Microscopy (SEM) instruments to understand the behaviour of atoms in photoluminescent materials when they interact with radiation. XRF analysed the chemical elements in concrete specimens based on the measurement of wavelengths and intensities of their spectral lines emitted [18]. The XRF imaged the cross-section of the concrete specimens and determined the concentrations of Strontium Aluminate and glass particles. X-rays interact weaker with matter than electrons and so they penetrate deeper into the sample.

This deeper penetration makes the results less sensitive to the surface roughness [19]. XRF has comparatively high detection limits due to high background produced by the absorption and scattering of X-ray beam by the sample and matrix [18]. Different calibration plots are required for different matrices to offset the matrix effects. The XRD technique was used to find the crystal structure properties of the Strontium-Aluminate and glass powder determining the average bulk composition. The XRD produces the patterns of peak positions and relative intensities that characterise the crystal structures and enable identifying the presence of Strontium-Aluminate and glass powder in the concrete specimens [20]. The SEM analysed the morphology of Strontium-Aluminate and glass powder and assessed the interlocking spatial variations in chemical compositions of particles. The elongated particles with long edges are prone to air voids and structural disintegrity.

2.2.2 Structural experiments

Several tests such as accelerated pavement testing and plate load test examine the pavement response and modulus of subgrade reaction under the controlled and accelerated accumulation of wheel loads in a compressive time period. Local streets in residential neighbourhood, mostly utilise for leisure and play, have slow speed light traffic, occasional delivery vans and walking and bicycle facilities. Since the neighbourhood streets are not subjected to accelerated accumulation of damage by heavy trucks and the concrete pavements are much stronger in compression than in tension, the compressive strength tests were experimented on the concrete specimens. Moreover, glass powder replaced sand as the filler in the concrete specimens and the durability and deformation of the specimens with glass powder were compared with the controlled specimens (Group 3).

2.3 Testbed at neighbourhood street in Peshawar, Pakistan

A test bed of 1 x 7.5 feet (30 cm x 230 cm) was constructed at the neighbourhood street in Peshawar, Pakistan to understand the impact of photoluminescent materials on lighting the neighbourhood street (Figure 4a). The photoluminescent concrete blocks were cured in the laboratory and placed on the neighbourhood road after for 3 days to avoid hindering the neighbourhood mobility. This neighbourhood street was selected as a testbed because there is no streetlight in this street and the emission rate and duration of light rays from the photoluminescent concrete pavement can be examined after the dawn (Figure 4b).

(a) The photoluminescent concrete blocks were placed after 3 days curing (b) The photoluminescent concrete pavement testbed

Figure 4. Testbed construction at the neighbourhood street in Peshawar Pakistan

3. Experimental Results

3.1 Chemical Experiments

The non-destructive analyses of photoluminescent materials using XRF, XRD and SEM tools produce the controlled populations of point defects and multiple defect clusters such as voids, stacking-fault tetrahedra and dislocation loops [21]. The XRF elemental analysis, independent of the chemical form of the elements, measures the concentration of different elements in the photoluminescent materials by the X-rays intensity that is proportional to the concentration of an element and the strength of the ionizing source [22]. This study applied the fundamental parameter method for quantitative XRF analysis that is based on the primary and secondary radiations to calculate the intensity of analyte radiation in the specimen of known composition [23]. Ten elements were identified in the Strontium-Aluminate materials such as Strontium (Sr), Chlorine (Cl), Calcium (Ca), Manganese (Mn), Dysprosium (Dy), Thulium (Tm), Copper (Cu), Tungsten (W), Rubidium (Rb) and Rhodium (Rh). The Dy^{3+} (1.14%) doped Sr has major contribution (86.62%) in the Strontium-Aluminate (Table 1). The Dy^{3+} doped Strontium-Aluminate has excellent emission characteristics, single luminescent centre and high absorption efficiency in the Ultraviolet region [24]. The Dy^{3+} doped Strontium-Aluminate has higher emission rate comparing to Eu^{2+} phosphor and can emit light 6 to 10 hours at night [8,10].

The recycled glass waste powder consists of Sr, Silicon (Si), Sulfur (S), Rh, Potassium (K), Ca, Titanium (Ti), Chromium (Cr), Mn, Iron (Fe), Cu and Zirconium (Zr). Major constituent of recycled glass waste is quartz (SiO_2) that is also the major constituent of the sand (Table 1). A typical brown sand in concrete application contains up to 80% quartz along with iron, carbonate, potassium and other elements. The recycled glass waste contains 87.31% silica that ensures the stronger binding of concrete mixture (Table 1). The recycled glass powder also contains 23.91% Ca and 11.41% calcium oxide (CaO) (Table 1). The CaO crystallises each other and other substances in the concrete through hydration process and the carbonation process of calcium hydroxide increases the mechanical strength of the concrete and decreases the pH level. There is no toxic element or compound in the recycled glass powder.

The qualitative analysis of XRD data based on relative peak intensities identifies the presence of any impurity phases of composition different from Strontium-Aluminate and glass materials [24]. The XRD data were compared with a known phase diagram available at online databases such as International Centre for Diffraction Data (ICDD) to find unknown minerals [25]. The highest peak in the XRD diffractogram of strontium aluminate is at 30^0 angles on 2Theta axis with a value of 2.98 (Figure 5a). The compound phases of strontium aluminate in the XRD diffractogram are $SrAl_2O_4$ and $SrAl_2O_8Si_2$

IOP Conf. Series: Materials Science and Engineering 1196 (2021) 012020 doi:10.1088/1757-899X/1196/1/012020

while the crystal system and space groups are cubic and monoclinic, and P1211 and Pm-3m, respectively.

The recycled glass waste powder has too many peaks and no sharp diffraction peak because glass is amorphous. Instead of peaks, the rising background is observed in the XRD diffractogram of glass that consists of one or more broad diffuse rings [26]. The intensity curve of recycled glass waste in the XRD diffractogram very closely follow the absolute basis (electron units per SiO_2) (Figure 5b). The compound phase, crystal system and space group of recycled glass waste powder in the XRD diffractogram are SiO_2, hexagonal and P6/mmm.

The XRD pattern of strontium aluminate and glass powders provides data on crystal structure but unable to provide the morphology of the surface [6]. The SEM images observe the surface morphology and crystalline sizes of strontium aluminate and glass powders (Figure 6 and 7). Figure 6 shows that the particles in the strontium aluminate range from sub-micron size to several tens of μm sizes and have well distinct boundaries that provide good interlocking between particles. The strontium aluminate powder has finer and round shaped particles than glass powder but also possesses needle like particles at very small scale ensuring interlocking ability (Figure 6). The interlocked and closely spaced particles in the strontium aluminate powder increase the compressive strength of concrete because the load transfers uniformly from particle to particle down to the subgrade soil.

The SEM images of recycled glass waste powder show that particles are irregular and angular with distinct boundaries which is an indication of good particle interlocking (Figure 7). The particle sizes of glass powder range from 10μm to hundreds of μm with little number of particles in sub-micron size (Figure 7). There are more surface area contacts between the particles in glass powder mitigating the crack development in the concrete. In addition, the irregular and angular shapes of glass powder particles increase the strength of hardened concrete but prejudice the workability of fresh concrete [27]. The XRF and XRD analyses determine the chemical composition and crystal structure of Strontium-Aluminate and recycled glass waste powders arguing the materials purity in the photoluminescent concrete pavement. The SEM analyses conclude that mixture of Strontium-Aluminate and recycled glass waste powders in the concrete will enhance the compressive strength and workability of concrete after assessing the morphological structure of the materials.

3.2 Compressive strength analysis of concrete specimens

The compressive strength tests of 3-group of concrete specimens were conducted to correlate with the flexural strength by calibrating the relationship between flexural and compressive strength in order to check compliance. The compressive strength of Group 1 specimens that included RWG instead of sand reduced by 5.6% in comparison with Group 3 specimens (fine aggregates are sand) (Table 2). The compressive strength of Group 2 specimens was reduced by 22.65% comparing with that of Group 3 specimens (Table 2). The reduction of compressive strength of group 1 and 2 specimens was due to the presence of Strontium-Aluminate materials in the upper 2-inch layer because the SEM images of RWG powder showed the good particle interlocking. The crack developed in the bottom layer (contains sand) of Group 2 specimens during the compression tests justify the above-mentioned argument. The average compressive strengths of target specimens of Group 1 and Group 2 are 3738 psi (25.77 N/mm^2) and 3323.78 psi (22.92 N/mm^2), respectively (Table 2). The recommended minimum compressive strength classes of road pavement (concrete cube) are within the range of C16/20 (20N/mm^2) and C35/45 (40N/mm^2) to achieve suitable durability, regardless of the structural strength requirements. Higher strengths are sometimes used but can lead to higher shrinkage and higher costs. This is usually only justifiable where extreme conditions are present such as highly abrasive operations or where high impact resistance is required. Otherwise the detrimental effects of higher shrinkage usually outweigh any benefits from higher strength. There is also often a drop off in correlation between high compressive strength and flexural strength which leads to diminishing returns.

Table 1. XRF Analysis of Strontium-Aluminate and RWG powder.

	Powder elements			**Powder oxide**		**Element [3-sigma]**		**Oxide [3-sigma]**		**Line**	**Element - Int (cps/uA)**		**Oxide - Int (cps/uA)**	
Analyte	Strontium-Aluminate	Glass	Analyte	Strontium-Aluminate	Glass	Strontium-Aluminate	Glass	Strontium-Aluminate	Glass		Strontium-Aluminate	Glass	Strontium-Aluminate	Glass
Si		73.37%	SiO_2		87.31%		[0.788]		[0.937]	SiKa		24.770		24.866
Sr	86.62%	0.06%	SrO	87.78%	0.02%	[0.116]	[0.002]	[0.118]	[0.001]	SrKa	11960.139	9.032	11959.416	9.943
Cl	11.64%		Cl	10.39%		[0.346]		[0.304]		ClKa	21.619		22.163	
Dy	1.14%		Dy_2O_3	1.13%		[0.028]		[0.028]		DyLa	27.263		27.082	
Ca	0.19%	23.91%	CaO	0.23%	11.41%	[0.034]	[0.117]	[0.042]	[0.056]	CaKa	1.331	123.245	1.329	121.307
Fe		1.55%	Fe_2O_3		0.65%		[0.011]	[0.008]	[0.005]	FeKa		48.279		49.809
Mn	0.17%	0.04%	MnO	0.19%	0.01%	[0.007]	[0.005]		[0.002]	Mn Ka	8.649	0.795	8.604	0.837
Rb	0.13%		Rb_2O	0.10%		[0.007]		[0.007]		RbKa	23.849		19.531	
Tm	0.07%		Tm_2O_3	0.06%		[0.023]		[0.023]		TmLa	2.307		2.137	
W	0.03%		WO_3	0.02%		[0.009]		[0.010]		W La	1.692		0.836	
Cu	0.02%	0.04%	CuO	0.03%	0.01%	[0.004]	[0.003]	[0.004]	[0.001]	CuKa	3.057	2.373	3.322	2.352
S		0.44%	SO_3		0.40%		[0.068]		[0.064]	SKa		0.415		0.409
Ti		0.32%	TiO_2	0.08%	0.16%		[0.009]	[0.021]	[0.004]	TiKa		2.618	0.913	2.616
K		0.16%	Ag_2O		0.02%		[0.025]		[0.001]	K Ka		0.483		2.387
Zr		0.06%					[0.002]			ZrKa		10.164		
Cr		0.04%	Cr_2O_3		0.02%		[0.006]		[0.002]	CrKa		0.650		0.709

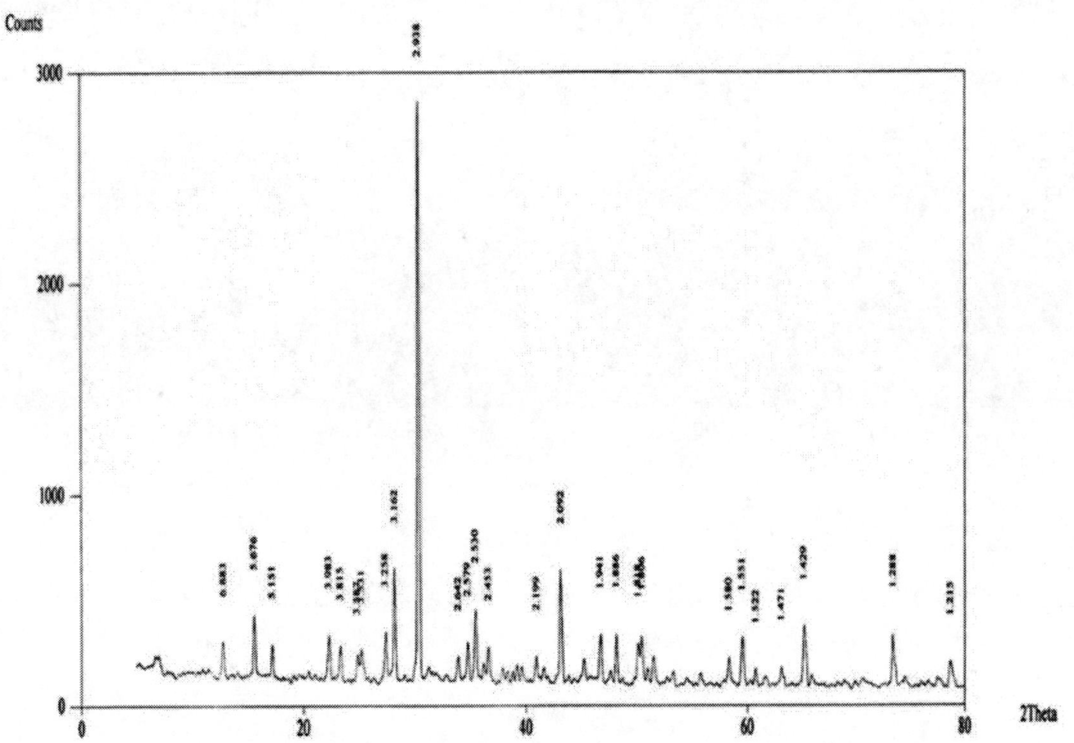

(a) Strontium Aluminate

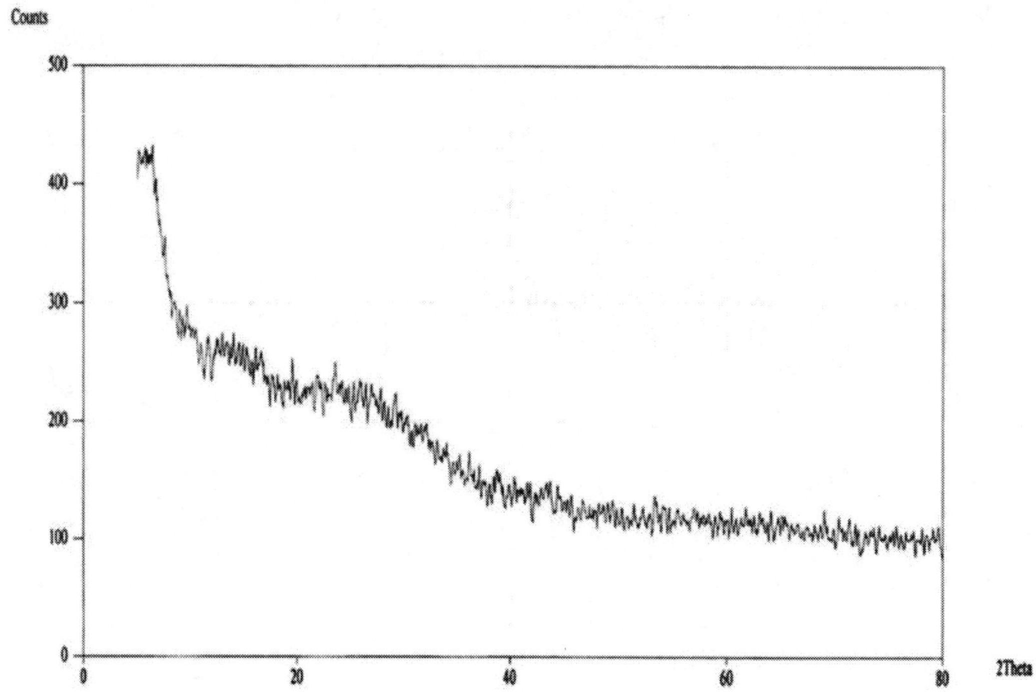

(b) Glass Powder

Figure 5. XRD diffractograms analysis

(a) x250 magnification level (100μm)

(b) X500 magnification level (50μm)

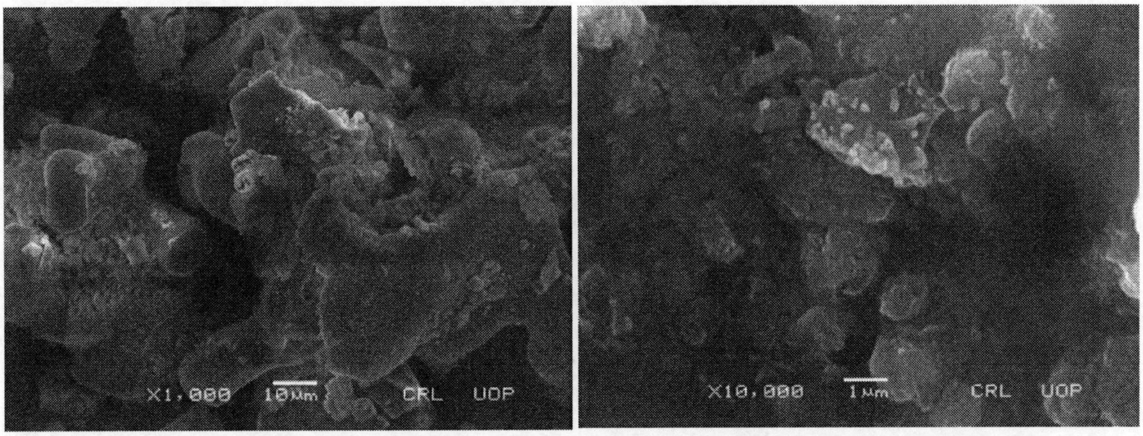

(c) X1000 magnification level (10μm) (d) X10000 magnification level (1μm)

Figure 6. SEM image micrographs of Strontium Aluminate

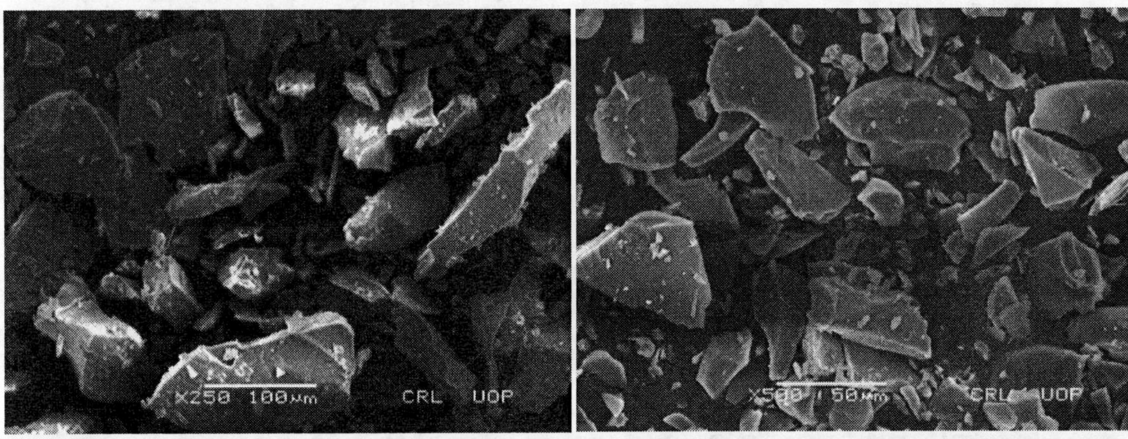

(a) x250 magnification level (100μm) (b) X500 magnification level (50μm)

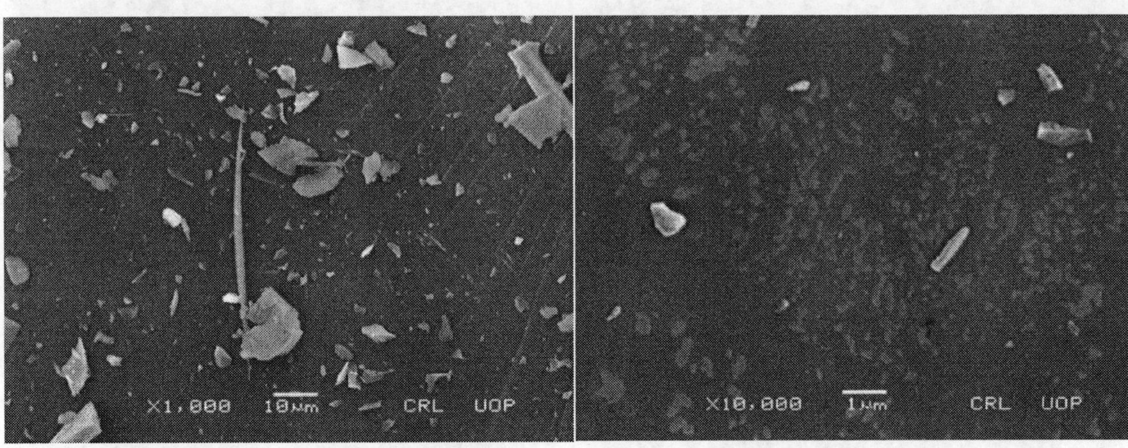

(c) X1000 magnification level (10μm) (d) X10000 magnification level (1μm)

Figure 7. SEM micrographs of recycled glass waste powder.

IOP Conf. Series: Materials Science and Engineering 1196 (2021) 012020 doi:10.1088/1757-899X/1196/1/012020

Table 2. Compression test results of Group 1, 2 and 3 specimens

	Sample No.	Load (tonnes)	Compressive Strength (psi)	Compressive Strength (N/mm^2)	Average compressive Strength (psi)	Average compressive Strength (N/mm^2)
GROUP-1	1	36	3654.95	25.20	3738.11	25.77
	2	38.7	3818.84	26.33		
	3	37.4	3740.52	25.79		
GROUP-2	1	30.5	3317.01	22.87	3323.78	22.92
	2	30.6	3324.26	22.92		
	3	30.7	3330.07	22.96		
GROUP-3	1	39.7	3881.21	26.76	3872.51	26.70
	2	39.6	3873.96	26.71		
	3	39.4	3862.36	26.63		

3.3 Photoluminescent concrete pavement at a neighbourhood street in Peshawar, Pakistan

The illuminance (lumen/m^2), also known as lux, of the photoluminescent concrete pavement was recorded at the testbed in the neighbourhood street from dusk to 9:00 o'clock at night (Figure 8). It was observed that during the time period between dusk to sunset, lux value was within the range of 0.5 to 2. The illuminance of pavement increases after the sunset (approximately 6:30 pm) within the range of 1-3 lux until 8:30 pm. The illuminance started decreasing after 8:30 pm but at deep twilight (approximately 1 lux) condition after two hours. The photoluminescent materials absorbs the light rays from the passing by vehicles and started glowing despite low lux level after 8:30 pm. The emission rate of light ray from the photoluminescent pavement strip was uniform and higher after 3-4 days of placement on the neighbourhood streets. There was no crack or structural damage of the testbed 28 days after the construction. The National Optical Astronomy Observatory recommends the illuminance level for local roads within the range of 3 to 6 lux. The observed illuminance of the 30 cm x 230 cm photoluminescent pavement strip at neighbourhood street was within the lower recommended illuminance level for local roads for a couple of hours after the sunset. The complete construction of neighbourhood cul-de-sac and streets enhances the attractiveness of evening walking, bicycling and social gathering.

(a) Day 1

(b) Day 2

Figure 8. The illuminance of photoluminescent concrete pavement

4. Conclusions

This study carried out an experiment mixing the Strontium-Aluminate and RWG powders in concrete to manufacture the photoluminescent concrete pavement for the neighbourhood street. The XRF test of Strontium-Aluminate and glass powder identified that Strontium, Chlorine and Dysprosium were the main components of Strontium-Aluminate sample and the Silicon, Calcium and Iron were the main components of RWG power passed through the 20-mesh sieve. The presence of Chlorine (11.64%) in the Strontium-Aluminate sample reduces the potential growth of bacteria and algae on the concrete pavement. The major constituents of Strontium-Aluminate and RWG powders were $SrAl_2O_4$ (85.56%) and SiO_2 (87.31%), respectively. The XRD results of Strontium-Aluminate and RWG powders didn't find any foreign substance justifying the purity of materials. The SEM image micrographs analysed the morphological structure of Strontium-Aluminate and RWG powder. The Strontium-Aluminate particles were rounded and finer shape like clay and had smooth edges, whereas the particles of RWG were irregular with distinct and sharp edges. The SEM images show the good interlocking capability of both materials.

The compressive strength tests of the specimens show the structural strength and durability of photoluminescent concrete cubes. The testbed experiments at the neighbourhood street observed the illuminance of the photoluminescent concrete pavement for a period of 6 to 8 hours with highest level of illuminance from the sunset to 8:30 pm. This study not only utilised the RWG in place of sand in the concrete pavement but also manufactured the energy efficient concrete pavement by mixing the Strontium-Aluminate material. The Strontium-Aluminate powder is usually used in painting and toys, however for the first time this study examined its effectiveness in the photoluminescent concrete pavement and saving the energy cost for streetlighting at the neighbourhood street. This study opens the window for potential study examining the further improvement in the photoluminescent materials and techniques for enhancing the illuminance of photoluminescent concrete pavement that would reduce the energy costs of city councils for streetlighting.

5. References

[1] Pagden M, Ngahane K and Amin M S R (2020). Changing the colour of night on urban streets - LED vs. part-night lighting system. *Socio-Economic Planning Sciences* **69**, 100692. https://doi.org/10.1016/j.seps.2019.02.007

[2] Escolar S, Carretero J, Marinescu M and Chessa S (2014). Estimating Energy Savings In Smart Street Lighting By Using An Adaptive Control System. *International Journal of Distributed Sensor Networks*. **10**(5), 971587. DOI: https://doi.org/10.1155/2014/971587

[3] Iveland J, Martinelli L, Peretti J, Speck J S, and Weisbuch C (2013). Direct measurement of Auger electrons emitted from a semiconductor light-emitting diode under electrical injection: identification of the dominant mechanism for efficiency droop. *Physical review letters*. **110**(17), 177406. DOI: https://doi.org/10.1103/PhysRevLett.110.177406

[4] Plainis S, Murray I J and Pallikaris I G (2006). Road traffic casualties: understanding the night-time death toll. *Injury Prevention*. **12**(2), 125-138. DOI: http://dx.doi.org/10.1136/ip.2005.011056

[5] Katsumata T, Sasajima K, Nabae T, Komuro S and Morikawa T (1998) Characteristics of Strontium-Aluminate crystals used for long-duration phosphors. *Journal of the American Ceramic Society*.

[6] Bone A N (2017). *Optimization of a Strontium-Aluminate* ". Oak Ridge National Lab. (ORNL), Oak Ridge, TN (United States).

[7] Gill R (2014) *Modern Analytical Geochemistry: an introduction to quantitative chemical analysis techniques for Earth, environmental and materials scientists*. Routledge.

[8] Luitel H, Watari T, Chand R, Torikai T and Yada M (2013) Giant Improvement on the Afterglow of Sr4Al14O25:Eu2+,Dy3+ Phosphor by Systematic Investigation on Various Parameters. *Journal of Materials*. 2013, pp.1-10.

[9] Egerton R F (2005) *Physical principles of electron microscopy*. New York: Springer.

[10] Praticò F, Vaiana R and Noto S (2018) Photoluminescent Road Coatings for Open-Graded and Dense-Graded Asphalts: Theoretical and Experimental Investigation. *Journal of Materials in Civil Engineering*. **30**(8), p.04018173.

[11] Colombo P, Brusatin G, Bernardo E and Scarinci G (2003). Inertization and reuse of waste materials by vitrification and fabrication of glass-based products. *Current Opinion in Solid State and Materials Science*. **7**, 225–239. https://doi.org/10.1016/j.cossms.2003.08.002

[12] Arulrajah A, Ali M Y, Disfani M M and Horpibulsuk S (2014). Recycled-Glass Blends in Pavement Base/Subbase Applications: Laboratory and Field Evaluation. *Journal of Materials in Civil Engineering*. **26**(7), 04014025. https://doi.org/10.1061/(ASCE)MT.1943-5533.0000966

[13] Arulrajah A, Piratheepan J, Disfani M M and Bo M W (2013). Geotechnical and geo-environmental properties of recycled construction and demolition materials in pavement subbase applications. *Journal of Materials in Civil Engineering* **25**(8), 1077-1088. https://doi.org/10.1061/(ASCE)MT.1943-5533.0000652

[14] Jamshidi A, Kurumisawa K and Nawa T (2016). Performance of pavements incorporating waste glass: The current state of the art. *Renewable and Sustainable Energy Reviews*. **64**, 211-236. https://doi.org/10.1016/j.rser.2016.06.012

[15] Demir I (2009). Reuse of waste glass in building brick production. Waste Management & Research: *The Journal for a Sustainable Circular Economy*. **27**(6), 572-577. https://doi.org/10.1177/0734242X08096528

[16] International Trade Administration (2018). Pakistan - Waste Management | Export.Gov (2018) [online] available from <https://www.export.gov/article?id=Pakistan-Waste-Management> [26 April 2019]

[17] Duan Y, Huo Y and Duan L (2017). Preparation of acrylic resins modified with epoxy resins and their behaviors as binders of waterborne printing ink on plastic film. *Colloids and Surfaces A: Physicochemical and Engineering Aspects*. **535**, 225-231. https://doi.org/10.1016/j.colsurfa.2017.09.041

[18] Misra N L and Mudher K D S (2002). Total Reflection X-ray Fluorescence: A Technique for Trace Element Analysis in Materials. *Progress in Crystal Growth and Characterization of Materials*. **45**(1-2), 65-74. https://doi.org/10.1016/S0960-8974(02)00029-3

[19] Moradllo M K, Sudbrink B, Hu Q, Aboustait M, Tabb B, Ley M T and Davis J M (2017). Using micro X-ray fluorescence to image chloride profiles in concrete. *Cement and Concrete Research*. **92**, 128-141. https://doi.org/10.1016/j.cemconres.2016.11.014

[20] Snellings R (2016). X-ray powder diffraction applied to cement. In: K. Scrivener, R. Snellings, and B. Lothenbach (eds.) *A Practical guide to microstructural analysis of cementitious materials*. CRC Press, New York. Pages 107-176.

[21] Stiegler J O and Mansur L K (1979). Radiation Effects in Structural Materials. *Annual Review of Materials Science*. **9**,405-454. https://doi.org/10.1146/annurev.ms.09.080179.002201

[22] El-Taher A (2012). Elemental analysis of granite by instrumental neutron activation analysis (INAA) and X-ray fluorescence analysis (XRF). *Applied Radiation and Isotopes*. **70**, 350–354. doi:10.1016/j.apradiso.2011.09.008

[23] Stiko R and Zawisza B (2012). *Quantification in X-Ray Fluorescence Spectrometry. In: S. K. Sharma (ed.) X-Ray Spectroscopy*, pp. 137-162. IntechOpen. Available from: http://www.intechopen.com/books/x-ray-spectroscopy/quantification-in-x-ray-fluorescence-spectrometry

[24] Karabulut Y, Canimoglu A, Kotan Z, Akyuz O and Ekdal E (2014). Luminescence of dysprosium doped strontium aluminate phosphors by codoping with manganese ion. *Journal of Alloys and Compounds*. **583**, 91–95. http://dx.doi.org/10.1016/j.jallcom.2013.08.172

[25] *ICDD Database Search - ICDD* (2019) [online] available from <http://www.icdd.com/index.php/pdfsearch> [22 April 2019]

[26] Warren B E (1940). X-ray Diffraction Study of the Structure of Glass. *Chemical Reviews.* 1940 **26** (2), 237-255. DOI: 10.1021/cr60084a007
[27] He Z H, Zhan P M, Du S G, Liu B J, Yuan W B. (2019). Creep behavior of concrete containing glass powder. *Composites Part B: Engineering*. **166**, 13-20. https://doi.org/10.1016/j.compositesb.2018.11.133.

Analysing the Sustainability of Cascade Recycling in Plastic Manufacturing

Karl Tanti[1], **Arif Rochman**[1] **and Paul Refalo**[1,*]

[1]Department of Industrial and Manufacturing Engineering, Faculty of Engineering, University of Malta, Msida, MSD 2080, Malta

* **E-mail:** paul.refalo@um.edu.mt

Abstract. This study ventures to further sustainable manufacturing by showing how a cascade recycling approach can be utilised by a plastic cosmetic packaging company. Three reprocessing cycles were assessed from product quality, environmental, financial and social points of view. It is shown how the cascade approach did not negatively affect the quality of the produced parts and when compared to a no-recycling scenario, the proposed three-cascade recycling approach results in 28.1% less costs and 29.3% less carbon footprint. The labour increases up to a maximum of 7 hours per 50,000 products. When compared to a conventional in-house recycling scenario, the proposed approach would reduce the costs by 11.7% and the carbon footprint by 12.9%.

Keywords: cascade recycling, regrind, plastic injection moulding, sustainable manufacturing, cosmetic packaging.

1. Introduction

Sustainable manufacturing in current businesses is no longer solely motivated by economic feasibility, rather it must take into consideration all three pillars of sustainability, that is; environmental protection, economic development and social development [1]. The manufacturing industry, being a large contributor to the use of natural resources and waste generation, results in significant environmental impacts. In fact, in recent years, sustainable manufacturing has stood out to be a comprehensive strategy for reducing environmental impacts whilst advancing the economic performance of manufacturing facilities. Materials used in manufacturing are themselves major sources of environmental impacts, due to the energy and resources required to extract and produce them. Hence, with better material management, minimising material consumption will help in achieving better sustainability performance of a manufacturing process [2].

This study aims to further sustainable manufacturing within the polymer processing industry, specifically the manufacturers of plastic packaging for the cosmetic sector. This analysis focuses on the sustainability of recycling plastic injection moulding waste, a process which consumes energy and requires financial and human resources. The objectives of this study are (i) to analyse the technical, financial, environmental and social sustainability of cascade recycling of plastic injection moulding waste, and (ii) to identify the optimum recycled plastic ratio and recycling approach that yields good quality injection moulded parts which can be produced in a sustainable manner.

2. Plastic Recycling

Today, several recycling approaches exist in relation to the plastic manufacturing industry and these are sustainably assessed and analysed in various studies. Whilst recycling can usually be done at different stages of the product's life cycle, to maximise resource efficiency, inner loop recycling is essential, given its inherent higher energy and value saving potential [3]. Outer loop recycling is also beneficial, as it has considerable advantages over incineration and landfilling [4][5]. In general, the highest environmental benefits are achieved through the substitution of virgin resin with recycled material, the production of which leaves a significantly lower impact on the environment [6].

2.1. Regrind

Regrind, as per definition, is a material that has undergone one or more processing cycles such as injection moulding, exposing it to at least one heating cycle. It is produced by grinding sprues, runners, flash and/or reject parts [7][8][9]. When compared to virgin material, regrind is irregularly shaped, resulting in a lower bulk density. Theoretically, this lowers the machine's specific throughput which in turn decreases the process efficiency. Therefore, an improved feeding process may be required [10]. Due to possible changes in mechanical and rheological properties, manufacturers and consumers still doubt regrind to be a reliable replacement of virgin resin, especially in high quality and surface finished parts. This has also created the perception that recycled resin has a lower economical value. Nonetheless, primary recycling, meaning reprocessing, is considered as the most feasible solution for recycling waste in the plastic industry [11].

2.2. Regrind and virgin resin mixing

Many manufacturers assume that a maximum of 25 percent regrind can be used in accordance with third party certification standards, despite the fact that these actually read that regrind can be used up to the 25 percent limit unless further testing is conducted. In fact, there is no limitation on how much regrind may be used in approved components if further testing proves that a material retains its properties and remain within specification limits. Such a misinterpretation can result in excessive and unmanageable regrind stockpile [12]. According to Bryce [9], as a general rule of thumb, regrind can safely be added to virgin material at a rate of between 1 percent and 20 percent with most materials. This statement though assumes that the regrind is of high quality and not degraded or contaminated. Therefore, if regrind has been circulating in a manufacturing facility for a prolonged period, its use would require more attention, as degradation and contamination issues may come into play.

Regrind usually holds a lower quality value compared to its virgin equivalent and can usually be acquired at half its price. Luckily, this relatively low-cost material, if not excessively used and abused, will retain up to 90 percent of its original properties [9], making it possible to use regrind at a 100 percent rate. This being said, it is important to have knowledge on the heating/reprocessing history of the acquired material and on how this was handled and stored due to possible contamination. Nonetheless, actual testing ultimately determines the acceptable level of regrind to be used. In a recent study concerning ABS, PLA and PP30GF in various virgin-to-recycled ratios, regrind content, even at a 100% rate, had no effect on part quality [13].

Tracking the level of regrind in parts may seem possible on paper but not that simple in practice, especially when regrind is used in different amounts and possibly for different components. While it is simple for the first processing cycle, it becomes more complicated with every subsequent cycle. This is because an amount of regrind from each previous cycle, is always present, although in a diminishing manner. Therefore, this calls for proper material handling and storing, which may not be the easiest thing to achieve in a manufacturing facility [7].

2.3. Cascading regrind approach

An alternative approach to mixing regrind with virgin resin in a continuous regrind loop is the cascading regrind system. This system is implemented by first using up all virgin resin and then move onto using subsequent generations of regrind in a 100 percent manner. Until one particular generation of 100 percent regrind is used up, the next generation regrind is stored. This process can continue until almost all the regrind in the system is entirely utilised. In a cascading regrind system no mixing occurs, hence eliminating related problems and issues. Chances of contamination are also reduced and in addition, such a strategy is considered as self-purging. Once regrind is used up, any contamination in the regrind stream is removed. In this way, only a small number of parts are affected [7][8][12]. Since this strategy makes the number of heat histories a known variable, polymer degradation is avoided by not allowing molecules to stay in the system for a long period of time, hence related problems are eliminated [12].

Such an approach raises no questions about the amount of regrind actually being used or if it is being properly mixed. Also, regrind tracking becomes much simpler as parts need only be labelled for the resin lot number and its number of passes through the machine, or better, its heat history. With this approach, further separation of regrind for future use is simplified, and even if not directly utilised, such regrind would hold more value, since it can be classified and tested. Hence, such reground material can be sold at a higher price as buyers can be provided with knowledge regarding how many heat cycles it had been through [12] From a financial point of view, this results in lower capital costs, since it eliminates the need to invest in mixing equipment [7][8]. Nonetheless, as pointed out by Bozzelli [7] and Solomon [14], one particular issue with the cascading regrind system may be colour matching.

In a laboratory scale cascading regrind study, Janicki et al. [12] illustrate the feasibility of this approach by testing two high impact polystyrenes (HIPS) from different suppliers, ignition resistant polystyrene (PS) and polycarbonate (PC) at each of the first three regrind generations. Results pertaining to the tensile and impact strength tests conducted in this study resulted positive. Furthermore, in view of the fact that average molecular weight is stated to be the most direct indicator of polymer degradation, Janicki et al. [12] noted no significant reduction in this property in any of the materials after three processing cycles, hence, negligible molecular degradation had occurred. Other conducted tests in this study showed the same general trend, hence, property retention was proved to be quite good up to several regrind generations. In another study, Vidakis et al. [15] experienced a 30% increase in mechanical properties when recycling ABS up to five processing cycles but then noticed a rapid degradation after the fifth cycle. When tackling this approach from a financial point of view, although testing of parts produced from 100 percent regrind may be costly, contaminated regrind inventories also come with a price. This is because they are either useless, hence taking up space, or will subsequently result in rejects [8].

3. Materials and methods

3.1. Case study

Being a world leader in the cosmetic packaging industry, Toly Products Ltd was chosen as the industrial partner for this case study. Within Toly's manufacturing facility in Malta, a wide variety of high quality packages for the beauty industry are continuously being manufactured in accordance with the high-quality requirements imposed by such an industry.

3.1.1. Product selection. Due to its frequent demand, colour and material, the Tivoli Large Compact (detailed in Table 1) was selected as a foundation for a sustainability analysis of injection moulding of recycled plastic at Toly Products Ltd. The base and lid components are manufactured entirely from ABS (Novodur® P2H-AT), while the platform component is produced from an ABS-SAN blend in a 1:3 ratio. For this study, all three components were manufactured separately as explained in section 3.2. Any regrind resulting from the manufacture of any part was solely used for the manufacturing of that same part.

Table 1. Tivoli Large Compact specifications.

Component	Material	Colour	Mass (g)	Outside Diameter (mm)
Base	ABS	Black	9.09	72.98
Lid	ABS	Black	9.86	72.98
Platform	1 part ABS 3 parts SAN	Black	2.06	69.00

3.1.2. New targets. Plastic recycling was already common practice at the facility, however it was limited to either customer specifications, material supplier recommendations or problems encountered due to material limitations or current recycling practices. Hence, using a new approach to recycling, new targets were set and this study aims to further plastic recycling at the facility to new levels. This was done by manufacturing parts at different regrind to virgin ratios containing higher levels of regrind than the company's present practices. The three material ratios used in this study were (i) 60% Regrind with 40% Virgin, (ii) 80% Regrind with 20% Virgin, (iii) 100% Regrind with no virgin material.

Although literature suggests that a cascading regrind approach, as explained in section 2.3 (with no mixing), is the way forward to recycling in the injection moulding industry, material properties are very dependent on the grade of the material, which includes different additives depending on its use, hence, generalisation is not suggested. This study, therefore, considered the other two recycled material ratios (60% and 80%), should the 100% regrind products fail to satisfy the necessary quality requirements.

Moreover, the reprocessing capability of the materials used was put to test for each of the three ratios. This was done by adapting the working principles of the cascading regrind approach to each. After manufacturing all the required products, product quality was subsequently analysed as explained in section 3.4. With regard to reprocessing, this study covered up to three reprocessing cycles.

The quality of the products produced using these new material ratios was compared to the quality of virgin equivalent products and also to product samples taken from a typical production run at the factory. Typically, platforms contained 20% regrind, while the base and the lid contained 40% regrind. Furthermore, these samples were manufactured from material of unknown heat history, as no information on reprocessing of material is typically recorded.

3.2. Production

In order to accumulate enough injection moulding waste of known heat history, it was required to first manufacture an adequate number of components from 100% virgin material using three injection moulding machines loaded with the three required moulds of the components making up the Tivoli Large Compact. It is important to note that for consistency reasons, only sprues and runners were ground and used as recycled material throughout this study. This process also yielded virgin products, which later acted as a benchmark during quality testing.

Once virgin production for every component was complete, and the resulting waste ground, the three mixtures embodying the three different ratios were produced for each component making up the Tivoli Large Compact. Furthermore, in order to test their reprocessing capability, these mixtures were successively processed for three times in total by grinding their subsequent production waste, mixing with virgin where necessary (excluding the 100% recycled material ratio) and loading them back into the injection moulding machine. This was done, separately, for all three components making up the compact. The whole process is portrayed in Table 2. R_0 refers to the virgin production, R_1 to a production containing first generation regrind, and similarly R_2 and R_3 refer to productions comprising of the second

IOP Conf. Series: Materials Science and Engineering 1196 (2021) 012021 doi:10.1088/1757-899X/1196/1/012021

and third generations of regrind respectively. Calculations pertaining to the planning of these productions allowed for the least waste possible at the final reprocessing cycle.

Table 2. Production process.

0% Regrind	60% Regrind	80% Regrind	100% Regrind
	R_1	R_1	R_1
R_0	R_2	R_2	R_2
	R_3	R_3	R_3

3.3. Energy metering system
In order to conduct an energy consumption analysis on the injection moulding process, energy consumption was metered for every process portrayed in Table 2 whilst manufacturing platform components to monitor whether this resulted in any trend or discrepancies in energy consumption. Furthermore, energy consumption was metered during the production of the lid and base components from virgin material, so that any variance noticed with the platform could be mirrored on these too. This study also looked into the energy consumed by the grinding process. Hence, energy was metered accordingly so as to allow for the holistic analysis presented in this study.

3.4. Quality testing
In order to verify whether or not the manufactured products were of good quality, validation tests were conducted on the assembled parts by the quality control lab at Toly Products Ltd. For such tests, around 50 samples from each of the ten production runs previously listed in Table 2 were needed, together with another 50 samples taken from a typical production run at the factory. Samples were taken from each mould impression/cavity for a more holistic representation. Given the fact that the only variation from previous validation tests was coming from the material, more precisely regrind content and reprocesses, and also in view of the lengthy process entailed, tests were limited to quality criteria affected by the material. Furthermore, tests were subdivided into three categories: package integrity, visual aspect and functional aspect.

4. Results and discussion

4.1. Quality results
4.1.1. Package integrity and visual aspect. With regard to package integrity and the visual aspect, all compacts passed the required tests. These included several different drop tests, open/close cyclic tests, stress cracking tests and other tests which measured the mismatch or alignment between the base and lid components and also the gap between the two components when the compact is closed. Additionally, a colour comparison test was carried out, in view of colour matching issues mentioned in section 2.3. This test showed that for the platform, colour was constant throughout, but with regard to the lid and base components, a slightly lighter finish was noticed at high regrind contents. However, for all recycled material ratios, lid and base components from R_3 were noticed to be slightly darker than those from R_1. Nonetheless, differences were very minimal and difficult to detect, hence these were of no particular issue.

4.1.2. Functional aspect. With regard to the functional aspect, three tests were conducted: clip opening, hinge breakage and welding resistance at the hinge area. Both the clip opening tests and the welding resistance tests resulted in positive results for all the different ratios and reprocessing cycles tested. Values related to these tests were all within specified control ranges.

Being the ones most likely to be affected by material changes, hinge breakage tests were required to test for initial bending of the hinges, which plastically deform with increasing force. Most values obtained fall within experimental error. Furthermore, compacts produced using virgin material seem to have bent prior to compacts containing regrind. Also, from the results obtained, components from the 80% R_1 production seem to have superior strength, while bending at the hinge was impeded for a longer time. The uncertainty related to these results varies between 1% and 2.7%, which is an acceptable range. Similar to the clip opening test, no particular trend was noticed, however, it is clear that regrind content increased the strength of ABS. All values obtained were in excess of 7.0 kg (force equivalent), hence all compacts clearly passed this test in view of the minimum specification of 2.7 kg (force equivalent).

To conclude on quality tests, in view of these results and validation reports generated for all samples, all compacts were validated by the quality control lab at Toly Products Ltd and hence, the Tivoli Large Compact is capable of being manufactured in the three new material ratios. The conducted tests more specifically prove that this particular compact can be manufactured with 100% regrind that has been reprocessed for three times, without any compromise on quality.

4.2. Injection moulding energy consumption
For each production run, injection moulding energy was metered for 10 consecutive cycles, with data readings starting after the 5th cycle from the start of each production run, in order to give chance for the injection moulding machine to reach a stable working state.

It is important to note that for each different production, the injection moulding machine was purged from any remaining material used for the prior batch and reloaded with the required material for the next. Furthermore, energy data was collected over a number of days and therefore some variances could have easily occurred due to possible noise such as changes in ambient temperature and fluctuations in supply voltage. Thus, injection moulding parameters may have varied slightly from one reading to the other. This being said, an effort was still made to keep injection moulding parameters as constant as possible.

In view of the results, most energy consumption values were similar and only vary within experimental error (130 Wh/cycle ± 3.8%). However, some readings were slightly out of this range, with the most energy consuming production being 80% R_2 (143 Wh/cycle) and the least being 100% R_3 (121 Wh/cycle). The first is attributed to a higher than usual power demand, whilst the second is attributed to a combination of low cycle time and low power demand. All in all, the outcome of this injection moulding energy analysis shows that with different regrind content and reprocessing cycles, variance in injection moulding energy is for the most part slight and erratic, and no trend is evident with increasing regrind content. Moreover, the maximum possible discrepancy noted between 80% R_2 and 100% R_3 is of approximately 19%. Although this cannot be deemed negligible, it is not substantial. In view of the outcomes of this analysis, variances in energy consumption for the injection moulding process were not accounted for in calculations and not included in the following sustainability analysis.

4.3. Sustainability analysis
In view of the promising results obtained and the conclusions deduced from available literature, this analysis focuses on a 100% regrind cascading approach and a breakdown of this is provided so as to analyse and compare the different recycling opportunities the manufacturer can opt for. Recycling opportunities include:

- Manufacturing with virgin material only (waste material is sold for recycling),
- Manufacturing with the recycling approach currently employed,
- Manufacturing with a 100% recycling rate and 1 reprocessing cycle,
- Manufacturing with a 100% recycling rate and 2 reprocessing cycles,
- Manufacturing with a 100% recycling rate and 3 reprocessing cycles.

Different scenarios are also possible for the grinding process, and moreover, different options exist with regard to the sale of post-production injection moulding waste, i.e. the remaining runners at the end of a production run:

G1. Automated and continuous grinding throughout all injection moulding cycles, where the grinder is left switched on during production.
G2. Batch grinding at the grinder's maximum throughput, either at the end of every reprocessing cycle, or alternatively, in smaller batches.
W1. Selling post-production injection moulding waste in runner form to a waste broker at the price of €0.10/kg for any material.
W2. Grinding post-production injection moulding waste and selling it in regrind form at a price of €1.38/kg for lid and base components and €1.28/kg for the platform component. These prices are 50% of the respective virgin material prices which is an assumption based on the prices of various material suppliers, [16].

The following analysis, summarised in Table 3, compares all possible combinations of the above to a virgin equivalent production utilising waste sale option W1 (values of which can be found in the first row), with this combination being the most basic and simplest recycling opportunity.

Table 3. Comparison of different recycling opportunities for a production run of 50,000 compacts.

Regrind %	Number of Reprocessing Cycles	W	G	Total Cost (€)	Cost Savings	Total CO_2 Emissions (kg)	CO_2 Savings	Grinding Labour (h)
0%	0	W1		4,133	-	5,201	-	0.0
		W2	G1	3,634	12.1%	5,377	-3.4%	0.0
			G2	3,581	13.4%	5,208	-0.1%	7.0
Current	Unknown	W1	G	3,365	18.6%	4,221	18.8%	2.8
100%	1	W1	G1	3,302	20.1%	4,140	20.4%	0.0
			G2	3,247	21.4%	4,015	22.8%	5.1
		W2	G1	3,154	23.7%	4,186	19.5%	0.0
			G2	3,102	24.9%	4,017	22.8%	7.0
	2	W1	G1	3,110	24.8%	3,913	24.8%	0.0
			G2	3,057	26.0%	3,760	27.7%	6.3
		W2	G1	3,055	26.1%	3,930	24.4%	0.0
			G2	3,003	27.3%	3,761	27.7%	7.0
	3	W1	G1	3,047	26.3%	3,839	26.2%	0.0
			G2	2,995	27.5%	3,677	29.3%	6.7
		W2	G1	3,022	26.9%	3,847	26.0%	0.0
			G2	2,970	28.1%	3,677	29.3%	7.0

In Table 3, CO_2 values include emissions from primary material production and emissions from energy required to power the grinding process. Also, grinding scenario G in the fourth row refers to a combination of G1 and G2 according to the current practice at the facility.

4.4. Economic analysis

Results show how grinding scenario G2 is always a cheaper option than G1, whatever the waste sale option. This holds true despite the introduction of labour cost with G2. Such an outcome is attributed to the shorter operating times of grinders, which in turn consume less energy although loaded at their maximum rating. Furthermore, the lower maintenance costs for the grinders in scenario G2 also have a substantial effect on total costs. Waste sale option W2 is always more cost-effective than W1 for similar grinding scenarios. This means that although more energy is consumed to grind the remaining post-production waste, the income related to the higher priced regrind outweighs the higher grinding costs.

Moreover, when comparing possibilities within the 100% regrind approach, manufacturing with more reprocessing cycles results in greater overall cost savings. This is due to less virgin material consumption and more efficient use of injection moulding waste, which in turn results in less post-production waste. Hence, instead of selling most of this waste at an inferior price, this same waste is replacing more virgin material within the same production. Additionally, savings are seen to increase with increasing regrind percentage, hence a 100% regrind approach is a more cost-effective way to deal with regrind. This analysis concludes that the most economical recycling opportunity is the one using a 100% regrind approach, three reprocessing cycles, waste sale option W2 and grinding scenario G2.

4.5. Environmental analysis

The environmental analysis is based on CO_2 emissions related to grinding energy consumption and primary material production. The CO_2 footprint data was obtained using CES EduPack™ [17]. This analysis shows how grinding scenario G2 results in approximately 2% to 3% reduction in total CO_2 emissions when compared to grinding scenario G1 within the same waste option. This reduction is attributed to the shorter operating times of grinders, which in turn result in lower electricity demands and hence, less CO_2 emissions.

Although, as deduced economically, waste sale option W2 is more cost-effective than W1, this produces more CO_2 emissions for the same grinding scenario due to the increase in energy consumption, as more material is being ground. However, this difference in CO_2 emissions at maximum, results in less than 1% reduction in total CO_2 emissions, and furthermore, this difference is seen to narrow down to a negligible percentage with three reprocessing cycles due to lower quantities of post-production waste. When manufacturing with more reprocessing cycles using a 100% regrind approach, apart from being more cost-effective as previously explained, CO_2 emissions decrease correspondingly. This reduction is attributed to the lower consumption of virgin material, which plays an important part in CO_2 emissions. Furthermore, CO_2 emissions were seen to decrease with increasing regrind percentage.

Recycling opportunities with a virgin production are only limited to external recycling of post-production waste, since, with such an approach, the facility does not make internal use of any of its production waste. Hence, no CO_2 emissions are reduced with regard to primary material production as none of the virgin material is replaced with regrind. To make matters worse, CO_2 emissions related to grinding are further added with waste option W2, hence resulting in more CO_2 emissions than the first and most basic recycling opportunity (0% W1). On top of the current recycling approach, a 100% regrind cascading approach is seen to be furthermore beneficial for the environment.

This environmental analysis concludes that the most environmentally benefitting recycling opportunity is the one using a 100% regrind approach, three reprocessing cycles and grinding scenario two (G2). With regard to the waste sale scenario, technically, W1 is less harmful to the environment than W2, but since the difference in CO_2 emissions is minimal, this is decided upon after an overall analysis.

4.6. Social analysis

Factors discussed in the environmental analysis, also affect society as part of the environment. Less CO_2 emissions, reduce the Global Warming Potential of the operations and indirectly result in a better quality of life. Furthermore, more labour is introduced with grinding scenario G2. With waste option W1, this added labour increases with increasing reprocessing cycles for the 100% regrind approach, whilst with waste option W2 for the same 100% regrind approach, this is constant throughout. When labour hours are compared, any recycling opportunity that makes use of grinding scenario G2 generates more work, specifically green jobs. For 50,000 products, a maximum of 7 hours of labour related to grinding of injection moulding waste is added with waste option W2 and grinding scenario G2. Hence, given the economic and environmental benefits of the 100% regrind approach with three reprocessing cycles, it would be wise to opt for waste sale option W2 and grinding scenario G2.

5. Conclusions

In view of the above analyses, it is clear that in-house post-production recycling affects positively all the three pillars of sustainability. Moreover, a 100% regrind approach with the greatest number of reprocessing cycles is the most benefitting for all the three pillars of sustainability. This study has tested up to three reprocessing cycles, however, more reprocessing cycles would mean less virgin material consumption and less post-production waste, hence more efficient use of regrind. In view of the economic, environmental and social analyses, the sustainable solution proposed concerns a 100% regrind approach with three reprocessing cycles (R_3), where grinding is carried out in batches (G2) and any post-production waste is sold in regrind form (W2). When compared to a no-recycling scenario, this will ultimately result in 28.1% less costs, 29.3% less CO_2 footprint, and labour is increased up to a maximum of 7 hours per 50,000 products. When compared to the typical in-house recycling and mixing scenario currently employed by the case study factory, the proposed cascade recycling approach would reduce the costs by 11.7% and the CO_2 emissions by 12.9%. This approach would increase the labour required by 4.2 hours per 50,000 products.

Cost savings are mainly attributed to the low cost of grinding when compared to the virgin material cost. In fact, with regard to ground material, electricity costs amount to less than €0.01/kg, labour costs amount to approximately €0.10/kg and maintenance costs also amount to less than €0.01/kg. When their total of €0.12/kg is compared to the virgin material prices of circa €2.75/kg for ABS and €2.50/kg for SAN, cost cutting opportunities are very evident.

With a simple shift in how regrind is being handled and used, the proposed system would bring remarkable benefits with respect to all three pillars of sustainability. The proposed recycling system produces the least waste possible, hence making the best use of the available material. Therefore, apart from being the most cost-effective system, this reduces the amount of non-biodegradable plastic waste that may in time end up in landfills, affecting positively both the environment and society. Furthermore, the cascading regrind approach on which the proposed system is based makes the material heat history a known factor. This in turn, allows the manufacturer to sell regrind at a higher price than what is normally attributed to post-production waste with an unknown heat history.

This study paves another way for manufacturing companies to further their commitment towards sustainable manufacturing. This is achieved through a reduction in the use of natural resources pertaining to both primary material and energy production, and also through better management and efficient use of inevitable injection moulding waste. As a result of the above, more green-job opportunities are created. Additionally, all this is accomplished with remarkable profits, hence deeming this development economically sound.

Acknowledgements

The authors would like to thank Mr Olaf Zahra, Chief Officer Technology and Sustainability at Toly Group, for his cooperation throughout the entire study.

References

[1] Stark R, Selinger G and Bonvoisin J 2017 *Sustainable manufacturing: Challenges, solutions and implementation perspectives*, Springer Nature

[2] Yuan C, Zhai Q and Dornfeld D 2012 A three-dimensional system approach for environmentally sustainable manufacturing *CIRP Annals* **61(1)** 39-42

[3] De Jong A and Mellquist, A 2021 The Potential of Plastic Reuse for Manufacturing: A Case Study into Circular Business Models for an On-Line Marketplace *Sustainability* **13(4)** 2007

[4] Chen Y, Cui Z, Cui X, Liu W, Wang X, Li X and Li S 2019 Life cycle assessment of end-of-life treatments of waste plastics in China. *Resources, Conservation and Recycling* **146** 348-357

[5] Hou P, Xu Y, Taiebat M, Lastoskie C, Miller S and Xu M 2018 Life cycle assessment of end-of-life treatments for plastic film waste *Journal of Cleaner Production* **201** 1052-1060

[6] Gu F, Guo J, Zhang W, Summers P and Hall P 2017 From waste plastics to industrial raw materials: A life cycle assessment of mechanical plastic recycling practice based on a real-world case study *Science of the Total Environment* **601-602** 1192-1207

[7] Bozzelli J. Regrinding Plastics | Prospector Knowledge Center [Internet]. Prospector Knowledge Center. 2021 [cited 5 March 2021]. Available from: https://knowledge.ulprospector.com/1468/pe- regrinding-plastics

[8] Bozzelli J 2015 Another way to deal with regrind *Plastics Technology* **61(1)** 38-41

[9] Bryce D M 1996 *Plastic injection molding: manufacturing process fundamentals*. Society of Manufacturing Engineers.

[10] Thieleke P and Bonten C 2021 Enhanced Processing of Regrind as Recycling Material in Single-Screw Extruders *Polymers* **13(10)** 1540

[11] Fei N C and a Kamaruddin N M M 2013 A Degradation Study of Virgin and Recycled ABS Blends Subjected to Multiple Processing *International Journal of Research in Mechanical Engineering* **1(1)** 93-102

[12] Janicki S L, Bain Jr M F and Groleau R J 1992 The material Cascade: An Alternative Form of Regrind Utilization *ANTEC 92--Shaping the Future* **1** 1201-3

[13] Vassallo C, Rochman A and Refalo P 2020 The impact of polymer selection and recycling on the sustainability of injection moulded parts *Procedia CIRP* **90** 504-9

[14] Solomon N 2021 Influence of repeated recycling on the quality of molded finished products *International Journal of Environmental Science and Technology*

[15] Vidakis N, Petousis M, Maniadi A, Koudoumas E, Vairis A and Kechagias J 2020 Sustainable Additive Manufacturing: Mechanical Response of Acrylonitrile-Butadiene-Styrene over Multiple Recycling Processes *Sustainability* **12(9)** 3568

[16] Plastics News [Internet]. Plasticsnews.com. 2021 [cited 4 August 2021]. Available from: http://www.plasticsnews.com/resin

[17] 2018 *Edupack*, C. E. S. Granta Design Limited

Life cycle assessment of recyclable, reusable and dematerialised plastic cosmetic packages

Isaac Jordan Gatt[1] **and Paul Refalo**[1,*]

[1] Department of Industrial and Manufacturing Engineering, Faculty of Engineering, University of Malta, Msida, MSD 2080, Malta

* E-mail: paul.refalo@um.edu.mt

Abstract. Design for Recycling and Design for Reuse are two different approaches which can be employed separately or concurrently. When designed for reuse, products are typically more robust in order to increase their probability to be used more than once. If reuse is not possible, it is essential that dematerialisation and recycling are applied. This study assessed the environmental impacts resulting from reusable, recyclable and dematerialised plastic cosmetic packages. Life cycle assessments of different versions were conducted, to identify what features are responsible for such impacts. Findings showed that removing components which are made from resourceful materials, and which render the package to be reusable, resulted in a 74% reduction in environmental impacts only when the packaging materials are fully recycled. Hence this study concludes that in such cases, reuse should be given prominence, as recycling would only depend on the user and the infrastructure in place.

Keywords: Life Cycle Analysis, Recyclability, Reusability, Cosmetic Packaging, Sustainability.

1. Introduction

The term sustainability has never been given as much importance and acknowledgment as it has been given in the last few years. It has become a trend to associate many aspects of our everyday life to sustainability, and awareness has increased significantly [1]. Additionally, when considering the drastic increase in human population over the last 50 years, from around 3 billion to today's 7.8 billion [2], an increase in demand for goods and services is also expected. This in turn leads to larger economies and increased prosperity around the world. However, this factor has also negatively affected the environment due to the increased rate of emissions, and increased depletion of resources [3]. The environmental issues worsen when we consider that packaging amounts for a large quantity of the generated waste, the most common of which is plastic. Packaging is necessary for protecting and distributing products worldwide, however, particularly secondary and tertiary packaging only serve their purpose until the product reaches its use phase, at which point such packaging turns into waste and is considered as an environmental issue [4].

1.1. The Cosmetics Industry

One of the industries which heavily uses packaging is the cosmetic industry. Globally, the cosmetic industry was valued at approximately USD 532 billion in 2017, with predicted forecasts expected to reach USD 863 billion by 2024 [5]. From this global value, approximately USD 25 billion are reflected in product packaging [6]. These values make the cosmetic industry extremely important and relevant. However, the drastic rise in this industry has also skyrocketed plastic use, and consequently has led to negative environmental impacts. With global plastic usage set to increase from 1.5 million tonnes/year in 1950 to 322 million tonnes/year in 2015, it is very evident to see just how dire the situation is [7]. Both primary packaging (the container holding the cosmetic product) and secondary packaging, are commonly found in significantly high quantities and sometimes even excessive quantities in such a luxury industry. Packaging helps create a more premium experience, such as by including thick ribbons, having individually packed items, or including lots of paper separation packaging. In 2018, more than 120 billion units of cosmetic packaging were recorded to be produced globally [8]. This is worsened when realising that studies show a significant lack of recycling. A recent EU investigation showed that in 2014 less than a third of the generated plastic waste was recycled in Europe, with the other two thirds being either deposited in landfills or being solely used to recover energy from them [9].

There are a number of existing case studies showcasing cosmetic companies which managed to improve their sustainability efforts by applying design changes and by carrying out sustainability assessments. One such company is LUSH cosmetics, which improved its environmental footprint by implementing design changes such as reducing product thickness and replacing the standard wrapping paper with reusable fabric wrapping, which comes from recycled plastic [10]. LUSH also claim that 35% of their physical outlet product range do not need any packaging whatsoever, whilst products purchased from their online outlet are shipped in biodegradable bags, cushioned by compostable renatured peanuts, and packaged in 100% recyclable cardboard boxes. When secondary packaging of the product itself is necessary, LUSH have switched to using reusable product packaging by using aesthetic and durable plastic boxes. These can be reused by either using them to resend another gift in them, or to organise jewellery or small items. Another study carried out by LCA Centre showed that if all cosmetic manufacturers worldwide adopted a reuse-refill system for their cosmetic products, the cosmetic industry overall would eliminate 70% of its total CO_2 emissions [11]. This is a very significant statistic and bodes well for companies who would like to increase their efforts towards sustainable product development.

2. Materials and methods

The main objective of this study is to quantify and compare the different environmental life cycle impacts resulting from recycling and reusing plastic cosmetic packages, and to identify the best solution for the manufacturing of such products. In order to determine these objectives, an extensive LCA study was carried out, based on a case study product produced by Toly Group.

2.1. LCA Methodology

2.1.1. Goal and Scope Definition. The main goal for carrying out this LCA, was to study the environmental impacts of different variations of plastic cosmetic packages and propose environmentally conscious possibilities. The scope of this study is to show how the defined goals can be met with the limitations that are in place. In summary the proposed main takeaways for this analysis were:

- To compare the total life cycle environmental impacts of reusable and recyclable cosmetic packages,
- To consider how the environmental footprint varies with regards to all phases of the plastic cosmetic product's life cycle, i.e. raw material extraction, manufacturing, transportation, use, and end of life,

IOP Conf. Series: Materials Science and Engineering 1196 (2021) 012022 doi:10.1088/1757-899X/1196/1/012022

- To create multiple scenarios by varying different parameters, in order to evaluate how different variables compare with each other, and decide what the most environmentally conscious option would be.

A case study cosmetic package (blush compact) which is shown in Figure 1, was selected and used for the analysis. This product is a small blush powder makeup compact, intended for portable everyday use. The design is relatively simple with the product possessing a circular shape and a pinned hinge design. The main parts that constitute the case are the lid, mirror, base, base plate, and pin. The circular mirror is glued to the inside of the case's lid, and the pin constrains the rest of the components together. The main materials include acrylonitrile butadiene styrene (ABS) for the main housing of the case (i.e. lid, base, and base plate), glass for the mirror, and stainless-steel for the pin that acts as the pivoting hinge. The cosmetic powder itself is contained in a circular aluminium container defined as the pan. This factor makes the product reusable, and users can easily purchase additional pans from retailers once their existing pan has been used up and swap out the old with the new refill whilst keeping the same compact case. An important clarification that shall be defined for the reader's ease of understanding, and that will be used throughout, is the difference between case and compact. The term case refers to the full assembly of the cosmetic case but without housing the pan (i.e. one empty case). On the other hand, the term compact refers to the final end product which is sold to consumers (i.e. one case + one pan). The cosmetic powder itself was excluded from the analysis as this study is only focussing on the packaging components.

Figure 1. *Tivoli cosmetic blush compact.*

2.1.2. Functional Unit. It is critical to establish a clear functional unit when carrying out an LCA study, as this is a quantitative measure of the function of the system being studied and allows for fair 'like with like' comparisons [12]. In this case, the cosmetic blush powder is estimated to last around 18 - 24 months from when it is first opened. Through surveying 20 cosmetic blush users, it was found that a 3.5g cosmetic pan approximately lasts for around 100 uses [13]. If the user uses the blush once a day (i.e. 1 day = 1 use), the user will require approximately four cosmetic pans in one year (365 days). Moreover, the probability that the user will not keep buying pan refills or may wish to change brand increases after 1 year. Therefore, the functional unit for this study was taken to be the usage of cosmetic powder in the period of one year, thus defining the functional unit as the usage of 14g (3.5g x 4 uses) of cosmetic blush powder. This will allow the output quantity of cosmetic powder to always remain constant, whilst being able to compare different versions of compact cases in a fair method.

2.1.3. Inventory Analysis. In the inventory analysis stage, all the required data necessary to model the LCA was researched and/or calculated. The full bill of materials of the existing blush compact is tabulated in Table 1, along with the relevant manufacturing processes utilised for each part. It should be clarified that the cosmetic blush powder was not assessed from an LCA point of view, as this was not

IOP Conf. Series: Materials Science and Engineering 1196 (2021) 012022 doi:10.1088/1757-899X/1196/1/012022

within the scope of this study. It was only inputted in the Bill of Materials to account for its weight during transportation and maintain a higher level of data accuracy.

Table 1. Compact Bill of Materials.

Part	Raw Material	Manufacturing Process	Mass (g)
Lid	ABS HI121H (25% regrind)	Injection Moulding	5.52
	Adhesive Resin	Stock Part	0.05
	Aluminium Foil	Hot Stamping	0.05
Mirror	Silver Coated Glass	CNC Diamond Cutting	5.94
	Adhesive Resin	Stock Part	0.5
Base	ABS P2HAT/792 (25% regrind)	Injection Moulding	2.47
Base Plate	ABS P2HAT/792 (25% regrind)	Injection Moulding	4.17
Pin	AISI304 Stainless Steel	Deep Drawing	0.1
Pan (Full)	Cosmetic Blush Powder	Stock Part	3.5
	Aluminium Alloy (Pan)	Stock Part	2.3
Total			**24.6**

2.1.4. Impact Assessment. The LCA software used in this study was SimaPro 8.4 [14]. The Life Cycle Inventory database used was Ecoinvent 3, and the chosen method was the ReCiPe 2016 endpoint method (Hierarchist version), which is the default ReCiPe endpoint method [15]. The main impacts which were assessed in this LCA study will resolve around two of the primary ReCiPe endpoints, which are the Human health and Ecosystems endpoints. To better understand what the chosen endpoints represent, it is important to understand how these endpoints are formulated [16]. The Human health endpoint covers the impacts associated with several diseases and malnutrition, while the Ecosystems endpoint covers damage associated to freshwater species, terrestrial species, and marine ecotoxicity amongst others.

2.1.5. Interpretation of Results. The generated results are shown and reviewed in detail in Section 3, along with a detailed discussion of the extracted takeaways from this LCA study.

2.2. Parameters and Scenarios Review

Three primary parameters were analysed in this study. The first parameter is the Number of Case Reuses which relates to the reusability aspect, as by continuing to reuse the existing case and utilising pan refills to reuse the cosmetic compact, the use phase is extended. In the default scenario (Reuse 3 in Table 2), it is assumed that the case detailed in Table 1 is used 4 times (first use and three reuses). In the Reuse 1 scenario, the case was dematerialised by 7% which led to a less robust product, and in turn allowed for less reusability. This concept was also applied to the reuse 0 scenario (single use product), the mass of which was reduced by 17% when compared to the default scenario. It should be clarified that the dematerialisation was applied to all plastic parts, i.e. lid, base, and base plate, and that the percentages were generated by assuming the lowest part thickness possible all around to be that of 1.4mm and 1.25mm respectively (as opposed to the benchmark thickness of 1.5mm). The other two parameters deal with the case and pan disposal scenarios respectively. To summarise, the proposed parameters along with the defined scenarios are listed in Table 2. For all the scenarios, the manufacturing country and

End-of-Life (EOL) country parameters were kept constant as Malta, and the European Union respectively. The terminology (a), (b) and (c) refer to the three different recycling rates of 0%, 50% and 100% recycling respectively. It is assumed that the pan which would be contaminated with cosmetic powder and the glass mirror are landfilled.

Table 2. List of Methodology Parameters and Scenarios.
(LF: Landfilling, REC: Recycling)

Scenario	Values
No. of Case Reuses	0 *(17% Dematerialisation)* 1 *(7% Dematerialisation)* 3 *(0% Dematerialisation)*
Case Disposal Scenarios	(a) 100% LF, 0% REC (b) 50% LF, 50% REC (c) 0% LF, 100% REC
Pan Disposal Scenarios	100% LF, 0% REC

2.3. Methodology Versions Review

The scenarios defined in Table 2, were grouped according to the different versions that they are related to. Version V1(a), is the existing cosmetic blush product with an assumed reuse capability of 3 times, i.e. 4 uses in total. This version shall be considered as the main benchmark against which other versions shall be compared. Version V1A, builds upon the existing product design, but applies 7% dematerialisation to all plastic ABS parts, i.e. the lid, base and baseplate. This reduces its maximum reuse capability, allowing only for the potential of one reuse. Hence, when considering the functional unit, 2 cases and 4 pans were required. Version V2 also builds upon the benchmark cosmetic blush product design, with the main difference being the 17% dematerialisation to all plastic ABS parts. Consequently, this completely eliminated all potential reuse capability, making it a single-use product that is neither reusable nor recyclable, since it houses a pin hinge mechanism making it difficult to disassemble, as well as utilising adhesive which is considered as a contaminant. Version V3, is the single use, fully recyclable scenario. This scenario differs significantly from the rest since this is the only scenario eliminating the use of the aluminium pan and allows blush powder to be added directly into the case. Moreover, this scenario can be recycled due to the assumed design changes taking place, such as by changing the steel pin to a plastic ABS pin, in order to ease sorting difficulties when disassembling. Finally, Version V3E can be considered as the eco-friendly version prioritising sustainability over build quality by trading the standard pin hinge mechanism to a clip-on hinge, which is deemed as poor quality from a consumer point of view. This version has the maximum dematerialisation rate of 17% applied with no reuse capability, as well as being recyclable. A summary of the proposed versions can be observed in Table 3.

Table 3. List of Methodology Versions.

	V1 (a)	V1 (b) (c)	V1A (a)	V1A (b) (c)	V2	V3 (a)	V3 (b) (c)	V3E (a)	V3E (b) (c)
Features	Reusable	Reusable + Recyclable	Reusable + Demat.	Reusable + Recyclable	Single Use + Demat.	Single Use	Single Use + Recyclable	Single Use	Single Use + Recyclable
Case Dematerialisation	/	/	-7%	-7%	-17%	/	/	-17%	-17%
Reuse Capability	3	3	1	1	0	0	0	0	0
Hinge	Steel Pin	Plastic Pin	Steel Pin	Plastic Pin	Steel Pin	Plastic Pin	Plastic Pin	Clip-on Hinge	Clip-on Hinge
% Plastic Recycle	0%	50%, 100%	0%	50%, 100%	0%	0%	50%, 100%	0%	50%, 100%
Pan Design	Reusable	Reusable	Reusable	Reusable	Glued Pan	No Pan	No Pan	Non-Reusable	Non-Reusable

3. Results and Discussion

This section will review the results generated from the LCA study, following the methodology above. It should be noted that the presented results all represent the environmental effects of the functional unit defined previously.

3.1. LCA Results

The resulting comparisons of all seven versions along with their respective recycling subcategories are shown in Figure 2. The results were normalised with respect to the existing version, V1(a), for both endpoints, as this was treated as the benchmark. It should also be noted that similar to other LCA studies carried out, there is a level of inaccuracy in the generated results. A 10% margin of error is overlayed on the results in Figure 2.

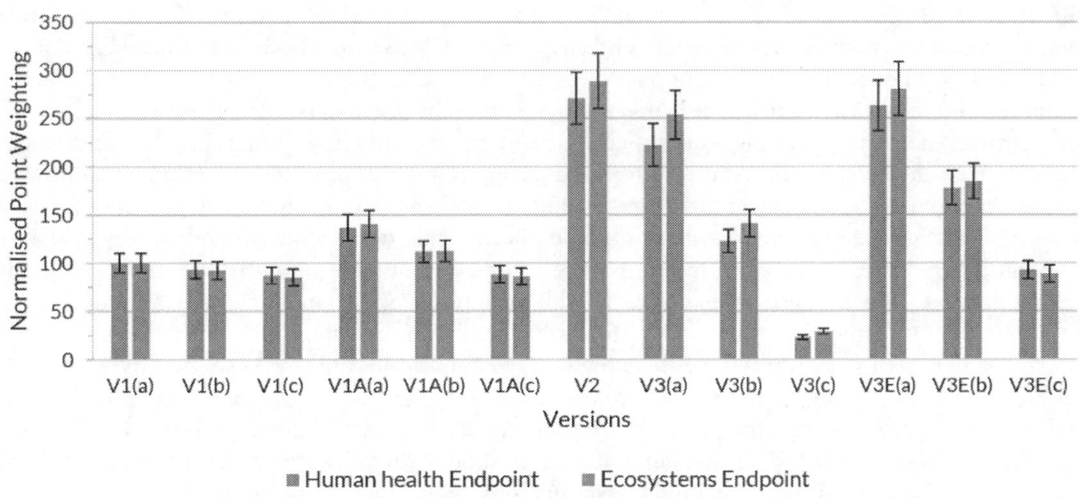

Figure 2. *Normalised endpoint life cycle impacts of the different versions (V1(a) = 100).*

One primary result that can be observed, is that the positive effect of reusability out ways by far the effects of dematerialisation by 171% when considering the human health endpoint (versions V1(a) vs V2). Another observation is that applying recyclability to a product that is already reusable, does not reduce the environmental impact by a significant amount, as indicated by only a 13% decrease when considering the human health endpoint (versions V1(a) vs V1(c)). In contrast, applying recycling to a single use product, resulted in very large environmental savings of 92% when considering the human health endpoint, and 90% when considering the Ecosystems endpoint (versions V2 vs V3(c)). It can therefore be concluded that the existing cosmetic blush compact performed extremely well when compared to the studied versions. The only argument that should be made when comparing the obtained results of V1(a) and V1(c), is that the recycling aspect throughout the entire study was considered only for plastic parts. This indicates that if the glass mirror and the aluminium pan could be recycled, the results would be far more significant than those registered in this study, especially when considering the large environmental discrepancy between ABS and aluminium, with aluminium having a 48% higher embodied energy and a 73% higher CO_2 footprint than ABS plastic [17]. The only version which performed significantly better than the current scenario (V1) is the recyclable single-use version V3(c). In this scenario, the aluminium pan which enables the case to be reused was eliminated. The environmental benefits resulting from the removed pan and the recycling counteract for the lack of reusability of the product. However, had the modified compact not been recycled entirely at its End-of-Life, the positive effects of the removed pan would be counteractive. In fact, V3(a-b) perform much worse than the benchmark scenarios V1(a-c).

4. Conclusions

There are a number of key takeaways which can be extracted from the LCA carried out in this study. The main takeaway is that the reusable version currently being manufactured, is very well suited when compared to the versions proposed in this research, even if the plastic waste is not recycled. Moreover, any potential improvements on the existing version, such as making the product recyclable in conjunction with it already being reusable, were found to yield very little benefits. However, it was found that if the existing product is not reused, the impacts increase by 171% and 189% for the two respective endpoints (versions V1(a) vs V2).

Only plastic recycling was considered in the study. If the other components (more specifically the aluminium pan) were to be recycled as well, the generated results are expected to be significantly different. Cosmetic packaging manufacturers should research this possibility, as well as study how realistic it is to setup an efficient circular economy recycling infrastructure. Moreover, there is a potential to increase environmental benefits by an average of 74% across both endpoints when compared to the existing product, should the compact design be altered in such a way that the use of the pan is eliminated altogether, i.e. V3(c). This is due to the fact that that the pan is made from aluminium, which has significant negative environmental impacts. The main difficulty here is that this change impacts the filler company the most, since with this proposed design it will not pack the cosmetic blush powder in the pan like previously, and instead will have to find a way how to insert the blush powder directly into the case or change the material to a more durable plastic. This raises strength and quality issues since the manufactured case would need to be extremely rigid and durable to withstand the cosmetic blush packing process. This scenario does not really fall within the plastic manufacturer's remit. However, this should be researched further and potentially trialled in collaboration with the powder filler companies, in order to identify if it would actually be viable and feasible to do so. Finally, it can also be concluded that attempting to recycle the plastic case, but not recycling the aluminium pan due to contamination, proves to be almost ineffective due to the fact that aluminium has a high embodied energy and carbon footprint. Therefore, manufacturers should attempt to use materials with less harmful impacts and move towards reusable products comprising recyclable mono-materials.

Acknowledgements
The authors would like to thank Mr Olaf Zahra, Chief Officer Technology and Sustainability at Toly Group, for his cooperation throughout the entire study.

References

[1] Lozano R 2008 Envisioning sustainability three-dimensionally *Journal of Cleaner Production* **16** 1838–46
[2] Worldometer 2020 World Population Projections *https://www.worldometers.info/world-population/world-population-projections/*
[3] EUROPEN 2011 *Green Paper Packaging and Sustainability: An open dialogue between stakeholders*
[4] Pongrácz E 2007 The environmental impacts of packaging *Environmentally Conscious Materials and Chemicals Processing* 237–78
[5] Zion Market Research 2018 Global Cosmetic Products Market Will Reach USD 863 Billion by 2024
[6] Alejandra Borunda 2019 The beauty industry relies on plastic. Can it change? *National Geographic*
[7] Andrea Cheng 2019 How Sustainable Are Refillable Beauty Products? *Elle*
[8] Kaleigh Moore 2019 New Ways The Beauty Industry Is Testing Sustainable Practices *Forbes*
[9] European Commission 2015 Plastic pollution: We can make things better *PlasticsEurope*
[10] Lush Cosmetics 2019 Reduce, Reuse, Recycle | Lush Fresh Handmade Cosmetics *Lush Cosmetics*
[11] Beth Owens 2019 3 Eco-friendly Initiatives For Sustainable Cosmetics Packaging *noissue.co*
[12] Pålsson A-C and Riise E 2011 Defining the goal and scope of the LCA study 1–5
[13] Gatt I-J 2020 Sustainability Assessment of Recyclable and Reusable Plastic Cosmetic Packages *B.Eng. (Hons) Dissertation (University of Malta)* 44–53
[14] SimaPro 2020 About SimaPro *simapro.com/about*
[15] SimaPro 2020 ecoinvent v3 | LCI Database *simapro.com/databases/ecoinvent*
[16] Ellen Meijer 2014 Consider Your Audience When Doing Impact Assessment *PRé Sustainability*
[17] Ansys (CES) 2021 GRANTA EduPack *ansys.com*

How Can China Achieve A Sustainable Development of Shale Gas? Implications of River Chief System to water governance for shale gas in China

Qianyu Dong[1,*] and Ling Zhou[1]

[1] School of Economics and Management, Southwest Petroleum University, Chengdu 610500, China.

*Email: 393299406@qq.com

Abstract. Along with China's ambitious strategy for shale gas development, water contamination risks with shale gas extraction has aroused great attention. Meanwhile, an innovative water management mechanism called as 'River Chief System' has been popularizing around China to improve the river water quality efficiently. Could this mechanism be available in shale gas water risk management? And how could it be adapted to improve the sustainable development of shale gas? Therefore, this paper starts by a discussion on the logic of River Chief System' success and worries; then, a comprehensive examination of current multi-channel governance system of shale gas in China is presented to explore the dilemmas in its water governance; and a causal loop model is built to clarify interrelations among these dilemmas. The result identifies that River Chief System's superiorities which can be learned are mainly reflected in clear responsibility, authority and multi-sectoral collaboration system. However, its temporary "campaign-style" character, over-reliant on the commitment and capability of river chief, blurred evaluation mechanism has hindered its expanding in the long run. Accordingly, 5 policy implications based on River Chief System for the sustainable improvement of shale gas is presented.

Keywords: water contamination risk, water management, shale gas, river chief system, governance.

1.Introduction

Recently, China pledged to have CO2 emissions peak before 2030 and lower the emissions per unit of GDP over 65% from the 2005 level in September and December 2020, separately [15]. To develop clean energy is a promising approach. China is estimated to have the largest technically recoverable shale gas resources in the world (1115 tcf), in addition to the third-largest shale oil reserves (32 billion barrels) [5]. As early as in China's 12th Five Year Plan, shale gas has been identified as a policy priority and set out to be 'promoted and accelerated' [7]. In the 13th Five Year Plan (2016-2020), shale gas production in China is expected to reach 30 bcm by 2020 – this represents 15% of total natural gas consumption in the country [7]. In June of 2020, 'Guidance on improving energy security in 2020' by National Development and Reform Commission (NDRC) and National Energy Administration (NEA) , continued to emphasize to accelerate the development of shale gas and keep growing in production. China's ambitious goal indicates rapid development for shale gas.

However, the environmental risks from fracking activities of shale gas extraction (particularly, flowback and produced water) have gained great environmental concerns. Such as the flowback water mainly contains extremely complex dissolved chemicals, hydrocarbons, heavy metals, naturally occurring radioactive material (NORM), bacteria and water-soluble salts, such as potassium, calcium, sodium, carbonate, and chloride [13]. Good evidence suggests these large volumes of contaminated water are difficult for the industry to manage on site [9], and able to cause adverse health effects to humans [1][2]. Water risk associated with shale gas extraction needs to be extremely careful controlled. However, these risks have not been appropriately managed yet [16].

Meanwhile, a new management system in China that called 'River Chief System' (RCS) has been proved which could improve water management efficiently in a short-term. The so-called RCS means that leading local officials are appointed as "He Zhang" (river chiefs) for a specific river section, and the RCS makes them responsible for water protection and management within their own jurisdictions [11]. Given the effectiveness and success of RCS, this policy had been fully implemented in 30 provinces, and more than one million officials had been assumed as the river chiefs at the provincial, municipal, county, township, and village levels by the end of 2018 [6]. And this RCS is being expanded to forest management already [8].

Therefore, since China's shale gas development produce more fragmented environmental problems, especially water contamination related problems, which beyond the capacity of shale gas companies even water supervision departments; meanwhile, RCS are proved efficient in water governance, therefore, could this RCS mechanism be applicable in shale gas water contamination management? and how could it promote the sustainable development of shale gas? This study could provide more insights into the mechanism of RCS and its feasibility in shale gas water governance. The remainder of this paper proceeds as follows. Section 2 describes the research methods and Section 3 discusses the logic of RCS's success and worries, as well as applicability of the RCS to water governance for shale gas; after a comprehensive examination of current multi-channel governance system and the dilemmas of shale gas in Section 4; a causal loop model is built to clarify interrelations among these dilemmas; and finally, 5 policy implications based on RCS for the sustainable improvement of shale gas are provided.

2. Materials and methods

This study adopted a literature review and field research as the main methods. The literature included domestic and foreign studies both on RCS and shale gas environmental and water governance in China and outside China.

To obtain a further understanding of the current status of water governance in China, in addition to the literature review, we conducted field research to several shale gas fields in both Sichuan and Chongqing (they are the largest shale gas fields in China). What's more, we also held several discussions with relevant scholars, corporate representatives of oil companies and government officials. What we discussed involved the development of shale gas, current situation and dilemmas of shale gas environmental and water governance, and third-party assessment indicators and mechanisms et al.

3. The success and worries of RCS

RCS is an innovative policy instrument in China which purposes to overcome fragmentation in water management and improve water quality effectively in a short time. The development of RCS can be divided into three phases: Emergency phase, Local extension phase, and Large-scale expansion phase:

Phase 1： Emergency phase. RCS was first innovated to address a blue algae outbreak in Taihu Lake as a policy instrument by the local government of Wuxi Municipality of Jiangsu Province in August of 2007.

Phase 2: Local extension phase. Since the water environment of Taihu Lake has rapidly improved, RCS was introduced to the entire Jiangsu Province, followed by Yunnan and Henan, and other provinces in 2012.

Phase 3: Large-scale expansion phase. RCS has been approved by the China's central government and began to be applied throughout the whole China in 2016 to reinforce river and lake protection.

Obviously, RCS has achieved a transformation from local trials to a national mechanism. It can be considered following a parallel, bottom–up and eventually top–down routine. And it also means that RCS can be available in various positions because of its performance generation mechanism.

3.1 The logic of RCS's success

As shown in Figure 1, the essence of RCS is the river chief, which follows Chinese traditional governance logic 'the most powerful man should in charge of the most important thing'. With a clear accountability system, effective multi-sectoral collaboration system, and an accessible and differentiated goals, the river chief has been involved, centralized and maximum use in RCS. And the whole RCS has become a positive feedback mechanism, and delicacy management has been achieved.

3.1.1 Clear accountability system. RCS sets the government leaders as river chiefs at several levels, making them the first responsible persons for the water environment within their respective geographical jurisdiction. River chiefs are responsible for the drinking water safety, water source dispatching, water saving management, and sewage treatment, which were once subordinate to the respective authorities of environment, agriculture, resources, and transportation [6]. Additionally, 'one vote veto power' has been generally introduced to RCS, which means if the river's water quality failed to the goal, the responsible river chief will also be failed to their annual government performance evaluation. Accordingly, an accountability system is feasible. To some extent, the administrative efficiency of river chiefs and officials has been improved.

3.1.2 Effective multi-sectoral collaboration system. As shown in Figure 2, a multi-sectoral collaboration system has been built for RCS. Vertically, a hierarchical system of four levels' river chief office is constituted (including provincial, municipal, country-level and township). Horizontally, RCS office coordinates the operation of various department agencies in charge of environmental protection, water affairs, land use, agriculture, industries etc. What's more, in order to promote the inter-agency collaboration, even the officials of RCS office are from various department agencies.

Consequently, with this multi-sectoral collaboration system, local authorities have been associated directly to environment affairs. Additionally, the senior government officials (also as the river chiefs) would have a comprehensive understanding to the complexity and systematicness of environmental issues as well.

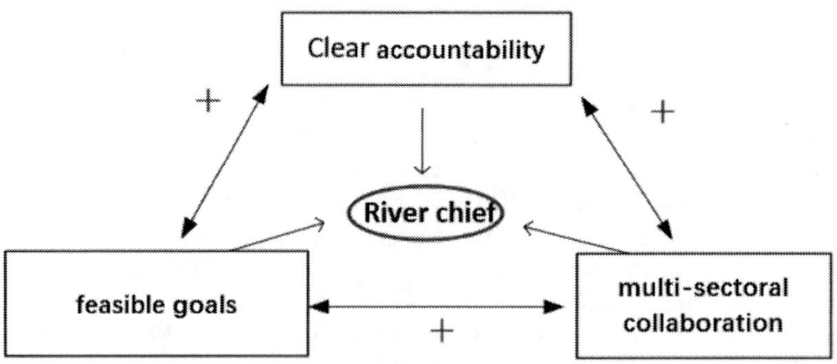

Figure 1. Performance generation mechanism of river chief.

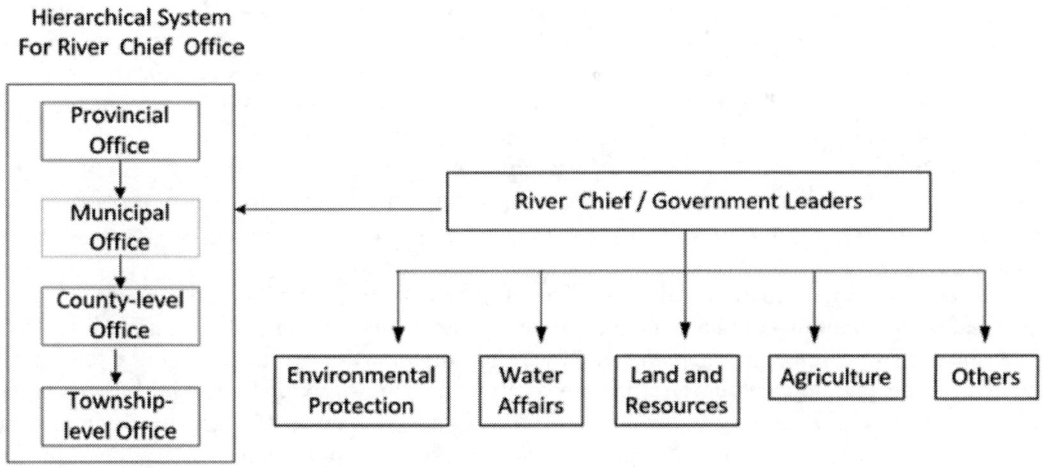

Figure 2. The collaborative system of administrative structure and RCS.

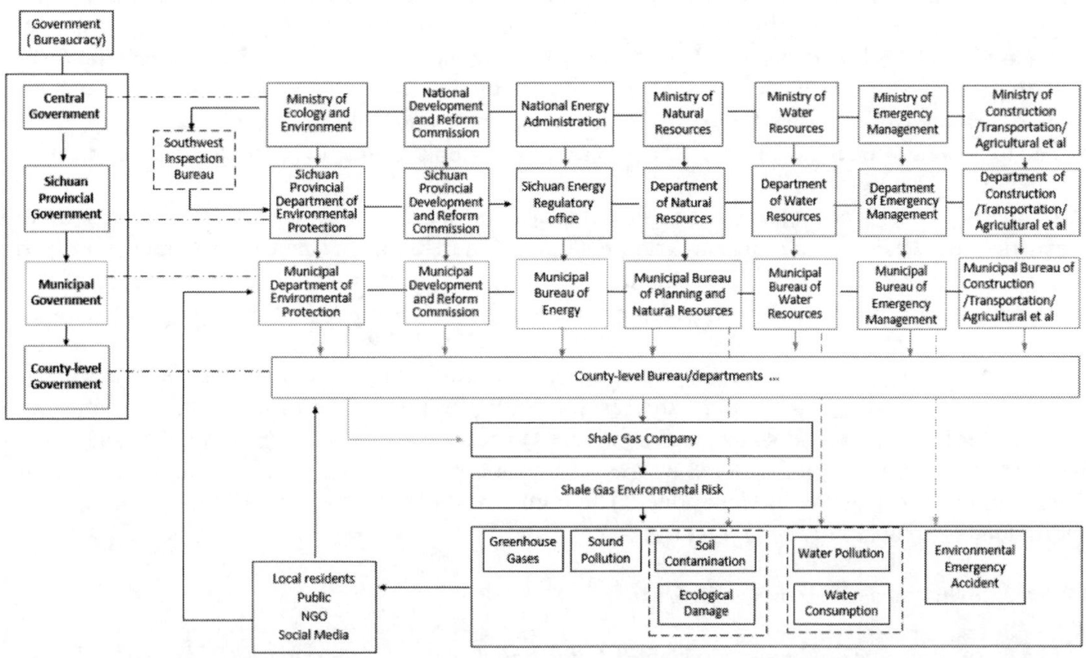

Figure 3. Multi-channel governance system of Shale gas in China-based on Sichuan.

3.1.3 Feasible implementation system. The features of each river diversified, and the challenges river chiefs face also different, for instance, existing water quality, enterprise quantity and industry type. Even just for water pollution control, the distinguished quality, composition, and spatial distribution of pollution sources and specific natural conditions of each river requires targeted control measures. Therefore, instead of traditional 'one-size-fits-all' environmental policy, RCS has conducted "one river, one file" and "one river, one plan" to promote differentiated measures on each river with achievable goals.

3.2. The worries on RCS

RCS has brought prominent improvement to China's river management for its innovation in accountability, collaboration and goal setting. However, based on literature review, field research as well as interview with related staffs, we found that the weaknesses of RCS remain in several aspects, which have already appeared in its implementation across the country, and impairing the effect and limiting this system's promotion in other environmental protection issues:

3.2.1 River chief-driven mechanism. The river chief is regarded as the center of the whole RCS (as shown in Figure 1and Figure 2), whose abilities, prestige and values can greatly determine the work efficiency. However, these could be influenced by many factors, including personal knowledge in pollution control, subjective initiative, values, individual attention distribution, local economic level, personal social network situation and so on. Therefore, the "rule of man" mode is flexible when facing diversified river conditions but hard to constant in efficiency and implementation more broadly.

3.2.2 Campaign-style governance model. RCS was initially innovated as an emergency setting for water security of a large-scale cyanobacteria bloom occurred in Taihu Lake. However, since RCS has been promoted to the whole country from 2016, it falling into the trap of 'Campaign-Style Governance' (CSG) in some places. Because CSG can concentrate resources to accomplish large undertakings and achieve the effect of governance quickly, CSG model has been adopted in China's governance system for decades, and getting further developed in recent years, especially in dealing with major social problems. For the river chiefs, in order to pass the achievement assessment, they tend to adopt CSG model in a short time (generally, half to 1 year before the assessment) to expect an immediate result. Similar as CSG, RCS is original an emergency policy without systematical design, which already breaks the conventional river administrative system, and it inevitably brings a series of problems, such as high resource dependency, non-institutionalization, less specialization and so on. Consequently, if RCS is adopted to be a new mechanism for water governance in whole China, to promote its regularization and constancy will be important.

3.2.3 Out of social participation. As plural subject system has been widely accepted in public governance, especially in solving environmental and other social problems, to attract more subjects involve in environmental governance should be a promising option. However, as the core of Chinese bureaucracy is considered as 'always responsible to a higher authority system' [17], the local policy-making process tends to be less systemic for chasing the immediate results, such as 'one-size-fits-all' environmental policy. A shorter policy-making process could imply not enough consideration for all stakeholders, especially, which will finally result in extra economic loss for the enterprises and farmers. Hence, fewer enterprises and farmers are motivated to participate in RCS. Additionally, since the river chief has been commonly considered as the responsible person for the river pollution affairs, which would lead to 'The Tragedy of Commons'. The public could form a concept of 'river pollution is only government's responsibility' in a long run.

4. Water governance of shale gas in China

4.1 The status of water governance system in shale gas

In the Sichuan basin of China, (mainly including Sichuan province and Chongqing region) the environmental governance system (including water governance) of shale gas can be classed into 4 levels bureaucracy hierarchies, every hierarchy has its corresponding supervision departments, as shown in Figure 3. Besides, the shale gas company, local residents, public, NGO and social medias are designed to be responsible in this system.

IOP Conf. Series: Materials Science and Engineering 1196 (2021) 012023 doi:10.1088/1757-899X/1196/1/012023

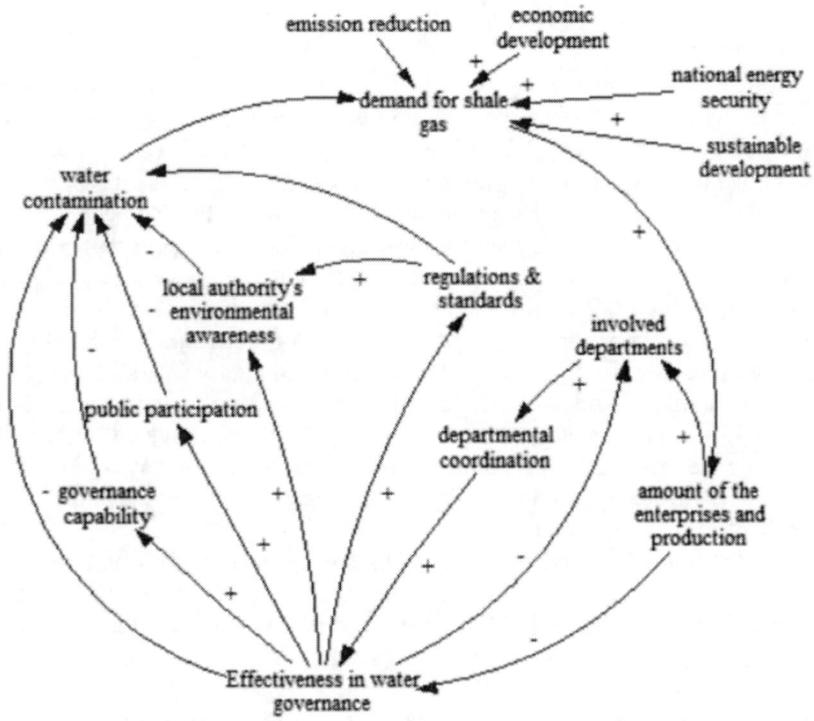

Figure 4. Causal loop diagram model of the water governance dilemma in Shale Gas.

4.2 The dilemmas of water governance in shale gas
RCS's success identifies that the effectiveness of water contamination control is directly positive correlated to environmental governance. According to the literature review, field research as well as discussions with officials at various government levels and shale gas companies, we find mainly 5 dilemmas in China's shale gas water governance system as following:
4.2.1 Dilemma of governance departmental coordination: China implements multi-channel management system for shale gas industry (as shown in Figure 3). However, without a unified governance model/authority, this system function complicated and fragmented. For instance, water issues have triggered high attentions in the world, but Ministry of Water Resources is not included in management system. Departments fight with each other for the departmental interests, especially between shale gas and environmental protection departments. And environmental protection is an unobvious political achievement which has been belittled involuntarily. Moreover, shale gas development in China is open to all types of firms now, firms even without any experience in oil or gas exploration are allowed to enter this industry. This institutional loophole could be exploited by some developers and sowing the seeds of the accidents.
4.2.2 Dilemma of governance incentive and capability: The environmental governance in China represents the characteristic of 'Inverted pyramid'. Which means, despite that the governance work is mostly carried out by the basic- hierarchy government/local government, the government with higher hierarchy owns not only more financial resources, but more advanced eco-environment technology and stronger sense of responsibility, and vice versa. In terms of shale gas industry, firstly, the exploration fields are mainly located in undeveloped cities/towns, where the most important assessment criteria for local government is GDP, and shale gas industry plays an irreplaceable role in advancing the local economic growth. Consequently, local government has no incentive to governance it. Additionally, local water specialized supervision institutions are required by Ministry of Environmental Protection as well as Department of Water Resources, however, these water supervision institutions are affiliated with local government, and their employees basically have neither college degree nor major in water or

environmental protection. Therefore, there is no incentive and capacity for local government to struggle with powerful oil enterprises.

4.2.3 Dilemma of governance criterion: over rely on the enterprises and lack of sector-specific standards : The applicable water regulations for shale gas activities can be only referred to oil and gas laws. Such as the 'Environmental Protection Law', 'Water and Land Conservation Law'. However, China's oil and gas related water protection regulations are not complete yet. Even the largest shale gas fields in China, such as Sichuan and Chongqing, they also do not have their own shale gas water

4.2.4 Dilemma of local authority environmental awareness: With the improvement of living standard, Chinese people and government put more emphasis on environment in recent years. However, start from the 50s-60s of last century, natural environment has once been regarded as an 'enemy' in China, people must fight with to survive. Based on this point of view, China has taken a development strategy of 'Pollution first, Treatment later' for a long time. Hence, governments at various levels and local residents have to change their traditional mentality of 'destruction first, repair later; emissions first, reduction later'. For example, 'Shale gas development plan (2016–2020)' promulgated by NEA targets of shale gas to 30 billion cubic meters by 2020 and Sichuan province is set as the key region. Despite of its high population density, complicated geologic conditions, seasonal water shortage and even near earthquake zone, mainly for the vast shale gas reserves.

4.2.5 Dilemma of public and NGO participation: Dislike a common facility, expertise is required in shale gas environmental governance. For instance, fracturing fluid contains extremely complex dissolved chemicals, heavy metals, NORM, bacteria and water-soluble salts etc., and the contents of each component various depend on properties of fracking shale reservoirs. Specialized knowledge is a precondition for public participation. What's more, since the shale gas well sites are located in the countryside and locked by gates, the only feasible approach to visit these places is to apply to Petro China Company officially. It's time-consuming, and even the application is accepted, information public can receive is limited and unintelligible.

4.3 Causal linkages among the dilemmas

In order to clarify interrelations among above dilemmas, we use system dynamics theory to build a causal loop diagram（CLD）model of the water governance in shale gas, as shown in Figure 4. In a CLD, variables are connected by an arrow and an assigned positive ('+') or negative ('-') polarity to represent causal relationships among the variables. A positive polarity indicates that both variables move in the same direction (increasing or decreasing) in response to a change in the variable, and vice versa (J Singh ,et al.,2019). In this study, the CO2 emission reduction pressure, sustainable development and national energy demand are exogenous variables which promote the amount of shale gas enterprises and production; meanwhile, if the governance is more effective, the environmental pollution should decrease, which will lead to a higher demand for shale gas again. However, departmental coordination, governance capability and driving force, sector-specific regulations and standards, local authority environmental awareness and public participation form negative feedback to shale gas pollution control.

5. The implications of RCS to China's shale gas water governance

Shale gas has been heavily backed to China's sustainable development. Simultaneously, since water contamination issues of shale gas involved in all stages of its product life cycle, which has also injected uncertainties into the development of its upstream and downstream enterprises and local resident living conditions. Considering the feasibility and effects of the RCS, as well as the challenges it facing, we argue that a reinvented RCS would play an important role in water governance of shale gas. Therefore, how to achieve the governance innovation and keep the advancement of RCS in shale gas water governance is our major concern.We find three keys to the success of RCS: firstly, in the initial stage, as local government leaders are appointed to be river chiefs, which indicates the authoritativeness and significance of water affairs, as well as ensuring the effectiveness of multi-sectoral system; then, since local government leaders are pushed to be river chiefs, a clear accountability system was established to ensure the responsible persons with clear tasks to ensure the efficiency of water management; the third is a feasible implementation system, which means, through scientific object design and effective implementation measures to ensure the accessibility of water management goal. Consequently, regarding the status quo and dilemmas of China's shale gas water governance, some policy implications based on the experience of RCS are discussed as following:

(1) Ensuring a seamless governance to explore and perfect the chain of responsibilities.

An integrated chain of responsibilities should include government, departments, enterprises, social organizations (third parties) and the public. The RCS emphasizes the "Administrative Subcontract" and clarifies the water management responsibilities of the local government leaders, but the participation of enterprises, social organizations and the public is insufficient and unclear. Moreover, due to the information asymmetry, the local governments are relatively incapable in the supervision of shale gas exploitation. Therefore, in a sense, China's shale gas water governance system can be regarded as a largely enterprise self-regulated system. Actually, in the largest shale gas development area, such as Sichuan province, there are many third-party organizations are capable of participating in the regulatory process (Southwest Petroleum University, Chengdu University of Technology, Sichuan University, Sichuan Academy of Environmental Protection Sciences, Chinese Academy of Sciences, etc.). What's more, different with the U.S., the shale gas fields in China are mostly separated over densely populated rural areas. Thus, the local public share more encompassing interests and could have greater enthusiasm to participate in the supervision.

Therefore, on the basis of defined functions of the supervision departments and the capabilities of the social subjects, water governance responsibility of shale gas could be disassembled seamlessly, so that the supervision responsibility shall be fully covered. As a result, the lack in professional supervision capabilities, over-reliant on self-regulatory of enterprises, imperfect disclosure system risk can be filled.

(2) Ensuring the authority of governance: establish an environmental issue-led governance system.

River chiefs are responsible for the drinking water safety, water source dispatching, water saving management, and sewage treatment, which were once subordinate to the respective authorities of environment, agriculture, resources, and transportation (Xia, B.). As discussed above, the leading organizations and leadership are the basis for the inter-departmental coordination of the RCS. And this leading organizations should also be authoritative. Leading organizations can be undertaken by different departments towards different environmental issues. For example, in water resource conflict areas, the shale gas environmental governance should be led by the water conservancy department; in the water pollution risky areas, the system could be dominant by environmental departments; and the complex environmental risks areas, the system can be borne by the government leaders.

In terms of organizational form, based on the distribution of shale gas production areas, similar as RCS, a four-level system of responsible persons at the provincial, municipal, county, and township levels can be established. For instance, in the provincial level, a provincial leader who should be the key person in charge of provincial shale gas affairs; and in the lower level, such as county, township, and village where the shale gas platform is located, or where the surface and groundwater flow through, the local government leader should be in charge of shale gas affairs.

(3) Ensuring the feasibility of implementation: refine the water governance dimensions and objectives.

A scientific and feasible implementation can be guaranteed from the following two aspects: firstly, refine the governance both from the time and the space dimension. In most cases, shale gas exploitations in China are only required environmental impact assessment（EIA）in two periods: one is in construction period and the other is in operation period. However, since shale gas production varies with each other from the length of drilling to operating life cycle (which could range from a few months to several decades). Apparently, the current EIA system is hard to prevent and control pollution risk. Therefore, if under a delicacy governance, in the time dimension, EIA can be clearly divided into 5 stages according the exposure level of risk factors, including exploration start-up period, early production period, mature production period, production decline period, and waste gas treatment period. Different stages should develop a regular and an emergency environmental governance system, respectively (such as daily monitoring and monthly review). Meanwhile, in the spatial dimension, surface water, groundwater and geologic changes supervision ought to be included, and the scope of single well and wellsite/platform affected/involved area also should be clarified. Accordingly, through the scientific and feasibility of goal setting both from the time and the space dimension, an effective implementation can be guaranteed. Secondly, Establish a differentiated management objective of "one platform one policy". Based on the Management by Objectives (MBO) theory, objectives are the ultimate goals towards which an organization's activities are directed. They should be specific, clear and concise enough understood by those who want to achieve them and to be used as standards for control purposes. Therefore, objectives should be measurable, specific, result oriented, realistic, attainable and time bounded [3][4][10]. The distribution of shale gas wells in China is mainly 4-5 wells to form a drilling platform. The platforms are significant different with each other from geological landforms, water, soil, climates to social conditions; additionally, surface water and groundwater pollution is latent regional, cumulative and lagging. Therefore, firstly, a comprehensive survey of the basic conditions of the utilization of shale gas resources, water environment, and ecological environment is needed. It is the basic for "one platform, one policy" policy. Then, according to the baseline survey results, "one platform, one policy" policy can be implemented. It means, the platforms which are classed as "excellent" should highlighting the "prevention" goal; the "light pollution/impact" class should highlight "protection" goal; and the "serious pollution/impact" class should highlight the goal of "control".

(4) Ensuring the effectiveness of implementation and the fairness of accountability: further improve water standards and technical specifications for shale gas.

Since the water issue is commonly regarded as the largest risk of shale gas exploration, the priority should be established environmental quantification standards for water resources utilization and wastewater treatment related to shale gas exploration, especially supervision rules focus on fracturing fluid and flowback water pollution. Moreover, based on geological condition, exploitation techniques, development processes and even social condition differences, provincial based water protection technical specification systems of shale gas are also needed to meet the practical conditions. And conventional oil and gas standards can be used as reference, and some shale gas specialized and normative enterprise standards which already be proved can also be considered as national standards.

(5) Ensuring the basis for achieving refined management and accountability: complete water quality grid monitoring network system to improve the informationization level of water governance .

Monitoring data is the core for refine management, and also the prerequisite for assessment as well as accountability. The water quality monitoring for shale gas mainly depends on samples taken from the upstream and downstream of the well site by the environment agencies and shale gas companies themselves. However, because of the low coverage, the automatic water monitoring network cannot meet the requirements of water environment and ecological management. More importantly, even now, the possible pollution elements produced by shale gas exploitation are not all included in this monitoring system. Therefore, based on the information technology such as big data, Internet of Things (IOT) technology and cloud computing, a real-time system for shale gas water monitoring is urgently needed. Based on this system, a national-wide information network of shale gas can be established and connect with relevant departments to communicate with each other, for decreasing the degree of information barrier. Consequently, information update and resource sharing throughout regions and departments can be guaranteed, the waste of information dissymmetry can be decreased as well.

Funding: This research was funded by Young scientific research innovation cultivation team program of Southwest Petroleum University, 2 grant number 2018CXTD14.

References

[1] Alley B, Beebe A, Rodgers J Jr and Castle J W Chemical and physical characterization of produced waters from conventional and unconventional fossil fuel resources. *Chemosphere*. 2011, **85**, 74-82.

[2] Colborn T, Kwiatkowski C, Schultz K and Bachran M Natural Gas Operations from a Public Health Perspective. *Human and ecological risk assessment*. 2011; pp.17, 1039-1056.
[3] Drucker P The practice of management. New York: Harper. 1954.
[4] Drucker P What results should you expect? A users' guide to MBO. Public Administration Review, 1976, 36, 12-19.
[5] EIA (Energy Information Administration) and ARI (Advanced Resources International) World Shale Gas and Shale Oil Resource Assessment. Washington, DC: EIA and ARI. Available online:http://www.advres.com/pdf/A_EIA_ARI_2013%20World%20Shale%20Gas%20and %20Shale%20Oil%20Resource%20Assessment.pdf. (accessed on 2013).
[6] Huang Q, Xu J, Rethinking Environmental Bureaucracies in River Chiefs System (RCS) in China: A Critical Literature Study. *Sustainability*. 2019, **11**.
[7] NEA (National Energy Administration). Shale gas development guidance (2016–20). Available online: http://www.gov.cn/xinwen/2016-09/30/content_5114313.htm. (accessed on 2016).
[8] NFGA（National Forestry and Grassland Administration）.全面推行林长制提升林业治理体系和治理能力现代化水平. Available online: http://www.forestry.gov.cn/main/5383/20200109/094542842127394.html. (accessed on 2020).
[9] Rahm B G, Riha S J Evolving shale gas management: water resource risks, impacts, and lessons learned. *Environmental ence Processes & Impacts*, 2014, **16**, 1400-1412.
[10] Rodgers R, Hunter J E, Impact of management by objectives on organizational productivity. *Journal of Applied Psychology*. 1991, **76**, 322-336.
[11] She Y, Liu Y, Jiang L, et al. Is China's River Chief Policy effective? Evidence from a quasi-natural experiment in the Yangtze River Economic Belt, China. *Journal of Cleaner Production*. 2019.
[12] Singh J, Sung K, Cooper T, et al. Challenges and opportunities for scaling up upcycling businesses – The case of textile and wood upcycling businesses in the UK. *Resources, Conservation and Recycling*. 2019, **150**:104439-.
[13] Fan WJ, Hayes K F, Ellis B R Quantitative survey and structural classification of hydraulic fracturing chemicals reported in unconventional gas production, *Environ. Sci. Technol.* 2018, **52**, 10839–10847.
[14] Xia B The reform of River Governor Mechanism and practice of integrated water management. *China Water Conserv*. 2017, **2**, 50–57.
[15] Xinhua. China announces further commitments to tackling climate challenge. Available online: http://chinaplus.cri.cn/recommended/1661/590983 (accessed on 12 December 2020).
[16] Zhang Z, Zhuang Y, Li J, et al. Feasibility evaluation of the treatment and recycling of shale gas produced water: a case study of the first shale gas field in the Eastern Sichuan Basin, China. *Environmental ence: Water Research & Technology*, 2019.
[17] Zhou Xueguang The Institutional Logic of Governance in China: An Organizational Approach. Beijing: SDX Joint Publishing Company, 2017.

Transition towards Solar-Powered built environment: spatial distributions and impacting factors of Australian solar installations

Yanquan Zhang[1,*], Ruidong Chang[2], Jian Zuo[2], Wei Emma Zhang[3]

[1]School of Civil and Environmental Engineering, University of Adelaide, Adelaide, Australia
[2] School of Architecture and Built Environment, University of Adelaide, Adelaide, Australia
[3] School of Computer Science, University of Adelaide, Adelaide, Australia
***E-mail**: yqzhang1120@gmail.com

Abstract. With the advantages of non-renewable energy conservation and environmental protection, solar energy technology has been widely used worldwide. Data shows that solar systems has been increasingly installed from 2006 to 2016 in Australia. However, impacting factors on Australian solar system installation remain unknown. Therefore, this paper provides a spatial regression analysis to identify the factors that influence households to install the solar panels. Solar installation maps of 2006, 2011, and 2016 are compared to investigate the changes and development of solar distribution. Besides, cluster maps are analyzed to investigate the cluster area distributions. Through analyzing statistical variables such as regression coefficient, p-value, z-value, and Moran's I value, it can be concluded that educational qualification, population, and household composition have substantial positive relevancies to the installation of solar systems. It also shows that the low solar installation cluster areas mainly distribute at the center of Australia.

Keywords. Solar power, solar energy, solar panel, Australia, spatial regression analysis, spatial distribution.

1. Introduction

The importance of adopting renewable energy technologies is growing rapidly for a sustainable future [8]. With the negative ecological impact of fossil energy, a severe contradiction among energy, environment and economy emerged due to the urgent energy requirement that the world is facing [12]. Therefore, renewable energy technology has been developed and widely adopted because of its higher energy efficiency, eco-friendliness and cost-saving features. According to Zhang et al. (2009), renewable energy can be referred to as the energy types that are recyclable and have little negative environmental impacts [12]. It contains solar energy, wind energy, biomass energy, small hydropower, geothermal energy and ocean energy. In Australia, the annual solar radiation is claimed to be approximately 10,000 times the annual energy consumption of the whole of Australia [4], and this provides favorable

IOP Conf. Series: Materials Science and Engineering 1196 (2021) 012024 doi:10.1088/1757-899X/1196/1/012024

conditions for developing the usage of solar power.

According to the Australian Government, the postcode data for small-scale solar installation of 2006, 2011 and 2016 shows that solar panel has been more commonly used. Chapman et al. (2016) claimed that due to the residential solar photovoltaics (PV) policy announced by the Australian government, the rapidly increased PV installation provided a huge amount of job positions across Australia between 2008 and 2012 [6]. Moreover, the solar installation price was reduced from $15 per watt in 2004 to a lower cost of $3 per watt in 2012. Meanwhile, the motivations for people to adopt residential solar electric technology also need to be considered. Schelly (2014) identified that motivations for adopting renewable energy technology are not only from environmental and economic aspects but also from the consideration of the expected lifestyle, household conditions as well as information communities [9].

However, the impacting factors of solar installation in Australia remain unknown. This paper focuses on the factors that may influence people to adopt residential solar technology across Australia based on the previous research. To identify the impacting factors of solar installation in Australia, a spatial regression analysis software named Geoda is applied on the Australian census data of 2006, 2011 and 2016. Results including impacting factors and the spatial distribution of solar installation are presented. It is expected that the solar installation shows a proportional relationship to the household population and personal qualification of people in economic and education aspects. Low-installation areas are found to cluster at the center of Australia.

2. Methodology

2.1. Data Collection

According to the aim of the research, the data required to be collected contains:

- Census data of 2006, 2011 and 2016
- Solar installation data of 2006, 2011 and 2016
- Australia postal area shapefile

Census data was downloaded from the Australian Bureau of Statistics (ABS) website. The corresponding solar installation datasets for the three specific years were downloaded from the Australian Government Clean Energy Regulator website. Besides, to reduce the error of spatial regression analysis, the Postal Areas ASGS Ed 2016 Digital Boundaries in ESRI Shapefile Format downloaded from the ABS website was used for all years. The raw datasets can then be cleaned and processed for the preparation of the spatial regression analysis.

2.2. Data Cleaning and Processing

Due to the content and order differences between postal area code and postcode, datasets need to be re-ordered for matching. Data order of shapefile was considered to be the standard. The missing postcode areas were added to the dataset with 0 solar system installation. Several important features were created as independent variables, which have been listed below:

- Percentage of males in the total population
- Percentage of people aged from 20 to 59 in all age groups
- Percentage of highly educated people in the total population (In this study, highly educated people is defined as people with at least a Bachelor degree)
- Percentage of married people in the total population
- Percentage of separated house in all dwelling types
- Percentage of families with high income in all families (According to Australian Bureau of Statistics (2021), the minimum high income (weekly) in 2015-2016 is $2083, $2036 in 2011-2012 and $1711 in 2005-2006 [3]. The statistics may not be accurate enough as only upper-income intervals were considered)

IOP Conf. Series: Materials Science and Engineering 1196 (2021) 012024 doi:10.1088/1757-899X/1196/1/012024

- Percentage of owned house in all tenure types
- Percentage of one family household in all households
- Percentage of households with more than 4 people in all households
- Percentage of households with high income in all households (Same standard as family income)

2.3. Data Analysis

To start with the Geoda analysis, weight files should be created first. According to Anselin (2020), spatial weights are a crucial element when constructing spatial autocorrelation statistics [2]. Besides, the quality of spatial data sources critically contributes to the construction of spatial weights. In this research, queen contiguity, which is better for disposing of potential inaccuracies in polygon files, was selected to create weight files.

With valid weight files, spatial regression analysis can be processed. Total Solar Installation Number was chosen as a dependent variable. Impacting factors for solar installation are aimed to be first analyzed. It is claimed that the regression coefficient represents the relationship between independent and dependent variables. According to Storey and Tibshirani (2003), an independent variable with a p-value less than 0.05 would be traditionally considered statistically significant [10]. Therefore, an independent variable with a positive regression coefficient and a p-value less than 0.05 would be considered a statistically significant factor proportional to solar installation. The solar installation distribution maps of the three specific years were compared to investigate the regional development of solar energy in Australia. Besides, to investigate the spatial clusters, Moran's I value and the associated p-value (statistical significance) and were mainly studied. A large positive Moran's I value indicates the studied location has similarly low or high values with the surrounding areas so that the areas are spatial clusters. For instance, high-high clusters, which can be considered as "regional hotspots", indicate that the studied location has high values in a high-value neighborhood [11]. Besides, Hazra (2017) claims that the z-scores of 1.65, 1.96 and 2.58 correspond to 90%, 95% and 99% confidence intervals, which also correlate with p-values of 0.1, 0.05 and 0.01 respectively [7]. This indicates that if the output z-score is larger than 2.58 and p-value is less than 0.01, the results should be fairly confident.

3. Results and Discussion

3.1. Impact Factors

Table 1 below demonstrates the variables significantly correlated with the solar installation, considering the regression coefficients and p-value of the independent variables. The variables including highly educated people, separated house, families with high income, one-family household and population in dwellings are notated as High_Edu, Sep_House, H_in_fam, 1fam_house and Dwe_pop correspondingly in Table 1. According to Table 1, the proportional variables are different between years. However, connections between the listed independent variables can still be identified. It can be summarized that the solar installation is mainly proportional to the factors including population, high education level and household combination. Variables such as the percentages of males and families with high-income levels are not significant.

Table 1. Proportional Variables to Solar Installation for each year.

		2006		
Variable %	Population	Male	High_Edu	Sep_House

Coefficient	9.71E-05	0.03869	0.04738	0.01207
p-value	0.00000	0.02604	0.00167	0.01227
		2011		
Variable %	Population	H_in_fam	Own_House	1fam_house
Coefficient	0.0066	6.08926	1.01179	3.37029
p-value	0.00000	0.00682	0.04102	0.00001
		2016		
Variable %	High_Edu	Dwe_pop	-	-
Coefficient	4.73137	0.00000	-	-
p-value	2.31155	0.02559	-	-

3.2. Spatial Distribution Analysis

Figure 1 shows the total solar installation maps for 2006, 2011 and 2016. Increasingly darker orange is used to indicate the increasing solar installation amount. Based on the comparison of the three maps, a noticeable increase in solar installations at coastal areas can be identified. In 2006, solar installation areas were relatively well-distributed. Installation amount differences between the inland to coastal areas are not obvious. However, solar installation difference has become wider in 2011 and 2016. Colors for inland Australia slowly changes over time, while most coastal areas grow to dark orange that indicates high solar installation amount.

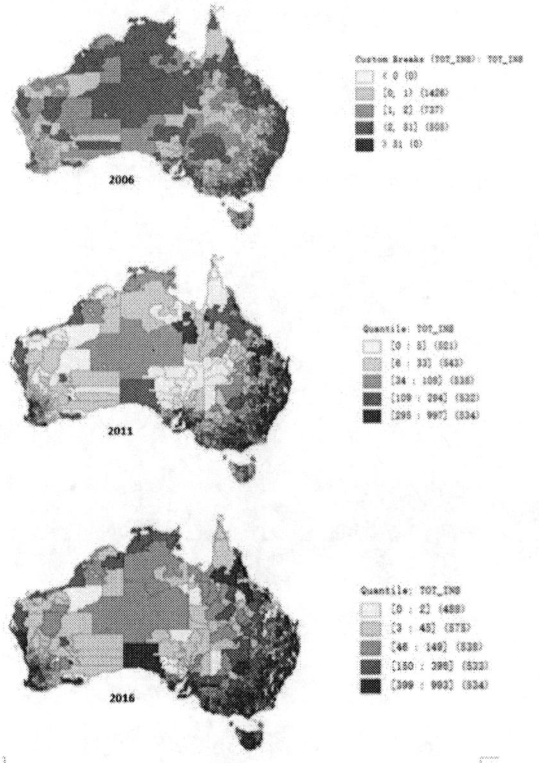

Figure 1. Solar Installation Distribution.

Statistical values including Moran's I value, p-values and z-values of the years from Geoda have been shown in Table 2. Cluster maps with color differentiation are shown in Figure 2 for spatial cluster

IOP Conf. Series: Materials Science and Engineering 1196 (2021) 012024 doi:10.1088/1757-899X/1196/1/012024

analysis. According to Anselin (2020), bright-colored areas, which are High-High and Low-Low areas, are considered clusters, while the shallow-colored areas represent spatial outliers [2]. Therefore, the highly solar-installed cluster area in 2006 mostly lays close to the edge of Australia, such as north Australia and east Australia, while the low solar-installed cluster areas of 2011 and 2016 mainly located at the center. Considering the migration movements and urban development in Australia, the presented result is considered rational.

Table 2. Statistical Values of Different Years.

Year	Moran's I	P	z
2006	0.312	0.001	25.9167
2011	0.254	0.001	19.7354
2016	0.135	0.001	10.8494

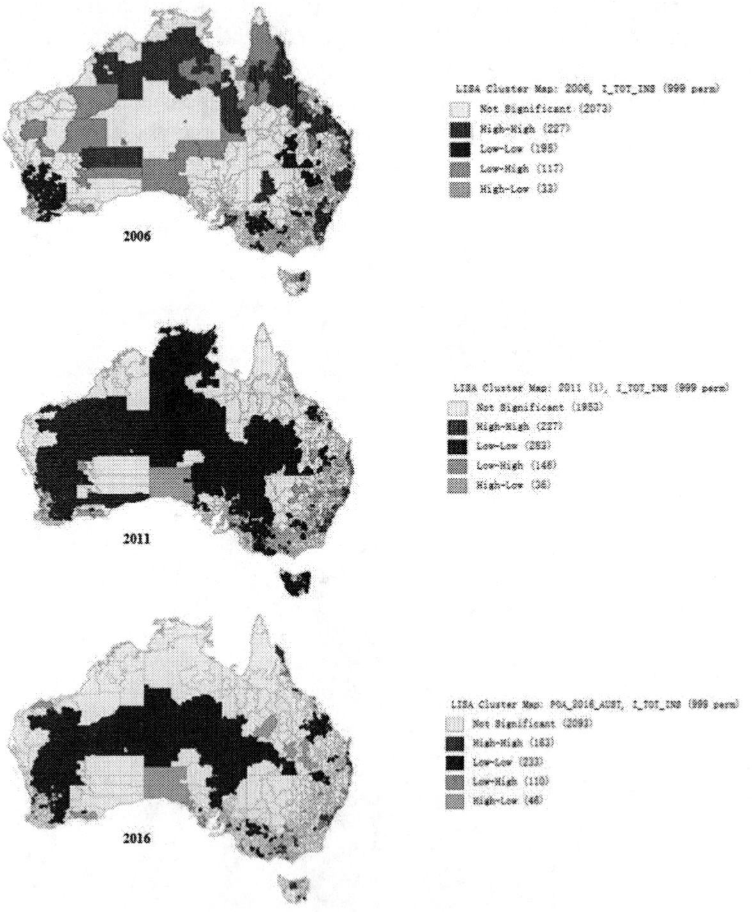

Figure 2. Cluster Map of 2006, 2011 and 2016.

4. Conclusion

After the spatial regression analysis of the 3-year census data, it can be identified that the statistically significant independent variables are different between years. In conclusion, personal qualifications, population and household composition are the three main factors that are proportional to the solar installation. Besides, the spatial cluster analysis shows that the cluster areas (Low-Low) aggregate to the center of Australia. In addition, by comparing the solar installation increases between inland and coastal areas of Australia, geographic location is determined to have a strong association with solar energy use. This study provides an important reference for future policy making regarding regional solar installations and incentives.

References

[1] Anselin L 2005 Exploring spatial data with Geoda: a workbook. *Cen. for Spat. Inte. Soc. Sci.*.
[2] Anselin L 2020 Local spatial autocorrelation (1): LISA and local Moran *GeoDa.*
[3] Australian Bureau of Statistics 2021 Household income and wealth, Australia *Australian Bur. of Stat.*
[4] Bahadori A and Nwaoha C 2013 A review on solar energy utilisation in Australia *Rene. and Sus. Ene. Re.* **18** 1-5
[5] Bao Q, Sinitskaya E, Gomez K, MacDonald E and Yang M 2020 A human-centered design approach to evaluating factors in residential solar PV adoption: A survey of homeowners in California and Massachusetts *Renew. Ene.* **151** 503-13
[6] Chapman A, McLellan B and Tezuka T 2016 Residential solar PV policy: An analysis of impacts, successes and failures in the Australian case *Renew. Ene.* **86** 1265-79
[7] Harza A 2017 Using the confidence interval confidently *J. Thor Dis* 4125-30
[8] Palm J 2018 Household installation of solar panels – Motives and barriers in a 10-year perspective *Ene. Pol.* **113** 1-8
[9] Schelly C 2014 Residential solar electricity adoption: What motivates, and what matters? A case study of early adopters. *Ene. Res. & Soc. Sci.* **2**, 183-91.
[10] Storey J and Tibshirani R 2003 Statistical significance for genomewide studies. *Proc. Nat. Aca. Sci Uni. Sta. Ame.* **100** 9440-45
[11] Zhang C, Luo L, Xu W and Ledwith V 2008 Use of local Moran's I and GIS to identify pollution hotspots of Pb in urban soils of Galway, Ireland. *Sci. Tot. Envir.* **398** 212-21.
[12] Zhang P, Yang Y, Shi J, Zhang Y, Wang L and Li X 2009 Opportunities and challenges for renewable energy policy in China. *Renew. Sus. Ene. Rev.* **13** 439-49

The evolution of China's new energy policy (1995-2021): an analysis based on policy tools

Yuechi Sun[1], Haiyan Liu[1,*] and Yaxin Su[1]

[1] School of Economics and Management, China University of Geosciences, Beijing, 100083, China.

* **E-mail:** liuhy@cugb.edu.cn

Abstract. Greenhouse gas emissions lead to global warming, which has been widely concerned by the international community. Further strengthening energy conservation and emission reduction is an urgent need to deal with global climate change. As a responsible big country, China has been formulating strong policies and measures to achieve the goal of energy conservation and emission reduction. Taking the evolution process of China's new energy policy as an example, based on the policy tool theory and using the content analysis method, this paper studies the structural evolution of China's new energy policy from 1995 to 2021. The results show that: (1) With the passage of time, the frequency and intensity of policy combination from the central government to local governments gradually increase; (2)In the use of new energy policy tools, Central government departments and local government departments mainly use mandatory command and control tools, supplemented by economic incentives and social autonomy tools (3) According to the different tasks of energy conservation and emission reduction, the eastern coastal provinces tend to use more applicable and flexible economic incentives, while the central and western provinces tend to use command and control tools. Based on the change trend of China's new energy policy from 1995 to 2021, this study predicts that in the next 10 years, the mandatory tools will be the first policy tools to be used, and the mixed tools and voluntary tools will be appropriately added to promote China's goal of carbon peak and carbon neutralization.

Keywords: new energy policy, policy tools, policy evolution, content analysis method, energy saving and emission reduction

1. Introduction

So far, nearly 200 parties to the United Nations Framework Convention on climate change (UNFCCC) have ratified the Paris Agreement and have begun to implement it in some countries[1]. The long-term goal of the agreement is to control the rise of the global average temperature within 2 degrees Celsius compared with the pre industrialization period, and strive to limit the rise of the temperature within 1.5 degrees Celsius[2]. China's carbon emissions account for about 28% of the global carbon emissions, and it is the largest carbon dioxide emission country in the world. China's carbon dioxide emissions are mainly generated by the consumption of a large number of fossil energy, and the combustion of fossil energy causes serious air pollution in China[3]. Air pollution will increase the risk of diseases[4]. Therefore, China's energy structure has begun to transform from traditional fossil energy to low-carbon and environment-friendly energy structure. The purpose is to further reduce carbon dioxide emissions and gradually increase the proportion of renewable energy in energy consumption, which plays an important role in alleviating global climate change, improving air quality and improving people's healthy living environment. New energy is an important strategic material in China and plays an irreplaceable role in the rapid development of China's social economy[5]. At the same time, the development of China's new energy industry is full of twists and turns. How can new energy continue to play an important role in energy conservation and emission reduction under the constraint of resources and environment? As a major energy consumer, the central government and local governments have implemented a series of policies in the field of new energy.

Public policy plays an unparalleled role in guiding and promoting the process of energy transformation,policy tools supporting new energy technologies are very important for energy transformation[6][7].At present, China's coal based energy system needs emission reduction technology to reduce carbon dioxide emissions. People's concerns about improving air quality and living environment also reflect a series of social expectations or interests. These fundamental pursuit or interests also promote the transformation of the traditional energy structure with coal as the main energy to the energy structure of sustainable development. In short, in order to promote the transition from the current energy system to a sustainable energy system, government agencies should design an energy policy portfolio suitable for future development, such as policies to support various renewable energy technologies, preferential policies to stimulate the development of new energy enterprises, etc[8].

The rapid expansion of new energy infrastructure will bring unexpected consequences. For example, hydropower is attracting global attention. China is a leading country in the development of small hydropower projects[9], and hydropower accounts for more than 16% of global power production. The advantage of hydropower generation lies in its low operation and maintenance costs. At the same time, it has low carbon dioxide emissions per unit of power generation, because of its flexibility in operation, high efficiency and profitability even in remote rural areas. However, the construction of hydropower infrastructure such as artificial dams and reservoirs has promoted the change of hydrological characteristics, resulting in adverse effects on the downstream river ecosystem. The government has promulgated a number of policy tools for governance and control, For example, the purpose of environmental impact assessment measures and watershed level management plans is to prevent the adverse impact of hydropower projects on the environment[9]. Therefore, we should develop new energy in a sustainable way and take policy measures to deal with the adverse consequences of the development process on the environment[10].The world's energy structure has been transformed into a green and low-carbon energy structure. The world energy structure has begun to change to a green and low-carbon energy structure, which has been transformed into a climate, ecological adaptability and friendly energy structure. The main feature is to ensure energy supply and realize the intelligent transformation of energy services[11].

In order to promote green development and realize the strategy of sustainable development, Chinese government departments and local government departments have promulgated a large number of policies to promote the development of new energy[12][13]. From now on to 2030, carbon emissions will reach the peak, and further to 2060, China will talk about "carbon neutral" of carbon emissions. China will practice the green development strategy and take the road of sustainable development for a long time. The Chinese government is adjusting and optimizing the energy consumption structure and choosing new energy with low energy consumption and low carbon emission as the main energy in the new era. Since the Chinese government implemented the reform and opening-up policy for more than 40 years, the growth in new energy production and consumption has been huge. The central and local government departments have issued a series of environmental policies for energy and new energy production and consumption, and used a variety of policy tools to correct the problems of high energy consumption and environmental pollution.

Although some scholars have made good achievements in the research of new energy policy, the research in this field needs to be further improved. Firstly, most of the previous studies focused on specific new energy policy tools, such as laws and regulations, environment, finance, taxation and government procurement, and rarely conducted comprehensive and systematic research on new energy policy; Second, there may be differences in the new energy policies issued by government agencies, but most studies do not clearly distinguish the policies issued by different government agencies; Third, previous studies mainly adopted qualitative research and lacked quantitative analysis.

Firstly, this paper uses the content analysis method to code and statistically analyze the new energy policy texts implemented by the central government and local government departments since China's implementation of the reform and opening-up policy[14][15].This paper constructs the analytical framework of China's new energy policy from the perspective of the use of policy tools by the central government and local governments departments, and draws lessons from the classification principles of Hollett and Ramish's policy tools. On this basis, based on the theory of policy tools, this paper studies the policy documents issued by the central government and local governments in the field of new energy, combs the structure of policy portfolio tools in the field of new energy, and excavates the problems existing in China's new energy policy tools; Finally, this paper forecasts the development trend of new energy policy in the next two five-year plans.

The rest of this paper is arranged as follows: Section 2 mainly introduces the data collection and research methods of this paper; The third section combs the historical development trend, policy tool structure and policy combination tool structure of the central and local governments; Section 4 discusses the evolution of China's new energy policy and the existing problems; Section 5 introduces the conclusions of this study and forecasts the trend of new energy policies and policy tools in the next 10 years.

2. Materials and methods

2.1. Data

This paper studies all the policy documents issued by the central government and local government departments in the field of new energy, and traces the evolution of China's environmental policies in the new energy industry from 1995 to 2021. The relevant policy documents in the field of new energy covered the regulatory documents related to the new energy industry issued by China's central government and local governments since the founding of the people's Republic of China. As of May

12, 2021, the central government has issued 329 laws and regulations, all of which are classified into administrative regulations and policy documents, departmental rules and policy documents and industry regulations and policy documents; There are 1778 laws and regulations issued by local government departments, which are classified into four categories: local laws and regulations, local norms, local working documents and administrative permission approval documents.

First, research data collection. Policy text data collection, this article in the "pkulaw" legal database(www.pkulaw.cn), the official websites of the central government departments and the local government departments searched the policy documents of China in the new energy industry. Secondly, preprocess the data, such as the supplement of lost data and the correction of abnormal data; The main task of policy text data processing is to eliminate repetitive and weak correlation text. Finally, we conduct in-depth mining on the new energy industry data and promulgated laws and regulations, analyze the change trend of the number of new energy industry data and policy promulgation over time, and sort out the evolution process of China's new energy industry policy framework and policy promulgation through the analysis and comparison of the documents issued by the central government and local government departments in the new energy industry, This paper analyzes the problems existing in the new energy policy promulgated by the central and local government departments, and the promulgation trend of China's new energy policy in the next 10 years.

2.2. Methods

From the perspective of the use and innovation of policy tools, this paper constructs an analysis framework according to the classification method of policy tools, and introduces statistical analysis technology and content analysis method[16][17] into policy research. Content based analysis is a widely used research method in the field of social science research, which can transform qualitative symbolic content into systematic quantitative data[18]. Carry out background analysis on the policies issued by the government to better understand the connotation of the policies for the policy formulation and implementation departments, so as to ensure the correct direction of these policies [19]. At present, content-based analysis method is applied to the analysis of policies in various fields, covering many industries, such as environmental protection industry[20], transportation industry[21], and mineral resources industry[22].

This study mainly refers to Howlett and Ramesh's classification of policy tools[23], and divides the categories according to the participation of central government and local government in each link of policy. According to the classification results of policy tools, the new energy policy tools adopted by China's central and local governments are divided into three categories: voluntary tools, mandatory tools and mixed tools. In addition, voluntary tools are divided into four strategic tools, mandatory tools are divided into three strategic tools and hybrid tools are divided into four strategic tools, a total of 11 specific strategic tools. Table 1 Types and structure of new energy policy tools.

Table 1. Types and structure of new energy policy instruments

Types of policy instruments	Policy tools
Voluntary tools	Publicity and education

	Non governmental organizations
	Market mechanism
	Family and community guidance
	Establish laws and regulations
Mandatory tools	Direct service delivery
	Public Enterprises
	Financial assistance
Hybrid tools	Information and persuasion
	Auction of property rights
	Taxes and usage fees

3. Results

3.1. Analysis of the development process of new energy policy

3.1.1. An analysis of the central government's new energy policy

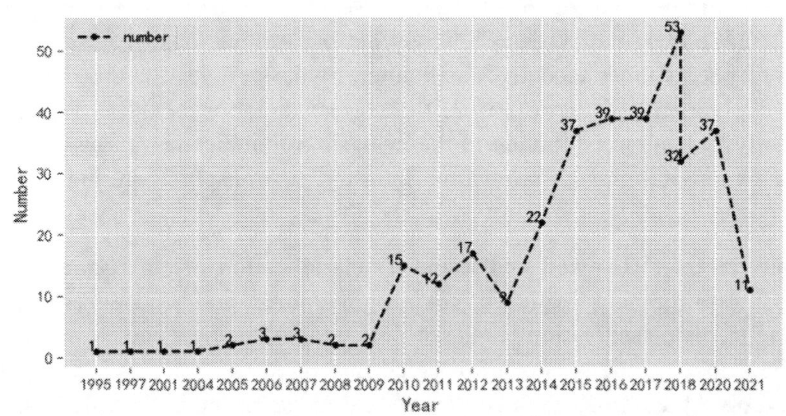

Figure 1. The number of policies issued by the central government departments each year

According to the statistical results of the number of policies implemented by the central government in the field of new energy every year, It can be concluded that from 1995, the first new energy policy was issued, and from 1995 to 2007, the central government implemented about two policies in the field of new energy every year; Since 2008, the number of policies implemented by the central government in the field of new energy has increased year by year, peaked in 2018, decreased in recent years, and more than 20 policies have been issued every year.

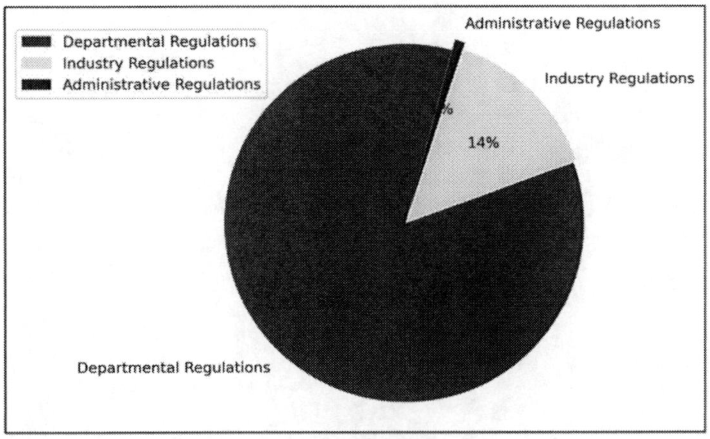

Figure 2. Distribution of effectiveness levels of policies issued by the central government departments

From the statistical analysis of the effectiveness level distribution of policies issued by central government departments, it can be concluded that 85% of the policies issued by central government departments are concentrated in departmental regulations, 14% of the policies issued by central government departments are industrial regulations, and 1% of the policies issued by central government departments are administrative regulations.

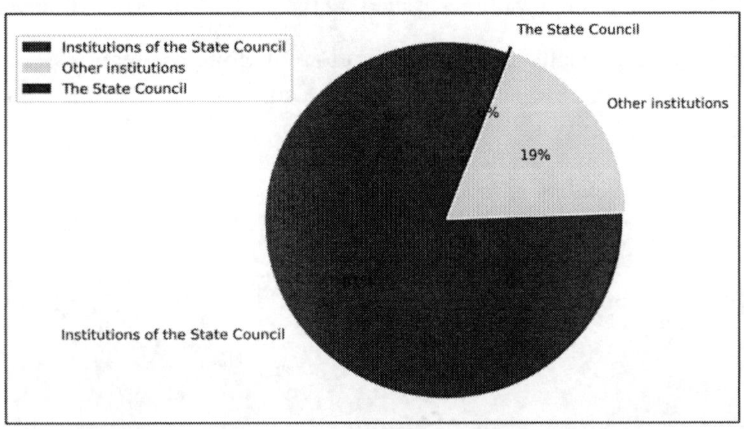

Figure 3. Distribution of policies issued by the central government departments

According to the statistical results of policy distribution issued by central government departments in Figure 3, it can be concluded that 82% of the policies issued by central government departments are distributed in various organs of the State Council and 18% in other organs.

IOP Conf. Series: Materials Science and Engineering 1196 (2021) 012025 doi:10.1088/1757-899X/1196/1/012025

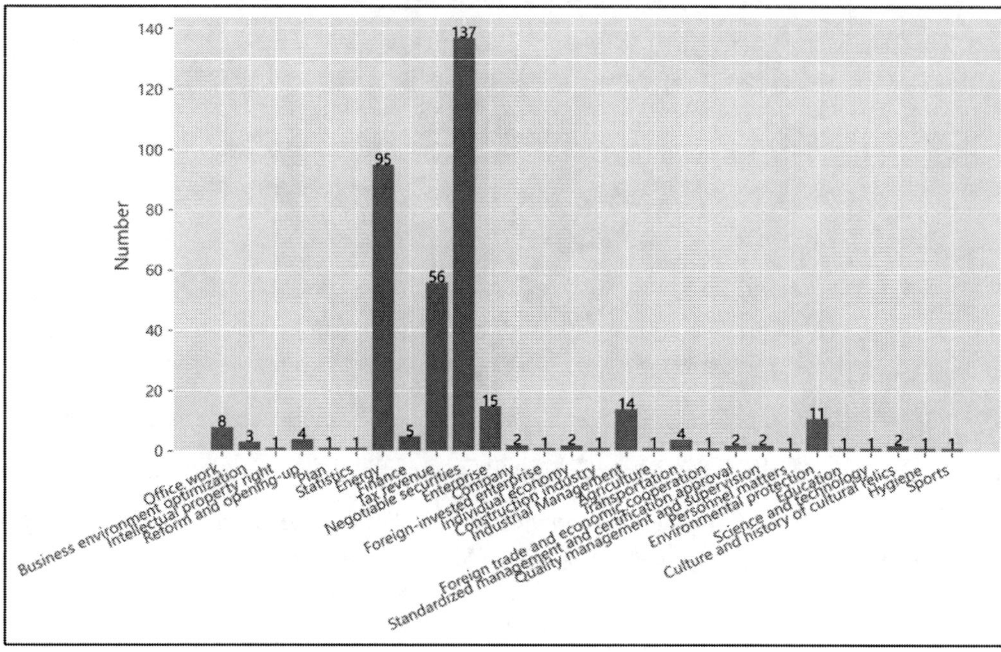

Figure 4. Distribution of policies issued by the central government departments

According to the statistical results of the distribution of policy categories issued by the central government in Figure 4, it can be concluded that the top policy categories issued by the central government are securities, energy, tax and other fields.

3.1.2. Analysis of new energy policy of local government

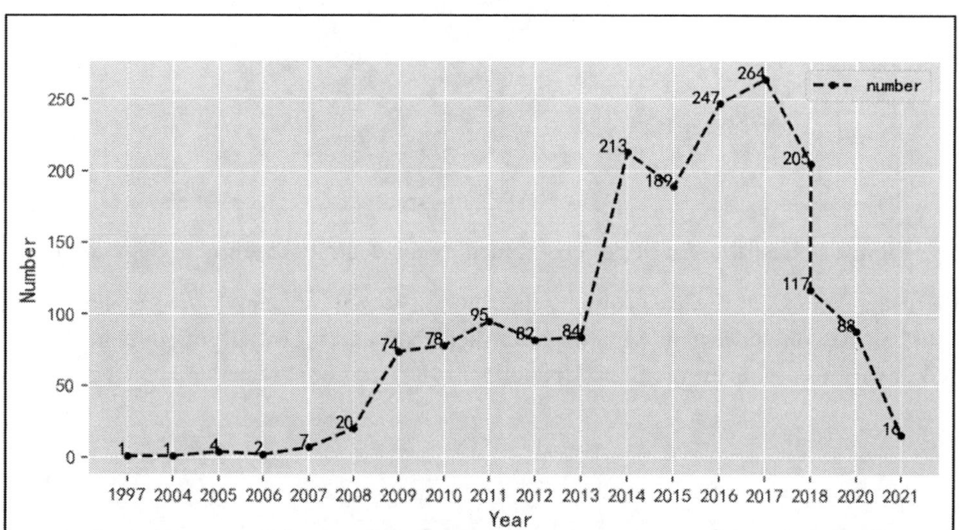

Figure 5. Annual number of policies issued by local governments

As can be seen from Figure 5 statistical chart of the number of policies issued by local governments every year, the number of new energy policies issued before 2007 is very small. Since 2007, the number of policies in the field of new energy has increased rapidly, and the number of policies issued each year from 2007 to 2020 will remain at about 80; Since 2014, the number of policies issued by local governments has increased greatly, reaching about 200.

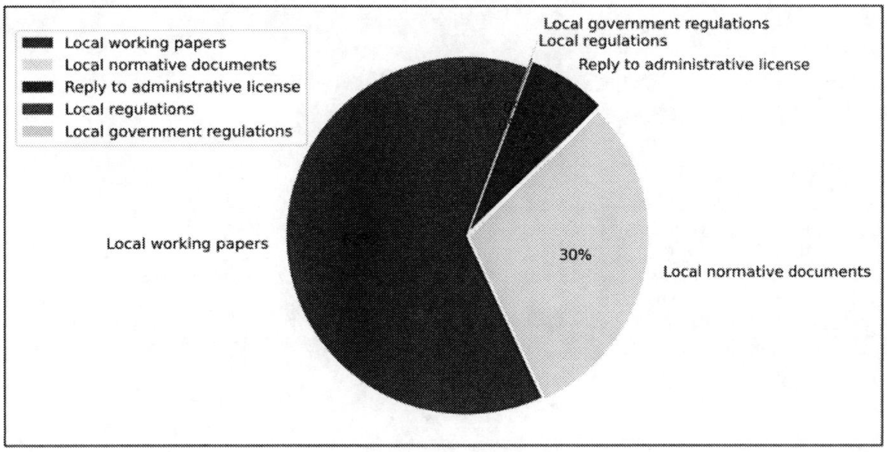

Figure 6. Distribution of the effectiveness level of policies issued by local governments

From Figure 6, we can see that 62% of the policies issued by local governments are local work, 30% are local normative documents, and 7% are administrative approval.

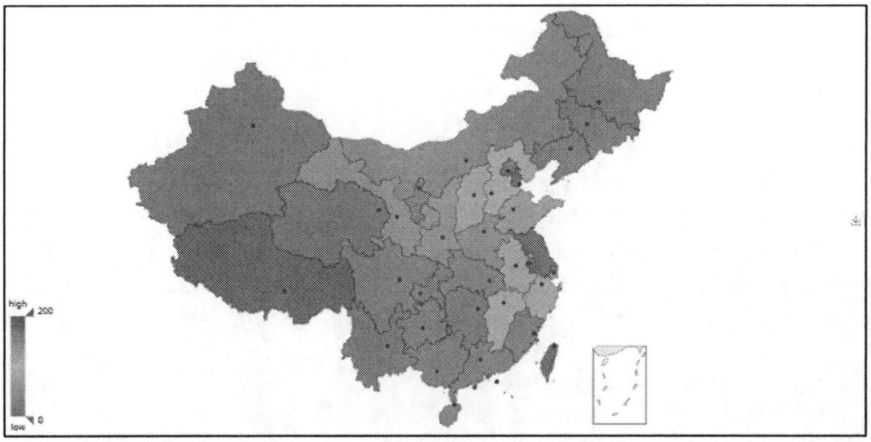

Figure 7. Comparison of the number of policies issued by provinces, municipalities directly under the central government and autonomous regions

According to statistics, the number of policies issued by government departments of provinces, municipalities and autonomous regions is shown in Figure 7, it can be concluded that the most issued policies are ranked in Jiangsu, Fujian, Guangdong, Hebei, Jiangxi and Zhejiang. Most provinces and cities are concentrated in the eastern coastal areas and the provinces with high economic development level.

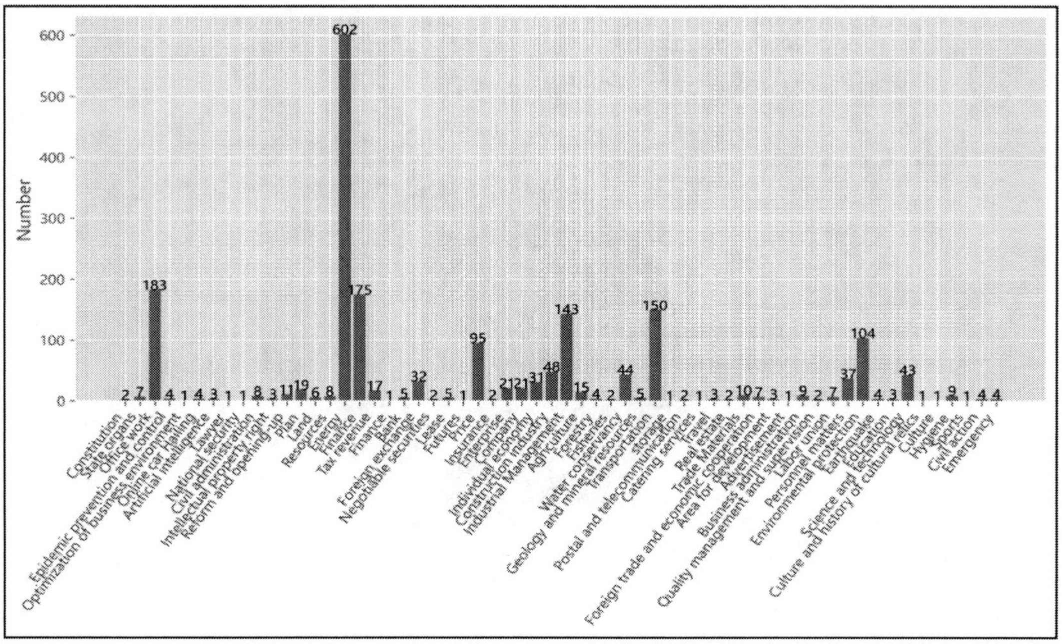

Figure 8. Comparison of policy categories issued by local governments

According to the comparative statistical chart of the data of policy categories issued by local governments in Figure 8, it can be seen that the policy categories issued by local governments are mainly distributed in the fields of energy and fiscal policies.

3.2. Analysis on the development process of new energy industry policy tools

By using content analysis and policy tool theory, this paper studies the development trend of policy tools in the field of new energy issued by China's central and local government departments. Referring to previous studies, this paper divides the types of new energy policy tools into three types, and further subdivides the three policy types into 11 policy tools. After the classification criteria have been established, we have made statistics on the frequency of using 11 specific policy tools corresponding to more than 2000 policy texts published by central and local governments.

3.2.1. The policy tools of the central government

This paper makes a statistical analysis of the central new energy policy tools from three types: voluntary tools, mandatory tools and hybrid tools. By analyzing the statistical data of new energy policy tools issued by central government departments (more than 329 policy texts from 1995 to 2021), mandatory policy tools account for the majority of the total policies, followed by hybrid tools and finally voluntary tools. As shown in Table 2, the statistical table of central new energy policy tools.

Table 2. Statistical data of central new energy policy instruments (1995-2021, more than 329 policy texts)

Types of policy instruments	Frequency of use	Policy tools	Frequency of use

Voluntary tools	11	Publicity and education	5
		Non governmental organizations	3
		Market mechanism	3
		Family and community guidance	0
Mandatory tools	384	Establish laws and regulations	333
		Direct service delivery	2
		Public Enterprises	49
Hybrid tools	221	Financial assistance	19
		Information and persuasion	142
		Auction of property rights	1
		Taxes and usage fees	60

3.2.2. Policy tools of local government

This paper makes a statistical analysis of local new energy policy tools from three types of new energy policy tools: voluntary tools, mandatory tools and hybrid tools. By analyzing the statistical data of new energy policy tools issued by local government departments (more than 1778 policy texts from 1995 to 2021), mandatory policy tools account for the majority of the total number of policies, followed by hybrid tools and finally voluntary tools. As shown in Table 3, the statistical data of new energy policy tools of local government departments are shown in the table.

Table 3. Statistics on new energy policy tools of local government departments (more than 1778 policy texts from 1995 to 2021)

Types of policy instruments	Frequency of use	Policy tools	Frequency of use
Voluntary tools	71	Publicity and education	65
		Non governmental organizations	2
		Market mechanism	4
		Family and community guidance	0

Mandatory tools	1629	Establish laws and regulations	1606
		Direct service delivery	21
		Public Enterprises	2
Hybrid tools	1001	Financial assistance	196
		Information and persuasion	730
		Auction of property rights	1
		Taxes and usage fees	74

3.3. Problems in China's new energy policy tools

3.3.1. Problems of voluntary policy tools

The market tool system has not yet been established, and China has been promoting the establishment of new energy market. However, due to the imperfection of economic policies such as price policy and financing policy, the new energy field has not formed a complete market; China's state-owned new energy enterprises account for a large proportion of China's state-owned enterprises. Compared with private enterprises, state-owned enterprises can obtain more resources. Therefore, the inherent management system of state-owned enterprises can easily lead to the difficulty in making full use of enterprise capital. At the same time, due to the relatively rigid management mode of state-owned enterprises, it is difficult to introduce a competitive mechanism, It's easier to miss out on competitive opportunities in the energy market.

The tool of voluntary organization is absent. China Electric Power Enterprise Federation is a national joint organization of enterprises and institutions in electric power industry approved by the State Council in 1988. It is also a non-profit social and economic organization. Its basic function is to study the current situation of new energy field and put forward corresponding policy suggestions. It does not play a complementary role to other policy tools.

The effect of family and community tools is not significant, and citizens' awareness and action on the use of new energy are still insufficient. In today's sharing economy, the improvement of citizens' awareness is very important to promote the use of new energy in all aspects, and the role of family and community in this process is very important. Only in families and communities do a lot of publicity, To guide citizens to understand and support the development of new energy can effectively improve the awareness of citizens.

3.3.2. Problems of mandatory policy instruments

The *Renewable energy law of the people's Republic of China* is the most perfect regulatory policy tool in the field of new energy. Although it has broken the blank of zero, it is only the general framework for the field of new energy, and its specific implementation needs a series of supporting documents such as detailed rules and technical specifications, The corresponding supporting documents are few

IOP Conf. Series: Materials Science and Engineering 1196 (2021) 012025 doi:10.1088/1757-899X/1196/1/012025

and lack of systematicness. In addition, the *Renewable energy law of the people's Republic of China* has not been revised since the second revision in 2009. The reality is constantly changing and making progress, and regulatory policy tools also need to be constantly adjusted and updated. Therefore, the new energy mandatory tools have not formed a system, and have not been continuously improved and revised.

Regulatory tools and direct service delivery tools are lack of enforcement basis, but also lead to insufficient enforcement. The lack of enforcement basis will lead to some wrong behaviors in the process of enforcement, and the enforcement strength will be greatly reduced.

Public enterprise tools lack competitiveness. In recent years, many large and small enterprises have emerged in the field of new energy, and they occupy a certain position in the international market, and some influential large-scale new energy enterprises have formed. However, there is still a big gap between the operation ability of large enterprises and that of International enterprises with excellent operation ability, The problem of management and operation level needs to be improved. At the same time, the technology of China's new energy state-owned enterprises can not reach the international advanced level, which is difficult to be compared with the United States, the European Union and Japan, which have strong technical strength, restricting the further development of new energy enterprises.

3.3.3. Problems of mixed policy instruments

The use of information tools is not comprehensive, especially in today's society, there is still a waste of resources. Fundamentally speaking, the cognition of protecting resources and environment is not enough, especially in the field of new energy. Improving the education level and strengthening the information policy tools are the problems to be solved. In daily life, there are always people and things who do not understand the current environmental situation, which also means that we do not have enough publicity on environmental protection topics and new energy advantages. We should strengthen the publicity on the use of new energy, so that more citizens can understand new energy, buy new energy vehicles and practice green travel.

The content of tax tools is not perfect, and tax incentives are attractive to many enterprises. A single financial subsidy tool is very unfavorable to promote the establishment of China's new energy market. Only by changing the way and content of financial subsidy tools, establishing a diversified financial subsidy tool, and increasing the intensity and amount of subsidies, can the policy effect be more significant. For the development of enterprises, the government's tax tool is an important tool to guide the progress of new energy industry. The establishment of guiding tax tools by government departments, guiding traditional enterprises to enter the field of new energy and promoting the development and progress of enterprises in the field of new energy are very important to the rapid development of new energy in China.

The role of financing tools is weak. For the development of enterprises, capital is very important, so a variety of financing methods are needed. The degree of financing is an important indicator to measure the healthy development of enterprises. At the same time, the development of new energy enterprises is the same. How to greatly improve the financing channels of enterprises in the field of new energy has become an urgent problem to be solved. Nowadays, the development of new energy is still in the primary stage, and has not formed a large scale. Compared with other advantageous enterprises, the ability of financing in the capital market is relatively low. Therefore, the ability of active financing is very weak, and the financing ability of new energy enterprises can only be improved through the support of national policies. Although China has adopted a series of financing tools, their role is limited and the effect is very little. The progress of enterprises in China's new energy field has not reached the expected level. Therefore, it is urgent to use financing tools to improve the financing ability of enterprises.

China has implemented the fiscal subsidy policy earlier in the renewable energy field, but the subsidy means are single, lack of strength, the amount of subsidy is small, and the source of funds is unstable, which leads to the policy effect is not significant. The most popular subsidies in the world include investor subsidies, producer subsidies and consumer subsidies, while the rare one in China is consumer subsidies, which is a policy tool with great effect for stimulating terminal consumption to expand the market of renewable energy products. A single financial subsidy tool is very unfavorable for promoting the establishment of China's new energy market, Only by changing the way and content of financial subsidy tools, establishing diversified financial subsidy tools, and increasing the intensity and amount of subsidies, can the policy effect be more significant.

4. Discussion

This paper analyzes the evolution of the new energy policy portfolio in the past 27 years by using the data of the new energy policy promulgated by the Chinese government based on time series, which is helpful for the international community to further understand the work of the Chinese government in the design of the new energy policy portfolio and the promulgation of the new energy policy, And support China's unremitting efforts in promoting sustainable energy development.

In the future, China's demand for new energy, the development of energy field and the change of energy structure will affect the choice of policy tools used by the Chinese government. In addition, the preference of government departments in the choice of policy types and the types of new energy will also affect the final use of policy tools. Generally, policies in the field of new energy also follow the self strengthening mechanism of "path dependence", which can usually explain the persistence of the system[24]. The policy tools issued by the government in the field of new energy and the types of policy tools used will continue to be used in the future on the basis of the existing policy tool structure. Based on the theoretical framework of "path dependence" in the next 10 years, China's new energy policy tools predict many factors affecting the evolution trend of new energy policy tools.

The type of mandatory policy will continue to be the main policy tool adopted by the Chinese government for future new energy issues. Mandatory policies can give full play to the power of the government, the cost of investment is relatively low, and achieve better policy objectives. Command and control tools show rapid response speed to reflect government orientation and economic interests. At the same time, it may affect the production profit of the enterprise. From the perspective of central government departments and local governments, the implementation goal of command and control tools is relatively high, and the government is more inclined to choose command and control tools. In practical application, we will further strengthen the supervision and inspection of industry access standards, industry regulations, administrative license approval and other links[25].When renewable energy companies exceed pollution emission standards, penalties for violations of laws and regulations will gradually increase[26].

In addition, financial subsidies and tax instruments will be strengthened. Due to the implementation of a dedicated environmental management fund system and environmental tax, the environmental costs of the new energy industry will be internalized continuously[27].The Chinese government will use more financial subsidy policy tools and tax policy tools to promote the production of enterprises to expand to the field of new energy, such as the application of new technologies and the production and application of new materials[28].

In addition, new energy policy tools should focus on guiding the improvement of pollutant emission trading mechanism of renewable energy in the future. At present, in order to achieve the goal of energy conservation and emission reduction, the Chinese government has established a carbon dioxide emission trading system. With the continuous improvement of carbon dioxide emission trading system and the maturity of pollutant trading market, the pollutant emission trading mechanism of renewable energy will be widely promoted and used in the near future [29][30].

The role of information and persuasion tools in the field of new energy is not as good as other types of tools. Therefore, this type of tool can be used for publicity in the cultivation of awareness of energy conservation and emission reduction, so that more people can understand and understand the benefits of new energy and increase the publicity role of information tools.

Finally, social autonomy policy tools, which will have outstanding performance in guiding and supervising the promotion and application of new energy. Secondly, in order to protect its ecological rights and health interests, the voluntary participation of social citizens will be improved, and residents will participate in the new energy supervision. There will be differences in the choice of new energy policy tools between provinces and cities, which will continue to be maintained. The new energy market in economically developed provinces and cities will be more mature than that in economically underdeveloped provinces.Therefore, in these provinces, more and more people adopt economic incentives and social autonomy policy tools that can provide more independent choices for the new energy industry to develop and adopt new technologies.

5. Conclusions

In the next 10 years, mandatory tools will become the preferred tool type for central government departments and local government departments to choose to use policy tools. The proportion of financial subsidy policy tools and tax policy tools will also gradually increase. In addition, with the adjustment of energy structure and the increase of energy conservation and emission reduction tasks, mandatory and direct tools are more effective than other types of policy tools to deal with environmental problems. However, the frequent use of this policy tool is bound to restrict the enthusiasm and independent innovation ability of enterprises.Therefore, the frequency of financial subsidies, taxes and property rights trading tools will continue to increase, which will further stimulate the vitality of the new energy industry and accelerate the high-quality and rapid development of the new energy industry. So far, voluntary and social monitoring tools are rarely used. However, as citizens' voluntary participation increases, the frequency of adoption of these policy tools will also be greatly increased. The implementation effect of information persuasion tools in the field of new energy is not obvious, however, this type of policy tool can play an important role in publicizing the promotion and use of new energy. According to the different tasks of energy conservation and emission reduction, the eastern coastal provinces tend to implement more applicable and flexible policy tools of economic incentives, while the central and western provinces tend to use command and control tools to achieve sustainable development and achieve the goals of carbon peak and carbon neutralization in advance.

Acknowledgements

We would like to thank Ming Xu, Pezhman Ghadimi and Felix Carl Schultz for helpful discussions and comments.This work was supported by the National Natural Science Foundation of China(Grant No. 72033005).

Reference:

[1] Chen-ke, Cheng, Hua, Liao, Zhong-ju. Towards Sustainable Growth in the Textile Industry: A Case Study of Environmental Policy in China. Polish Journal of Environmental Studies. 2018;27(5):2325-36.

[2] Li L, Taeihagh A. An in-depth analysis of the evolution of the policy mix for the sustainable energy transition in China from 1981 to 2020. Applied Energy. 2020;263.

[3] Limin C, MingLiang X. The impact of industry cluster and environmental policies on residents' health risk evaluation using big data. Work. 2021(1):1-10.

[4] Tilt B. China's air pollution crisis: Science and policy perspectives. Environmental Science & Policy. 2019;92:275-80.

[5] Qiuyun D. Discussion on Strategic Transformation Situation and Motivation for Enterprises in a Low-carbon Economy. Science and Technology Management Research. 2013;33(24):113-7.

[6] Boqiang L, Yufang C. Impacts of policies on innovation in wind power technologies in China. Applied Energy. 2019;247(AUG.1):682-91.

[7] Polzin F, Egli F, Steffen B, Schmidt TS. How do policies mobilize private finance for renewable energy?—A systematic review with an investor perspective. Applied Energy. 2019;236(FEB.15):1249-68.

[8] Seto KC, Davis SJ, Mitchell R, Stokes EC, Unruh G, Rge-Vorsatz D. Carbon Lock-In: Types, Causes, and Policy Implications. Social Science Electronic Publishing. 2016;41:425-52.

[9] Couto TB, Olden JD. Global proliferation of small hydropower plants – science and policy. Frontiers in Ecology & the Environment. 2018;16(12):91-100.

[10] Kuriqi A, Pinheiro AN, Sordo-Ward A, Garrote L. Flow regime aspects in determining environmental flows and maximising energy production at run-of-river hydropower plants. Applied Energy. 2019;256:113980-.

[11] Bo Z, Xudong S, Ying L, Suping P. Development Trends and Strategic Countermeasures of China's Emerging Energy Technology Industry Toward 2035. Chinese Journal of Engineering Science. 2020;22(2):38.

[12] Jian W, Xu H, Ke H, Zhouquan C, Xin L. An exploration on corporate-community relationship in mining sector in China – Lessons from Yunnan Phosphate Chemical Group Co., Ltd. Resources Policy. 2017;52:54-64.

[13] Xiaohong L, Tong Z, Ching-Ter C, James FC. China's Renewable Energy Strategy and Industrial Adjustment Policy. Renewable Energy. 2021;170(6).

[14] Zhijun Y, Xu G, Ruoxue W. Tool Preference and Path Optimization of Environmental Governance Policies——Based on the Content Analysis of 43 Policy Texts. Journal of Northeastern University. 2017;19(03):276-83.

[15] Shu W. Transformation of Chinese Government Functions from the Approach of Policy Typology——An Empirical Study on Higher Education Policy. Chinese Public Administration. 2016(07):115-21.

[16] Zhongju L. The evolution of wind energy policies in China (1995–2014): An analysis based on policy instruments. Renewable & Sustainable Energy Reviews. 2016;56:464-72.

[17] Huatao P, Yang L. A comprehensive analysis of cleaner production policies in China. Journal of Cleaner Production. 2016;135(nov.1):1138-49.

[18] Yang L, Feiyan L, Jianfang Z, Jianbo G. Insights into the nature of food safety issues in Beijing through content analysis of an Internet database of food safety incidents in China. Food Control. 2015;51:206-11.

[19] Tixier JP, Hallowell MR, Rajagopalan B, owman DB. Automated content analysis for construction safety: A natural language processing system to extract precursors and outcomes from unstructured injury reports. Automation in Construction. 2016;62(Feb.):45-56.

[20] Huiming Z, Qunwei W, R. MS. Waste cooking oil as an energy resource: Review of Chinese policies. Renewable & Sustainable Energy Reviews. 2012;16(7):5225-31.

[21] Inhaber H. Shortfall in on-road fuel economy: Implications for public policy. Energy Policy. 1982;10(4):356-9.

[22] Chai S, Zhicong Z, Jianping G. Evolution of environmental policy for China's rare earths: Comparing central and local government policies. Resources Policy. 2020;68.

[23] Howlett M, Ramesh M. Studying Public Policy: Policy Cycles and Policy Subsystems. American Political ence Association. 2009;91(2):548-80.

[24] Moncada JA, Lukszo Z, Junginger M, Faaij A, Weijnen M. A conceptual framework for the analysis of the effect of institutions on biofuel supply chains. Applied Energy. 2017;185:895-915.

[25] Aiping H, Jianping G, Yalin L. Vertical vs. horizontal integration: Game analysis for the rare earth industrial integration in China. Resources Policy. 2016;50:149-59.

[26] Powell-Turner J, Antill PD. Will future resource demand cause significant and unpredictable dislocations for the UK Ministry of Defence? Resources Policy. 2015;45:217-26.

[27] Jianping G, Yalin L. Resource tax on rare earths in China: Policy evolution and market responses. Resources Policy. 2018;59(C):291-7.

[28] Zhongju L. Environmental policy instruments, environmental innovation and the reputation of enterprises. Journal of Cleaner Production. 2018;171(pt.2):1111-7.

[29] Qingqing W, Xu H. A review of China's carbon trading market. Renewable and Sustainable Energy Reviews. 2018;91(aug.):613-9.

[30] Zhonghua W, Yao H. Research on China's Carbon Emissions Trading Market Price Distort Based on Shadow Price Model. Ecological Economy. 2019;35(05):13-20.

The changes and influencing factors of urban LUCC and EF footprint in energy resource areas: An empirical evidence from China using HLM model

Miao He[1,2,3], Wei Xiao[1*], Hang Zhang[1], Chao Liu[1], Donghui Xie[1], Mouyuan Fan[1], Lei Shen[3,4]

[1] School of Economics, Hebei University, Baoding 071002, PR China;

[2] Research Center of Resource Utilization and Environmental Protection of Hebei University, Baoding 071002,PR China;

[3] Institute of Geographic Sciences and Natural Resources Research, Chinese Academy of Sciences, Beijing 100101,PR China;

[4] Key Laboratory of Resource and Environmental Carrying Capacity Assessment, Ministry of Natural Resources, Beijing 100101,PR China)

*** E-mail: x1324210139@163.com**

Abstract. Analyzing urban land use and cover change and its influence factors are of great significance for understanding the situation of regional development and environmental constraints and for evaluating the ability of sustainable development. This paper calculates the three-dimensional ecological footprint and total factor productivity of research cities, these cities have significant individual differences, Yinchuan, Shizuishan, Yan 'an, Yulin, Ganzhou, Shaoguan etc. are energy resource areas and comparing to those Guangzhou, Dongguan, Wuzhong, Qingyuan etc. are non-energy resource leading cities. The HLM model is used to decompose the influencing factors of three-dimensional ecological footprint. The study shows the important role of total factor productivity in regional and urban sustainable development. Also, the study shows the inter-group differences of three-dimensional ecological footprint produced by cities with different population sizes. And the inter-group difference of urban three-dimensional ecological footprint under the influence of total factor productivity. Result shows that land, population and technology factors have different impacts on three-dimensional ecological footprint. Also, relying mainly on mining areas of energy resources and other industries dominated areas of mining land proportion and industrial structure differences did not lead to the influence of different ecological footprint, energy resource area in affecting factors, the ecological footprint is still a total factor plays the leading role of ecological rate, population and real GDP, There is no obvious negative externality effect of mining activities on the value of regional ecological products, and the regional ecological carrying capacity keeps improving.

Keywords: land use and cover change, three-dimensional ecological footprint, energy resource area, total factor productivity, influencing factors.

1. Introduction

Energy resources area which is important in the development of the national geographic units and this paper focuses on the influence factors of land use and cover change and human ecological occupation and its impact on human sustainable development in cities characterized by energy and resource industries. With the change of the way of energy utilization, and the change of the regional leading industry, a number of these cities have taken place in transition, there are still some cites belong to the type of energy and mineral resources production and processing core area. The traditional belief is that energy resources industries has certain environmental negative externalities, however, because of the green mine construction, More and more energy and mineral enterprises are taking measures to control the adverse impact on resources and environment. These measures may have a profound impact on the economy, society and environment of the cities they belong to. From an economical point of view, the output value and efficiency characteristics of energy resources market demand and regional green economic development will be taken into account in the industry production, from the social level, regional population change plays an important role in the development of cities and regions. From the perspective of environment, land use and cover change plays an important role in ensuring the sustainable development of cities and regions.

Therefore, this paper mainly did the following job: measured the biological account of three-dimensional ecological footprint changes of the ten cities; Measured the total factor productivity (TFP) of the ten cities; analyzed the influencing factors of 3d ecological footprint and its relationship with regional TFP and industrial industries. Since there are few studies analysis of the perspective that urban development and the correlation effect of natural resources for regional industry characteristics areas. So the contribution of this paper mainly lies in the fact that first we studied the features and characteristics of the land use cover change and industrial structure change in energy resources areas. Second, combined with the city's population and the agricultural production to calculate the three-dimensional ecological footprint, calculated the total factor productivity of the city. Thirdly, this paper analyzed the influencing factors of land use and cover change from the perspectives of population, land, Gross regional product and economic development efficiency, and then obtained the interactive relationship among regional economic growth, industry and human natural imprint, and the influence of regional ecological product value and regional development change on regional ecological carrying capacity.

2. Literature review and research hypothesis

Energy and resource areas play an important role in the national development, their industrial structure, human-land relationship and resource utilization are all paid highly attention. Compared with the general area, there are obvious differences in their development modes. Regional natural resource structure is one of the important factors affecting industrial structure [24]. Ecosystems are related to people's well-being [3]. Land elements are the most direct influencing factors of human activities. Land is the influence of human activities on the natural earth's surface, the most direct signal from a global scale, some study are focus on global land use/land cover change and the impact of ecological system, including influence on runoff, the impact on carbon emissions, the impact on regional climate, etc. [1, 2]. It is generally believed that land use/land cover change is one of the key factors affecting land use/land cover change. It is the pioneer of sustainable development and the core of all related issues. Productive land is a good proxy for natural capital and many of the resource flows and essential life support services that this capital provides [8]. So there is also a close relationship between LUCC and nature resource capital.

As for the relationship of land use cover and regional economy, some studies have focused on the impact of agricultural activities in the Amazon region of South America such as Barbieri Alisson Flávio [5], some studies focused on African areas Qualitative analysis of economic and social influencing factors of land use and cover change such as Emmanuel Daata kullo [6]. For the Local studies of China did by Chinese scholars are mainly focus on the temporal and spatial characteristics of land use and cover change, land use intensity and land use type transfer, or the relationship between other driving natural factors, such as Shaokun Liu [4], Xiaowei Shi et al. [12], Yang Yang et al. [22], Yali Liet al. [23]. These researches is less concerned about the land use and land cover (LULC) and its change will be affected by what regional factors or economic development factors. As a matter of fact, human activities are essentially a kind of social activity and have a strong relationship with economy. What kind of policy should be adopted to achieve the goals of sustainable development? It should not only consider the effect of ecological service system. Moreover, the influence of economic and social factors should be considered in a systematical way. For energy and resource concentrated areas, this will involve resource utilization policy, industrial regional development policy, ecological and environmental protection policy, urban sprawl and other issues. Since REES E W [9] proposed the ecological footprint method to measure the natural imprint of urban economy, the ecological footprint method has been widely used. The ecological footprint method and the improved three-dimensional ecological footprint method are useful for understanding the carrying capacity of a certain resource and to understanding the regional natural resource capitals [10]

When technological progress has become the mainstream of social development, researches on technological progress and its far-reaching impact is of great significance for people to better understand the influencing factors of regional development. Related literatures such as Miao, C.L. [16] argue that technologies from different sources will have different impacts on economic activities. For example, Bengoa and Roman [17], Wang Shengyun et al. [18] measured the welfare efficiency and ecological efficiency of economic growth. From the methodology Farrell [12] applied data envelopment analysis (DEA) to the study of input factors and their interaction, and proposed the data envelopment analysis model to handle technological efficiency. Charnes et al. [13], and Simar & Wilson [14, 15] all make efforts to improve the DEA analysis method and appy it to energy efficiency, economic development efficiency, industrial development efficiency, resource input-output efficiency and so on. In this study, the DEA method and the stochastic Frontier Function Analysis method (SFA) are used to measure the total factor productivity of 10 cities from 2009 to 2019 in order to find the technological progress impact.

To summarize, this study took the leading industry of energy resources city in China as samples, using land inputs variable as the breakthrough point, applying of three-dimensional ecological footprint method and the rate of total factor of ecological measure method to characterization the status of land, total factor productivity and related resources of the research city. And using three dimensions of the ecological footprint of biological account refers to land use cover change in the interaction relationship between human activities.. The systematic analysis is used to analyse the influencing factors is of reference significance for understanding and evaluating the influencing factors of regional sustainable development from the micro scale and have reasonable reference meaning to formulating rational energy, industry and regional development policies.

According to the above research purposes and literature review, based on theory of environmental Kuznets curve, which believes that the impact of industrialization on the environment has a trend of t rising and decreasing trend at different developing stage, this paper proposed the following three hypothesis:

Hypothesis 1: Ecological footprint is a significant confirmation of human occupation of resources. The biological account in ecological footprint measures the mark of human activities in non-built-up areas, and the correlation between the scope of non-built-up areas and human activities can directly reflect the ecological environment nature resources capital carrying capacity of the whole regions.

Hypothesis 2: Different population sizes and different lands input factors in different regions will have different impacts on the ecological footprint occupation and ecological product value of these cities.

Hypothesis 3: The improvement of regional total factor productivity will reduce human's dependence on natural resources and reduce the occupation of regional and urban ecological footprint.

3. Empirical process of influencing factors of land use and cover change and results

3.1 Study design

Based on research hypothesis 1, Since the biological account of three-dimensional ecological footprint is closely related to land use and cover change, that it can be regarded as one of the measurement standards of land use and cover change. In addition, three-dimensional ecological footprint can better reflect the carrying capacity of regional resources and environment from multiple dimensions. At the same time, three-dimensional ecological footprints are closely related to land classes variation, and the three-dimensional ecological footprints of different accounts can show the changes of each land class variation .Therefore, this study designed and calculated the biological account of the three-dimensional ecological footprint of the study city, and the calculation process referring to the design ideas of Niccolucci V et al., and some scholars' previous studies and revised it for actual.Based on research hypothesis 2,3, the study takes the local permanent population, capital stock, energy consumption and lands input as input variables, and local real GDP as output variable. DEA method is used to estimate the change of total factor productivity of these study cities. The study shows that there are obvious urban differences in total factor productivity among the studied cities, which is closely related to urban industrial structure, population scale and urban construction situation. These difference will be used to further explain the mechanism of related to local land use and cover changes.

Table 1 shows the descriptive statistical analysis of HLM regression. Panel data of Yinchuan and other cities arranged from year 2009 to 2019 are used for analysis. Variables include three dimensional ecological footprint biological account, local real GDP, local nominal GDP, lands areas and scale efficiency. Cities are divided into four categories based on population size.

Table 1. Eigenvalues of HLM regression variables

Variable	Obs	Mean	Std. Dev.	Min	Max
ef3d	110	6.150727	1.01919	4.127868	7.646883
ip	110	2.7	0.7845995	1	4
realgdp	110	7.198384	1.157705	5.225155	9.859392
land	110	9.448637	0.9855921	7.783724	10.6613
lncrs	110	-0.0628283	0.0473741	-0.17304	0
gdp	110	7.345362	1.132486	5.225155	10.07021

3.2 Null model results

According to the size of urban population, these 10 cities are divided into four groups: 1.Guangzhou; 2. Dongguan, Ganzhou; 3. Yinchuan, Wuzhong, Yulin, Shaoguan, Qingyuan, Yan 'an;4.Shizuishan. Ip Group1 for Guangzhou city u1 is 0.0502136 and u0 is 0.0702613 ;Ip Group2 for Dongguan city and Ganzhou city u1 is0.04064 and u0 is 0.6378765; Ip Group3 for Yulin city, Shao guan city ,Qingyuan city and Yanan city u1 is 0.0502136 and u0 is 0.4851803; Ip Group4 forShizuishan city u1 is 0.0145514 and u0 is 0.2229575,where u1 represents the slope and u0 represents the intercept, which indicate the rationality of grouping.

And there are four model underlying HLM, respectively are model 1(null model), model 2(random intercept and fixed slope), model 3(Random intercept and random slope) and model 4 (full model). we compare the results of the full models.

3.3 Regression results analysis

For the estimation of the model Akaike information criterion, the reduction of Akaike information value can improve the model was better fitted. The AIC criterion in this study respectively is 293.2603, -155.8727, -164.1787 and -169.1141 under the null model, random intercept model, random intercept slope model and full model. The results show that the multilayer linear model has a better fitness than the general linear regression model and is more suitable for this study. The estimation results and fitting test results of the random intercept model, the random intercept slope model and the full model are reported respectively.

The full model fitting results are as follows:

$$ef3d_{ij} = \beta_0 + \beta_1 land_{ij} + \beta_2 real\lg dp_{ij} + e_{ij} \tag{1}$$

$$\beta_0 = \gamma_{00} + \gamma_{01} \ln crs_{ij} + \mu_{0j} \tag{2}$$

$$\beta_1 = \gamma_{10} + \gamma_{11} \ln crs_{ij} + \mu_{1j} \tag{3}$$

$$ef3d_{ij} = \gamma_{00} + \gamma_{01} \ln crs_{ij} + \mu_{0j} + \gamma_{10} land_{ij} + \gamma_{11} \ln crs_{ij} land_{ij} + \mu_{1j} land_{ij} + \beta_2 real\lg dp_{ij} + e_{ij} \tag{4}$$

IOP Conf. Series: Materials Science and Engineering 1196 (2021) 012026 doi:10.1088/1757-899X/1196/1/012026

Full model indicates three-dimensional urban ecological footprint of research city is affected by the land inputs and real GDP. There are different influences due to different population groups bring the intercept and slope change. In this model lands input is the transmission elements which bring between-group variance among different groups. Land input in the grouping variable is affected by the city scale efficiency of total factor productivity. Effects of land elements and total factor productivity on the biological accounts of three-dimensional ecological footprints in urban groups of different population sizes estimated by random intercept model, the random intercept slope model and the complete model are reported respectively.

3.3.1 Random intercept and fixed slope model and Random intercept and random slope model

The estimations of the random intercept model and the random intercept slope model are model 3 and mode 4 respectively. Table 2 shows the regression results of the random intercept model and the random intercept slope model. The regression results shows that the coefficients of land and real GDP are both positive. The coefficient of land random intercept model is 1.068521, and the coefficient of random intercept slope model is 1.066046. There is no obvious difference between the two slope. The coefficient of the random intercept model of real GDP is 0.0680272, and the coefficient of the random intercept slope model is 0.122755. It shows that the effect of GDP is enhanced when the slope of the land varied following with the size of the population in different cities. The random intercept model and the random intercept slope model show that the increase of real GDP and the increase of land input will lead to the increase of the biological account of three-dimensional ecological footprint, and thus indicating the deepening of human's occupation of natural resources.

$$ef3d_{ij} = \beta_0 + \beta_1 land_{ij} + \beta_2 rea\lg dp_{ij} + \mu_{0j} + e_{ij} \quad \text{(model 3)}$$

$$ef3d_{ij} = \beta_0 + \beta_1 land_{ij} + \beta_2 rea\lg dp_{ij} + \mu_{0j} + \mu_{1j} land_{ij} + e_{ij} \quad \text{(model 4)}$$

Table 2. Regression results (1).

Random intercept and fixed slope model		**Random intercept and random slope model**	
ef3d	Estimate	ef3d	Estimate
land	1.068521***	land	1.066046***
	(-0.0129211)		(-0.0389942)
realgdp	0.0680272***	realgdp	0.122755***
	(-0.0169601)		(-0.018311)
_cons	-4.395902***	_cons	-4.76729***
	(-0.2132267)		(-0.4137256)
ip:		ip:	
Var(_cons)	0.0414733	Var(land)	0.0033913
	(-0.0307632)		(-0.0031477)
Var(Residual)	0.0110621	Var(_cons)	0.4064371
	(-0.0015214)		(-0.3458117)
AIC	-155.8727	Var(Residual)	0.0086817
			(-0.0012179)
		AIC	-164.1787

(Standard errors in parentheses:* p<0.05,** p<0.01,*** p<0.001)

3.3.2 Regression results of the full model

The full model regression results in Table 3-3left column show that the increase of total factor productivity will reduce the occupation of natural resources by human activities, while the increase of real GDP and land input will increase the occupation of natural resources by human activities. The interaction term indicates that the improvement of technical efficiency will strengthen the positive effect of land input on EF3D, and has a significant moderating effect. At the same time, by comparing the random intercept fixed slope model, intercept and random slope model and the full model, we can also find that, in a situation of different city group, because of the inter-group difference of the land input factors influence degree , the influence of real GDP are enhanced. That is to say, on a premise of considering the groups of different population sizes, due to the inter-group difference in the influence degree of land input factors, the impact of real GDP on three-dimensional ecological footprint has a significant strengthening trend. This indicates that the mediating effect of land input and the moderating effect of total factor productivity have an intensification difference on the positive impact intensity of real GDP towards three-dimensional ecological footprint. Robustness test results show in Table 3 right column are all significant at the 1% level, indicating the robustness of the relationship among variables.

Table 3. Regression results (2).

ef3d(1)	**Coef.(1)**	**ef3d(2)**	**Coef.(2)**
lncrs	-12.01274**	lncrs	-12.01274***
	(-5.249063)		(-4.632525)
realgdp	0.1018642***	realgdp	0.1018642***
	(-0.0253987)		(-0.0392958)
land	1.115881***	land	1.115881***
	(-0.0358677)		(-0.0470166)
c.lncrs#c.land	1.291742**	c.lncrs#c.land	1.291742***
	(-0.5327014)		(-0.4546412)
_cons	-5.056307***	_cons	-5.056307***
	(-0.4387745)		(-0.3842426)
ip		ip	
Var(land)	0.0017711	Var(land)	0.0017711
	(-0.0019607)		(-0.0010279)
Var(_cons)	0.2244214	Var(_cons)	0.2244214
	-0.2205939		(-0.1728338)
Var(Residual)	0.008242	Var(Residual)	0.008242
	-0.0011775		(-0.0032408)
AIC	-169.1141		

(Standard errors in parentheses:* p<0.05,** p<0.01,*** p<0.001)

3.3.3 Robustness test of model

Bootstrap method and substitution variable method were used to test the robustness of the full model. The Z value of Bootstrap method was 14.64, and P>Z was significant at the level of 0.05%. Bootstrap Std.err is 0.0095521 indicates that model relationship is stable under the sample. In addition, for the final form estimate of the model, real GDP is replaced by nominal GDP, and the results are basically the same as the original results (Table 4), indicating the robustness of the model.

Table 4. Test results of substitution variable method.

ef3d	
gdp	0.09204*
	(-0.0542767)
lncrs	-12.27842**
	(-5.364671)
c.lncrs#c.land	1.327302***
	(-0.5179832)
_cons	-4.984711***
	(-0.4885845)
ip:	
Var(land)	0.0012055
	(-0.0008573)
Var(_cons)	0.1635858
	(-0.1460749)
Var(Residual)	0.0079013
	(-0.0026253)

(Standard errors in parentheses:* $p<0.05$,** $p<0.01$,*** $p<0.001$)

4. Conclusions and policy recommendations

Based on empirical evidence and hypothesis of environmental kuznets, this paper applied the improved three-dimensional ecological footprint method and multiple linear regression analysis for the empirical analysis.This paper further improved prior research of Dongguan city's land use cover and trend of the ecological footprint and total factor productivity evaluation, and further explores cities of different urban population scales, The hierarchical difference of influencing factors of its ecological footprint. This paper measured the three-dimensional ecological footprint and total factor productivity, and constructed a multi-layer linear model to verify the three hypotheses previously proposed. The study shows that total factor productivity plays an important role in regional and urban sustainable development. From the evidences of 10 cities of China, it can be seen that when GDP growth to a certain stage, China has entered a new stage of sustainable development, green development concept of the ecological environment improvement and transformation and upgrading of traditional industries, bring a change to ecological footprint, which means compare with before it has a positive impact, the original population and land use model bring a negative effect for ecological footprint and for regional development. However, with the changes brought by industry and new development concept, some interactive transmission effect has been produced to reverse the original trend.

This study quantitatively analyzes the different characteristics and influencing factors of ecological footprint and ecological product value of cities characterized by land, natural resources and their derived industries, and their different effects on urban and regional development. The study also shows that the multi-layer linear regression model has obvious advantages in solving the transmission mechanism of resources, economy and environment. In this study, multiple linear model, a general linear regression model fitting effect is much better. It can describe and reflect the different population size of the three-dimensional urban ecological footprint of the differences between groups, and when the slope of land inputs varies, the conductive follow-up effect and the impact of total factor productivity in three-dimensional urban ecological footprint have obivous differences between groups. It shows that land, population and technology factors have different impacts on three-dimensional ecological footprint. This plays an important role in formulating and evaluating the effect of policy implementation and formulating differentiated and targeted development plans.

The depth and size of three-dimensional ecological footprint is an objective standard to reflect whether ecological product value can meet the demand of future and current ecological carrying capacity and human life's need for the environment. It has a same mean that whether local ecological account can satisfy people pursuit for the value of ecological products. In the study,most cities have a positive development trend of ecological footprint, And these cities are mostly belong to energy resources city, mining land proportion is higher than the national average. It also reflects the Yinchuan, Yan'an, Shizuishan, Ganzhou, Qingyuan, Yulin such kind of resource-based industries as the leading industry of local has not so much inverse refect to local ecological footprint, Local resources and environment protection appears positive momentum, The economic and ecological achievements of green mine construction can be evaluated through this aspect. These positive changes are of great significance for us to continue to implement the concepts of sustainable development and green development, and to properly handle the relationship between population, resources, environment and economic development.

The implications of this study suggest that the increase of TFP plays an important role in the value of ecological products and the improvement of ecological footprint, and China should continue to improve TFP at both the urban and industrial levels, especially to evaluate the benefits of technology and economies of scale. At the same time, the research also shows that the urban population size has an impact on the three-dimensional ecological footprint, and a reasonable urban population size is of great significance to achieve sustainable development.

Acknowledgments

This paper was supported by Open Project of Key Laboratory of Carrying Capacity Assessment for Resource and Environment, Ministry of Natural Resources (No: CCA2019.20), Surface Project of National Natural Science Foundation of China (No.41771566), High-level Talent Introduction Project of Hebei University (No. 521000981365), Industrial distribution and Regional Economic Development Innovation Team Project of Hebei University.

5. References

[1] Gaur Abhishek,Lacasse Michael,Armstrong Marianne,Lu Henry,Shu Chang,Fields Allan,Palou Francisco Salamanca,Zhang Yujia. *Effects of using different urban parametrization schemes and land-cover datasets on the accuracy of WRF model over the City of Ottawa*[J]. Urban Climate,2021,35.

[2] Zhang Fan,Xu Ning,Wang Chao,Wu Feng,Chu Xi. *Effects of land use and land cover change on carbon sequestration and adaptive management in Shanghai, China*[J]. Physics and Chemistry of the Earth, Parts A/B/C,2020(prepublish).

[3] Walter V Reid,Harold A Mooney. *The Millennium Ecosystem Assessment: testing the limits of interdisciplinary and multi-scale science*[J]. Current Opinion in Environmental Sustainability,2016,19:

[4] Hou Jun,Qin Tianlin,Liu Shanshan,Wang Jianwei,Dong Biqiong,Yan Sheng,Nie Hanjiang. *Analysis and Prediction of Ecosystem Service Values Based on Land Use/Cover Change in the Yiluo River Basin*[J]. Sustainability,2021,13(11):

[5]Barbieri Alisson Flávio,Guedes Gilvan Ramalho,Onofre dos Santos Reinaldo. *Land use systems and livelihoods in demographically heterogeneous frontier stages in the amazon*[J]. Environmental Development,2020(prepublish):

[6] Emmanuel Daata kullo, Ernest Biney, Emmanuel Harris, *Eric Kwabena Forkuo,The impact of land use and land cover changes on socioeconomic factors and livelihood in the Atwima Nwabiagya district of the Ashanti region, Ghana.,Environmental Challenges*,2021,100226,ISSN 2667-0100

[7] Hua Wenjian, Chen Haishan, Li Xing. *China climate effect of land use/cover change and its research review* [J]. Journal of earth science, 2014, 29 (9) : 1025-1036.

[8] Mathis Wackernagel,Larry Onisto,Patricia Bello,Alejandro Callejas Linares,Ina Susana López Falfán,Jesus Méndez García,Ana Isabel Suárez Guerrero,Ma Guadalupe Suárez Guerrero. *National natural capital accounting with the ecological footprint concept*[J]. Ecological Economics,1999,29(3).

[9] REES E W. *Ecological footprints and appropriated carrying capacity: what urban economics leaves out* [J] . Environment and Urbanization, 1992, 4(2): 121 - 130.

[10] Fang Xiaojuan, TAO Qiangqiang, *Ma Zhaodi. Sustainable utilization of natural capital in the Yangtze River Economic Belt—Based on three-dimensional ecological footprint model* [J]. Environmental Ecology, 2013(05):32-37.
[11] Li Shao-kun, WANG Shao-jun, ZHANG Zhi, Chen Xiao-qian. *Evolution and driving factors of sustainable use of natural capital in Jiangxi Province from the perspective of three-dimensional ecological footprint* [J/OL]. Agricultural Resources and Regional Planning in China :1-14[2021-08-01]
[12] Farrell M J. *The measurement of productive efficiency* [J] . Journal of the Royal Statistical Society, 1957, 120(3) : 253-290.
[13] Charnes A, Cooper W W, R hodes E. *A Data Envelopment Analysis Approach to Evaluation of the Program Follow through Experiment in US Public School Education* [R]. Carnegie — mellon Univ Pittsburgh Pa Management Sciences Research Group, 1978.
[14] Simar L, Wilson P W. *Sensitivity analysis of efficiency scores: How to bootstrap in nonparametric frontier models* [J] . Management science, 1998, 44(1) : 49 — 61.
[15] Simar L, Wilson P W. *Statistical inference in nonparametric frontier models: The state of the art* [J] . Journal of productivity analysis, 2000, 13(1) : 49—78.
[16] Miao, C.L., 2017. *Natural resources utilization efficiency under the influence of green technological innovation*. Resour. Conserv. Recy. 126, 153–161
[17] Bengoa, M., Román, M., 2017. *Do R & D activities matter for productivity? A regional spatial approach assessing the role of human and social capital.Econ. Model*. 60, 448–461.
[18] Wang Shengyun, HAN Yajie, Ren Huimin, Li Jing. *China's provincial ecological welfare performance evaluation and its driving effect decomposition* [J]. Resources science,2020,42(05):840-855.
[19] V. Niccolucci,S. Bastianoni,E.B.P. Tiezzi,M. Wackernagel,N. Marchettini. *How deep is the footprint? A 3D representation*[J]. Ecological Modelling,2009,220(20):
[20] Li Shao-kun, WANG Shao-jun, ZHANG Zhi, Chen Xiao-qian. *Evolution and driving factors of sustainable use of natural capital in Jiangxi Province from the perspective of three-dimensional ecological footprint* [J/OL]. Chinese agricultural resources and division: 1-14 [2021-08-01]. http://kns.cnki.net/kcms/detail/11.3513.s.20210607.1759.044.html.
[21] Shi Xiao-wei, FENG Guang-jing, Su Pei-tian, HE Gai-li, ZOU Yi-jiang, Wang Xiao-feng. *Metropolis suburb land use temporal and spatial evolution characteristics and habitat quality evaluation* [J]. Journal of agricultural engineering, 2021, 5 (4) : 275-284.
[22] Yang Yang, Tang Xiaolan, LI Zhehui, Jia Yanyan. *Spatial-temporal evolution and driving factors of land use landscape pattern in the Yangtze River Basin: A case study of 2008-2018* [J]. Journal of Northwest Forestry College, 201,36(02):220-230.
[23] Li Yali, Yang Fenli, Yang Lian 'an, SHANG Xiaoqing, HU Guigui, Jia Lujing. *Spatial pattern change of land use and its influencing factors in Yulin city in recent 40 years* [J]. Arid land geography,2021,44(04):1011-1021.
[24] Fan J. *The macro role of natural resources structure in the evolution of regional industrial structure: a case study in bohai sea area, for example* [J]. Journal of Natural Resources,1992(02):97-105.

Heavy metal and fertility in a Tropical Oxisol amended with sewage sludge under *Eucalyptus* plantation

[1]Paulo H.S. Cardoso*, [1]Thays S. Mandu, [2]Antonio L. Florentino, [1]Rosana B. Oliveira, [2]Luís R.F. Alleoni, [3]Clayton A. Alvares, [4]Thiago A.R. Nogueira, [5]Arun D. Jani, [6]Gian F. Capra, [1]Cassio H. Abreu-Junior

[1]Center of Nuclear Energy in Agriculture, Universidade de São Paulo, Brazil
[2]College of Agriculture "Luiz de Queiroz", Universidade de São Paulo, Brazil
[3]Suzano SA Company; Department of Forest Science, São Paulo State University, Brazil
[4]Department of Plant Protection, Rural Engineering and Soils, São Paulo State University, Brazil
[5]USDA-NRCS Ecological Sciences Division Portland, OR USA
[6]Università degli Studi di Sassari, Italy
*Correspondent author: paulohscardoso@usp.br

Abstract. The objectives of our study were to evaluate i) the soil fertility and fractionation of Ba, Cd, Cu, and Zn in the topsoil layer (0-0.20 m depth); and ii) production and concentration and accumulation of Ba, Cd, Cu, and Zn in the components of *Eucalyptus* trees at 36 months after sewage sludge (SS) application, with or without mineral P fertilizer, compared to mineral fertilization. Application of SS (at N criteria) with P increased soil organic matter and heavy metal concentrations, which were mostly bound to the oxidic and organic matter fractions. SS provided *Eucalyptus* production and heavy metal concentrations and accumulation in the trunk, branches, and leaves similar to mineral fertilization for high wood production. The application of SS (at N criteria) supplied with P increased soil heavy metal, fertility, and *Eucalyptus* production, without risk of environmental contamination.

Keywords: Biosolid, Forest management, Organic fertilizer, Potentially toxic elements, Sequential extraction.

1. Introduction

The presence of heavy metals in sewage sludge (SS) can prevent or restrict its use in agriculture and forestry. However, the potential for using SS in *Eucalyptus* plantations is great [2] since the products of these operations are not intended for human or animal consumption. After its application, SS interacts with soil particles and properties. During this process, heavy metals can bind to different soil components, such as organic matter, clay minerals, and/or remain in soil solution, changing their availability. Thus, sequential extraction methods can assist in the interpretation of the behavior of heavy metals in soil [8].

In *Eucalyptus* plantations, an increase in heavy metal concentrations and soil mobility have been observed but without toxic effects [1]. There may be a residual effect of heavy metals of up to 17 years after SS application [10]. In this context, *Eucalyptus* components have a high capacity to accumulate elements, which presents potential phytoremediation of soil in contaminated areas, with a significant advantage when cultivated for wood production since heavy metal return to the soil is limited [13].

We hypothesized that when following recommended plantation management criteria, SS application (i) does not contaminate soils with heavy metals; (ii) increases soil fertility and *Eucalyptus* wood production; (iii) and that there is a greater accumulation of metals in the wood in comparison to canopy. Our objectives were to evaluate heavy metal concentrations and fractionation in the topsoil layer (0-0.20 m depth); heavy metal accumulation and concentration in different *Eucalyptus* components at 36 months after SS application with or without mineral P fertilizer in comparison to mineral fertilizer application for high wood production.

2. Materials and methods

2.1. Experimental area and materials characterization

The experiment was installed in February 2015 in a commercial Eucalyptus plantation (Boa Esperança do Sul, State of São Paulo; 21°57' S and 48°31' W), on a soil classified as Typic Hapludox [18]. Eucalyptus production represents the only land use during the last 40 years and SS was never applied before the experiment. The climate is classified as Cwa, i.e., humid subtropical zone with hot summer and dry winter [4]. Class B SS was obtained from the Jundiaí wastewater treatment plant [7]. The detailed soil and SS physical-chemical attributes are presented in Abreu-Junior et al. (2020) [3].

2.2. Field procedures

Minimum tillage started in January 2015, which consisted of subsoiling to 0.40 m depth between the lines of the previous rotation over the entire area. After tillage, lime was uniformly broadcasted on the soil surface at 1.8 Mg ha^{-1} to supply Ca and Mg. Eucalyptus seedlings were manually planted in February 2015. The genetic material used was a hybrid, produced via vegetative clonal propagation, from the cross between Eucalyptus grandis and Eucalyptus urophylla (E. urograndis, clone SP5727), spaced 3.00 x 2.25 m apart, totaling 1,481 trees per hectare.

The experiment was installed in a randomized complete block design with six treatments and four replications, totaling 24 experimental plots, each consisting of 100 plants (10 x 10 plants; 675 m^2), with the 36 central trees corresponding to the data collection area (243 m^2). Treatments consisted of the following: 1. Control, without fertilization (C); 2. Mineral fertilization for high wood production, as recommended by Suzano S.A. Company (MF); 3. 14.5 Mg ha^{-1} of SS + 22 kg ha^{-1} of P (S1P1); 4. 29 Mg ha^{-1} of SS (S2); 5. 29 Mg ha^{-1} of SS + 17.5 kg ha^{-1} of P (S2P2); and 6. 43.5 Mg ha^{-1} of SS (S3).

Triple superphosphate was applied to the planting pit next to the seedling. Seven months after planting, SS was applied in a continuous 0.6 m wide band, without incorporation into the soil, 0.2 m aside the planting rows. Sludge doses were calculated according to N criteria, following CONAMA Resolution No. 375 [6], to provide 192 kg ha^{-1} of N, as recommended by the Suzano S.A. Company. Thus, the doses of 14.5, 29.0, and 43.5 Mg ha^{-1} of SS, on a dry weight basis, were equivalent to 50, 100, and 150% of the recommended N. Doses of 22 and 17.5 kg ha^{-1} of P were applied to provide 83 and 66% of mineral P recommendation (26 kg ha^{-1}), respectively. All treatments, except C, were supplemented with mineral K (165 kg ha^{-1} K_2O) and B (6.5 kg ha^{-1} B). Detailed nutrient application is presented in Abreu-Junior et al. (2020) [3].

2.3. Soil sampling, preparation, and analysis

Soil samples were collected in September 2018, 36 months after SS application (42 months after planting), at the depth of 0-0.20 m, considering superficial SS application. Each soil sample consisted of four subsamples collected per plot in the planting rows and interrow, after carefully removing fallen organic material. In the laboratory, samples were air-dried and sieved at 0.5 mm prior to analysis.
2.3.1. Soil fertility. Soil fertility analyses were performed according to Raij et al. (2001) [17]. Soil pH was measured in a soil/solution mixture of 1:2.5 1 mol L^{-1} $CaCl_2$. Soil organic matter (SOM) concentration was estimated by the Walkley–Black method. Available soil P (P-resin) was extracted by an ion-exchange resin procedure and determined by the colorimetric method. Cation exchange capacity (CEC) was equal to $Ca^{2+} + Mg^{2+} + K^+ + (H+Al)$.

IOP Conf. Series: Materials Science and Engineering 1196 (2021) 012027 doi:10.1088/1757-899X/1196/1/012027

2.3.2. Soil heavy metal. The levels of heavy metals were extracted using the 3051A method [20], with concentrated nitric acid (HNO_3) and hydrogen peroxide (H_2O_2) in a closed microwave system. Element (Ba, Cd, Cu, and Zn) determinations were performed by mass spectrometry with inductively coupled plasma (ICP-MS).
2.3.3. Soil sequential extraction. Soil fractionation was conducted as described by Colzato et al. (2018) [8]. The fractions evaluated were: F1 - soluble and/or exchangeable fraction (extracted with $CaCl_2$); F2 - carbonated fraction (extracted with NaOAC at pH 5); F3 - organic matter fraction (extracted with NaOCl at pH 8.5); F4 - oxides fraction (extracted with ammonium oxalate and oxalic acid at pH 3.0); F5 - residual fraction, determined by the 3050B method [19]. The elementary (Ba, Cd, Cu, and Zn) determinations were performed by optical emission spectrometry with inductively coupled plasma (ICP-OES).

2.4. Wood production assessment and plant analysis

Circumference at breast height (CBH) of all trees in the inner plot was measured at 42 months after planting. These data were used to estimate the volume of wood and biomass of wood, branches, and leaves according to methods presented by Abreu-Junior et al. (2020) [3]. The CBH data were also used to select one tree per plot for a non-destructive sampling of trunk, branches, and leaves to assess heavy metal concentrations.

Plant samples were washed with distilled water and then dried in a forced air oven at 60 °C to a constant weight. Then, the samples were ground in a Willey knife mill and digested with concentrated nitric acid (HNO_3) and hydrogen peroxide (H_2O_2) in a closed microwave system. Element (Ba, Cd, Cu, and Zn) determinations were performed by mass spectrometry with inductively coupled plasma (ICP-MS).

2.5. Statistic analysis

Statistics were conducted using the software R [16]. Differences among treatments were compared using a one-way analysis of variance by the F test on each measured variable and, when a significant effect was observed, multiple comparisons were assessed by a Duncan's post hoc test ($p < 0.05$).

3. Results

Soil pH, P-resin, and CEC did not differ among treatments, ranging from 4.7 to 5.4, 3.1 to 8.1 mg dm-3, and 26 to 34 mmolc dm-3, respectively (Table 1). After 36 months of SS application, soil organic matter (SOM) was 27% higher in the S2P2 treatment (10.0 g dm-3), relative to other treatments. Wood volume production and wood, branch, and foliar biomass production increased after 36 months of SS application (Table 1). Wood volume production was higher in the S2 (203 m3 ha-1), S2P2 (210 m3 ha-1), and S3 (206 m3 ha-1) treatments than the control (188 m3 ha-1), but similar to mineral fertilization (199 m3 ha-1). In general, wood, branches, and foliar biomass production were higher in the S2P2 treatment than the control and S2, but similar to mineral fertilization.

Table 1. Topsoil fertility (0-0.20 m depth) and *Eucalyptus* production after 36 months of sewage sludge application.

Treat.	pH	SOM	P-resin	CEC	Volume	Trunk	Branches	Foliar
	-	g dm^{-3}	mg dm^{-3}	mmol$_c$ dm^{-3}	m^3 ha^{-1}	Mg ha^{-1}	Mg ha^{-1}	Mg ha^{-1}
C	4.8 a	8.0 b	3.1 a	26.7 a	188 b	69.8 c	4.7 c	7.6 c
MF	5.4 a	8.1 b	6.4 a	26.7 a	199 ab	74.4 ab	5.0 bc	8.2 ab
S1P1	4.7 a	7.8 b	4.1 a	26.0 a	200 ab	73.4 bc	5.0 ab	8.2 ab
S2	5.1 a	7.8 b	4.2 a	29.0 a	203 a	73.3 bc	5.0 bc	8.2 b
S2P2	4.9 a	10.0 a	5.3 a	34.5 a	210 a	78.4 a	5.3 a	8.7 a
S3	5.4 a	7.6 b	8.1 a	34.3 a	206 a	76.5 ab	5.2 ab	8.5 ab

CV (%)	12.4	12.4	41.6	19.2	4.1	3.7	3.7	4.1

Note: C – Control; MF – Mineral fertilizers; S1P1 – SS (50%) + P (83%) + B + K; S2 – SS (100%) - P + B + K; S2P2 – SS (100%) + P (66%) + B + K; S3 – SS (150%) - P + B + K; SOM – Soil organic matter; CEC – Cation exchange capacity. Means followed by the same letter did not differ by Duncan test ($p < 0.05$).

Soil concentration of heavy metals increased with SS application, but there was no difference for Cu among the treatments, ranging from 2.81 to 4.27 mg kg^{-1} (Table 2). Soil Ba concentration was 43% higher in the S3 (4.65 mg kg^{-1}) treatment in relation to C, S1P1, and S2 (3.25 mg kg^{-1}), but similar to MF and S2P2 (4.02 mg kg^{-1}). The highest Cd concentration was observed in the S3 treatment (0.016 mg kg^{-1}), but similar to S2P2 (0.014 mg kg^{-1}) and higher than the control, MF, S1P1, and S2 (0.009 mg kg^{-1}). Soil Zn concentration was higher in the S3 (5.64 mg kg^{-1}) treatment than the C, S1P1, and S2 (3.43 mg kg^{-1}), but similar to MF and S2P2 (4.88 mg kg^{-1}).

Table 2. Heavy metals concentration in the 0-0.20 m soil depth cultivated with *Eucalyptus* after 36 months of sewage sludge application obtained by the means of acid digestion and sequential extraction.

Treat.	Acid Digestion	Exchangeable	Carbonated	Organic Matter	Oxides	Residual
	Ba (mg kg^{-1})					
C	3.13 c	1.93 a	0.85 a	1.19 b	20.7 a	1.26 a
MF	4.13 ab	2.77 a	0.99 a	1.60 b	20.3 a	1.49 a
S1P1	3.36 bc	2.23 a	0.94 a	1.15 b	20.0 a	1.25 a
S2	3.26 bc	2.10 a	0.96 a	1.27 b	20.0 a	1.63 a
S2P2	3.91 abc	2.75 a	1.10 a	1.92 ab	19.6 a	1.37 a
S3	4.65 a	2.75 a	1.12 a	2.52 a	20.4 a	1.56 a
CV (%)	16.0	21.4	13.1	32.4	2.4	29.5
	Cd (mg kg^{-1})					
C	0.009 bc	0.025 a	< 0.01	0.039 a	0.340 a	< 0.01
MF	0.010 bc	0.020 a	< 0.01	0.031 a	0.303 a	0.050
S1P1	0.006 c	0.036 a	< 0.01	0.023 a	0.303 a	< 0.01
S2	0.010 bc	0.019 a	< 0.01	0.040 a	0.325 a	< 0.01
S2P2	0.014 ab	0.017 a	< 0.01	0.038 a	0.323 a	< 0.01
S3	0.016 a	0.026 a	< 0.01	0.036 a	0.294 a	< 0.01
CV (%)	33.1	56.5	-	48.5	11.4	-
	Cu (mg kg^{-1})					
C	2.81 a	0.84 a	0.74 a	3.56 c	0.48 a	3.43 a
MF	3.43 a	0.81 a	0.55 a	4.46 abc	0.72 a	3.11 a
S1P1	2.66 a	0.84 a	0.55 a	3.90 bc	0.49 a	3.80 a
S2	3.01 a	0.92 a	0.58 a	3.72 a	0.54 a	4.05 a
S2P2	4.27 a	1.02 a	0.65 a	5.31 a	1.00 a	2.98 a
S3	3.95 a	1.02 a	0.53 a	5.51 a	0.98 a	3.15 a
CV (%)	27.9	15.4	25.4	21.1	49.8	25.2
	Zn (mg kg^{-1})					
C	3.34 c	0.98 a	0.25 a	4.18 bc	1.91 b	3.61 a
MF	4.90 ab	1.14 a	0.52 a	5.33 ab	2.01 b	2.92 a
S1P1	3.22 c	0.93 a	0.42 a	2.99 c	1.84 b	2.91 a
S2	3.72 bc	0.65 a	0.24 a	3.41 bc	2.08 b	2.58 a
S2P2	4.86 ab	0.54 a	0.40 a	4.40 bc	2.21 ab	4.03 a
S3	5.64 a	1.19 a	0.26 a	7.26 a	2.47 a	3.59 a
CV (%)	18.1	36.6	82.0	30.4	11.6	20.7

Note: C – Control; MF – Mineral fertilizers; S1P1 – SS (50%) + P (83%) + B + K; S2 – SS (100%) - P + B + K; S2P2 – SS (100%) + P (66%) + B + K; S3 – SS (150%) - P + B + K; SOM – Soil organic matter; CEC – Cation exchange capacity. Means followed by the same letter did not differ by Duncan test ($p < 0.05$).

Sequential extraction of Ba in the soil showed statistical differences were limited to the organic matter fraction, where the highest levels were obtained in the S2P2 and S3 treatments, which were 71% higher than the other (Table 2). Cadmium was not detected by the analytic method used (< 0.01 mg kg-

[1]) in the carbonated and residual fractions, and there were no differences among treatments in the exchangeable, organic matter, and oxides fractions. There were no statistical differences among treatments for Cu in the exchangeable, carbonated, oxides and residual fractions. However, in the organic matter fraction, Cu concentration was 77% higher in the S2, S2P2, and S3 treatments in relation to the control and S1P1. Soil Zn bound to the organic matter and oxides fraction had the highest level in the S3 treatment and was similar to the MF and S2P2 treatments, respectively.

There were no differences among treatments for Ba, Cd, Cu, and Zn concentration and accumulation in the trunk, branches, and leaves of Eucalyptus treated with SS, except for Zn accumulation in leaves, which was 23% higher in the S2 and S2P2 treatments in relation to the other (Table 3).

Table 3. Heavy metals concentration and accumulation in *Eucalyptus* components after 36 months of sewage sludge application

Treat.	Ba mg kg^{-1}	Cd mg kg^{-1}	Cu mg kg^{-1}	Zn mg kg^{-1}	Ba g ha^{-1}	Cd g ha^{-1}	Cu g ha^{-1}	Zn g ha^{-1}
	Concentration				*Accumulation*			
				Trunk				
C	2.39 a	0.0027 a	0.41 a	0.88 a	167.6 a	0.180 a	27.6 a	58.2 a
MF	1.74 a	0.0018 a	0.37 a	0.53 a	134.6 a	0.132 a	26.4 a	41.4 a
S1P1	1.68 a	0.0017 a	0.33 a	1.62 a	123.3 a	0.122 a	24.4 a	123.4 a
S2	1.58 a	0.0018 a	0.43 a	0.98 a	115.2 a	0.128 a	31.2 a	68.8 a
S2P2	2.03 a	0.0024 a	0.51 a	2.29 a	162.8 a	0.191 a	29.5 a	190.7 a
S3	1.82 a	0.0022 a	0.33 a	1.05 a	138.5 a	0.160 a	25.2 a	83.2 a
CV (%)	30.1	38.2	36.3	82.8	32.5	38.8	25.7	90.1
				Branches				
C	6.14 a	0.0103 a	4.79 a	5.90 a	28.9 a	0.048 a	22.6 a	27.8 a
MF	6.87 a	0.0060 a	3.96 a	5.24 a	34.6 a	0.030 a	20.0 a	26.1 a
S1P1	7.21 a	0.0082 a	3.99 a	5.67 a	36.4 a	0.042 a	20.2 a	28.9 a
S2	6.31 a	0.0104 a	3.50 a	6.34 a	31.6 a	0.052 a	17.5 a	31.8 a
S2P2	6.66 a	0.0079 a	3.71 a	5.84 a	35.4 a	0.042 a	19.7 a	31.0 a
S3	6.67 a	0.0074 a	3.72 a	6.38 a	34.5 a	0.038 a	19.1 a	32.9 a
CV (%)	15.6	57.3	25.2	24.0	18.3	56.8	25.9	25.8
				Leaves				
C	21.6 a	0.0047 a	2.70 a	11.43 a	164.5 a	0.035 a	20.6 a	87.3 c
MF	15.8 a	0.0042 a	2.62 a	12.74 a	130.5 a	0.034 a	21.5 a	103.8 bc
S1P1	16.9 a	0.0054 a	2.80 a	12.60 a	140.2 a	0.044 a	23.0 a	103.3 bc
S2	12.6 a	0.0055 a	2.71 a	14.21 a	103.3 a	0.045 a	22.2 a	116.2 ab
S2P2	13.1 a	0.0048 a	3.03 a	15.03 a	114.1 a	0.042 a	26.5 a	131.4 a
S3	17.5 a	0.0056 a	2.62 a	12.65 a	149.5 a	0.048 a	22.2 a	107.1 bc
CV (%)	30.3	26.8	10.3	12.6	32.5	26.9	11.9	13.3

Note: C – Control; MF – Mineral fertilizers; S1P1 – SS (50%) + P (83%) + B + K; S2 – SS (100%) - P + B + K; S2P2 – SS (100%) + P (66%) + B + K; S3 – SS (150%) - P + B + K; SOM – Soil organic matter; CEC – Cation exchange capacity. Means followed by the same letter did not differ by Duncan test ($p < 0.05$).

4. Discussion

After 36 months of SS application, there was an increase in SOM and P-resin, although this last one was not significant, with SS doses, without changing the soil pH and CEC (Table 1). With the improvement of soil fertility, the production of Eucalyptus components increased with the SS doses in relation to the control, mainly that one providing 100 and 66% of recommended N and P (S2P2), respectively, and similar production to the mineral fertilization (Table 1), corroborating the results obtained by Abreu-Junior et al. (2017, 2020) [2].

However, together with the input of macro and micronutrients, SS application also furnishes heavy metals to the soil, which can be absorbed by plants, possibly affecting crop development [10]. Despite the increased soil heavy metal concentrations provided by the application of the highest dose of SS (S3; Figure 1), concentrations of Ba, Cd, Cu, and Zn obtained in the soil were below the reference values of soil quality recommended by the Environmental Company of the State of São Paulo [7], which are 75, <0.5, 35, and 60 mg kg^{-1}, respectively. The levels, obtained in this study, were also within the natural concentration of heavy metals in the soils of the state of São Paulo, which are from 2.9 to 548 mg kg^{-1} for Ba; 0.01 to 0.48 mg kg^{-1} for Cd; and 4 to 68 mg kg^{-1} for Zn [15]. These authors also suggest updated levels for the reference values of soil quality, being 54, 0.10, and 22 mg kg^{-1} for Ba, Cd, and Zn, respectively (Cu was not evaluated). Even with the restriction of the values, the levels obtained in the S3 treatment are below indicated levels, showing the lack of environmental contamination of the soil by these elements through the application of the SS.

The sequential extraction of heavy metals in the soil shows that the fraction bounded to organic matter was the most influenced by the LE application, except for Cd (Table 2). The sorption of Zn and, mainly, Cu is dependent on its interaction with organic matter [5]. Because of this, higher levels of these elements were observed in the fraction bound to organic matter in treatments with greater SS application. Furthermore, the proportion in the fractions showed the following order: organic matter (44%) > residual (34%) > exchangeable (9%) > oxides (7%) > carbonated (6%) for Cu; and organic matter (40%) > residual (30%) > oxides (19%) > exchangeable (8%) > carbonated (3%) for Zn. This demonstrates the greater interaction of Cu and Zn with SOM.

For the Ba and Cd fractionation, despite the Ba increase in the fraction bounded to organic matter with SS doses, the oxide fraction was the one with the highest proportion of these elements, representing 76 and 85% of the quantified Ba and Cd (Table 2), respectively. These results are important because these elements are more harmful to plants in relation to Cu and Zn, which are micronutrients. Barium is associated with feldspar minerals and when free it can precipitate and adsorb to clays or oxy and hydroxides (Lu et al., 2019), as observed in this study. Cadmium has high mobility in the soil and around 55 to 90% is in the ionic form (Cd^{2+}). The forms of Cd in the soil are dependent on pH, which controls the complexation of the metal by the SOM, influencing its ionic form [12].

In plants, the lack of differences in the heavy metal concentrations in different components (Table 2) may suggest the use of some mechanism to protect Eucalyptus, either complexing the elements in organic acids released by plants, and/or storing the elements in the cell walls or vacuoles of root cells (Fan et al., 2020). On leaves, where the main metabolic processes for plant development occur, there was a higher Ba, Cu, and Zn concentration, with no symptoms of Ba phytotoxicity and the levels of Cu and Zn were below and within the recommended values of 7-10 and 10-18 mg kg^{-1} [11], respectively, for the purpose of assessing the nutritional status of plants. Ba, Cd, and Cu accumulation were higher in the Eucalyptus trunk relative to canopy (branches and leaves), as evidenced by a high exportation of these elements by the trunk. However, a high proportion of Ba and Cu still remain in the field after harvest from the branches and foliar residues. In general, Zn accumulation was higher in the leaves in relation to the trunk, with this nutrient remaining in the field after the harvest.

5. Conclusions

The application of SS to supply 100 and 150% of recommend N for the cultivation of Eucalyptus provided an increase heavy metal concentrations in the soil, however, at levels lower than those established by current regulation. By sequential extraction, the levels of heavy metals were higher in the fractions: oxides > organic matter > residual > exchangeable > carbonated, showing low availability for plants. On the plant, heavy metal accumulation was similar in the wood and canopy. Sewage sludge application at a rate to provide 100% of recommended N, supplemented with P, also increased soil fertility, which provided Eucalyptus production at levels similar to mineral fertilization for high wood production. Therefore, SS can be a sustainable source of nutrients for Eucalyptus cultivation, without risks of environmental contamination by heavy metals.

References

[1] Abreu-Junior CH, Brossi M J L, Monteiro RT, Cardoso P H S, Mandu TS, Nogueira T A R, Ganga A, Filzmoser P, Oliveira F C, Firme L P He Z and Capra G F (2019). Effects of sewage sludge application on unfertile tropical soils evaluated by multiple approaches: A field experiment in a commercial Eucalyptus plantation. *Science of the Total Environment*. **655**, 1457-1467. https://doi.org/10.1016/j.scitotenv.2018.11.334

[2] Abreu-Junior C H, Firme L P, Maldonado C A B, Moraes Neto S P, Alves M C, Muraoka T, Boaretto A E, Gava J L, He Z, Nogueira T A R and Capra G F (2017). Fertilization using sewage sludge in unfertile tropical soils increased wood production in Eucalyptus plantations. *Journal of Environmental Management*. **203**, 51-58. https://doi.org/10.1016/j.jenvman.2017.07.074

[3] Abreu-Junior C H, Oliveira M G, Cardoso P H S, Mandu T S, Florentino A L, Oliveira F C, Reis J V, Alvares C A , Stape J L, Nogueira T A R, Capra G F and He Z (2020). Sewage Sludge Application in Eucalyptus urograndis Plantation: Availability of Phosphorus in Soil and Wood Production. *Frontiers in Environmental Science*. **8**, Article 116. https://doi.org/10.3389/fenvs.2020.00116

[4] Alvares CA, Stape J L, Sentelhas P C, Gonçalves J L M and Sparovek G (2013). Köppen's climate classification map for Brazil. *Meteorologische Zeitschrift*. **22**, 711-728. https://doi.org/10.1127/0941-2948/2013/0507

[5] Brady NC and Weil R R (2013). Elementos da natureza e propriedades dos solos. 3. ed. Porto Alegre: Bookman.

[6] Brasil 2006. Conselho Nacional Do Meio Ambiente - CONAMA. Resolução 375. Brasília, 32p. Available in: <http://www2.mma.gov.br/port/conama/res/res06/res37506.pdf>. Access in: July 2017.

[7] CETESB 2016. Companhia Ambiental do Estado de São Paulo – CETESB. Valores orientadores para solo e água subterrânea no Estado de São Paulo. 2016. Available in: <https://cetesb.sp.gov.br/aguas-subterraneas/wp-content/uploads/sites/13/2013/11/tabela_vos_2016_site.pdf>. Access in: October 2020.

[8] Colzato M, Alleoni L R F and Kamogawa M Y (2018). Cadmium sorption and extractability in tropical soils with variable charge. *Environmental Monitoring and Assessment*. **190**, Article 345. https://doi.org/10.1007/s10661-018-6666-7

[9] Fan W J, Feng Y X, Li Y H, Lin Y J, Yu X Z (2020). Unraveling genes promoting ROS metabolism in subcellular organelles of Oryza sativa in response to trivalent and hexavalent chromium. *Science of the Total Environment*. **744**, Article 140951. https://doi.org/10.1016/j.scitotenv.2020.140951

[10] Florentino A L, Ferraz A V, Gonçalves J L M, Asensio V, Muraoka T, Dias C T S, Nogueira T A R, Capra G F and Abreu-Junior C H (2019). Long-term effects of residual sewage sludge application in tropical soils under Eucalyptus plantations. *Journal of Cleaner Production*. **220**, 177-187. https://doi.org/10.1016/j.jclepro.2019.02.065

[11] Gonçalves J L M (2010). Eucalipto. In: Prochnow, L.I., Casarin, V., & Stipp, S.R. Boas práticas para uso eficiente de fertilizantes: Culturas. Piracicaba: IPNI – International Plant Nutrition Institute.

[12] Kabata-Pendias A, Mukherjee AB, 2007. Trace elements from soil to human. Berlin: Springer-Verlag.

[13] Kaur B, Singh B, Kaur N and Singh D (2018). Phytoremediation of cadmium-contaminated soil through multipurpose tree species. *Agroforestry Systems*. **92**, 473-483. https://doi.org/10.1007/s10457-017-0141-2

[14] Lu Q, Xu X, Liang L, Xu Z, Shang L, Guo J, Xiao D and Qiu G (2019). Barium concentration, phytoavailability, and risk assessment in soil-rice systems from an active barium mining region. *Applied Geochemistry*. **106**, 142-148. https://doi.org/10.1016/j.apgeochem.2019.05.010

[15] Nogueira T A R, Abreu-Junior C H, Alleoni L R F, He Z, Soares M R, Vieira CS, Lessa L G F and Capra G F (2018). Background concentrations and quality reference values for some potentially toxic elements in soils of São Paulo State, Brazil. *Journal of Environmental Management*. **221**, 10-19. https://doi.org/10.1016/j.jenvman.2018.05.048
[16] R Core Team 2013. R: A Language and Environment for Statistical Computing. Vienna: R Foundation for Statistical Computing.
[17] Raij B, Andrade JC, Cantarella H and Quaggio J A (2001). Análise química para avaliação da fertilidade de solos tropicais. Campinas: Instituto Agronômico de Campinas.
[18] Soil Survey Staff (2014). Keys to Soil Taxonomy, 12th, Edn. Washington, DC: USDA-NRCS. https://doi.org/10.1109/TIP.2005.854494
[19] USEPA (1996). Method 3050B: Acid digestion of sediments, sludges, and soils. Hazardous Waste Test Methods, SW-846.
[20] USEPA (2007). Method 3051A: Microwave assisted acid digestion of sediments, sludges, soils and oils. Washington, DC.

Acknowledgment

The authors would like to thank the São Paulo State Research Foundation – FAPESP (grant # 2004/15097-0; # 2017/26375-1) and the National Council for Scientific and Technological Development - CNPq (grant # 485205/2012-2; # 312728/2017-4) for financial support; to the Coordenação de Aperfeiçoamento de Pessoal de Nível Superior - Brasil (CAPES) - Finance Code 001; to Suzano S.A. Company for funding and technical-scientific support, especially for Dileto Benedito Bau; and to the Jundiaí wastewater treatment plant, for the supply of SS.

Integration of Artificial Intelligence and Life Cycle Assessment Methods

José Oduque de Jesus[1,*], Karla Oliveira-Esquerre[1] and Diego Lima Medeiros[1]

[1]Federal University of Bahia (UFBA), Growing with Applied Metrics and Mindful Analysis (GAMMA), Industrial Engineering Program (PEI). Professor Aristides Novis Street, 02, Federação, 40210-630, Salvador, Bahia, Brazil.
***E-mail:** jose.oduque@hotmail.com

Abstract. Artificial Intelligence (AI) techniques support environmental tools based on the growing availability of data and information, aligning the concepts of data modeling and analysis. The Life Cycle Assessment (LCA) is an environmental tool that requires a large volume of data to measure the performance of a product and to simulate the proposed scenarios to improve its performance. This research reviewed studies using AI techniques and their intersection with LCA from data mining. This study identified some AI techniques used in LCA studies. However, there is a lack of LCA literature using AI techniques, despite the benefits of integrated modeling. The results show that different AI algorithms are used to build LCA models. The AI algorithms of the studies act from the identification of the problem to the solution stage, therefore the integration between AI and LCA makes it possible to build predictive models of machine learning to enable assertiveness in decision making.

Keywords: Artificial Intelligence (AI), Life Cycle Assessment (LCA), Integrated Assessment Models.

1. Introduction

Currently, the growing environmental problem has caused the awakening of the sustainable awareness of society and provided a greater movement in the effort to guarantee harmony between the economy growth and environmental conservation. In this context, sustainable development plays an important role in the balance between society's demands and the preservation of natural resources[18]. The predominant socioeconomic development model worldwide puts environmental sustainability at risk, therefore science-based policy is required to support a societal transformation towards a sustainable well-being. Different methods are used to support decision making towards environmental sustainability, such as Life Cycle Thinking (LCT) and Life Cycle Assessment (LCA) [10]. LCA measures the environmental performance (aspects and impacts) of products and processes [9], supporting the identification of optimization opportunities. LCA is used as an environmental management tool that supports the understanding and quantification of the environmental performance related to different production and consumption systems for comparative and improvement purposes [7]. However, LCA studies are data intensive. The inventory analysis is the most laborious phase in LCA due to data collection and treatment that are time-consuming. Despite the increasing availability of data and information to date, the available data requires collection and treatment before using in

IOP Conf. Series: Materials Science and Engineering 1196 (2021) 012028 doi:10.1088/1757-899X/1196/1/012028

such studies. Hence, a combination of methods enables a synergic evaluation with more robust results, such as the integrated assessment models (IAM), increasing the assertiveness in decision making.
Data science is an emerging research field that contributes to synthesizing current theories and models to integrate data collection, analysis and decision making. The amount of data and information processed daily in commercial transactions, logistics, sensors, social networks, users and analysis toolscan be optimized by Artificial Intelligence (AI) systems. AI collects accurate and valuable information, analyze data, perform tasks and solve problems. The use of AI contributes to problem-solving in different areas of society, especially those regarding environmental issues such as real-time monitoring, environmental impact accounting, environmental mapping, parameter estimation and scenario simulation, providing useful information for decision making. However, the user community needs to understand the AI techniques in data science and form a comprehensive view of what can be done from these techniques in a specific context to make useful applications. LCA studies can be integrated with AI techniques to diagnose potential failures in data modeling and analysis, however, there is a lack of literature combining these methods. Therefore, this research reviewed studies using AI techniques and its intersection with LCA from data mining.

2. Materials and methods

A classic data mining approach was used to search the relevant articles regarding AI and LCA in the literature, considering three stages as shown in figure 1.

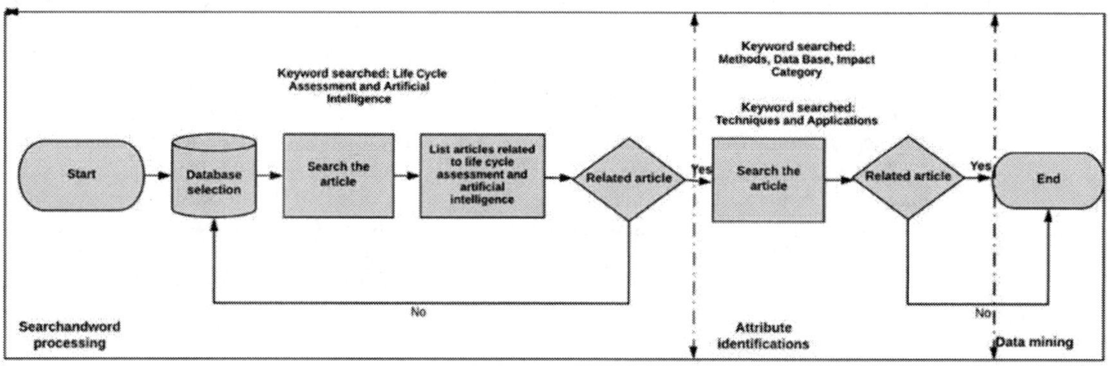

Figure 1. Data mining steps used in AI and LCA literature searching.

The Search and word processing stage considered the insertion of the keywords (Life Cycle Assessment (LCA) and Artificial Intelligence (AI)) in the searching engine of the available databases. Thereafter, the articles found were grouped in two categories: related articles (A); and others (B). In the Identification of attributes stage, two sets of keywords were defined as attributes, based on their frequency and meaning, and chosen as a centroid for mining. The sets of keywords were the following: Methods, Data Base and Impact Category in LCA; and Techniques and Application sin AI. The articles were selected from semantic similarities of the defined centroids and the keywords of each article found. The selected articles were grouped and used as a reference for this work.

3. Results and Discussion

The main characteristics of AI techniques used in LCA studies are presented in Table 1.

Table 1. Main characteristics of Artificial Intelligence(AI) techniques used in Life Cycle Assessment (LCA) studies.

Studies	Typesof AI technologies	Characteristics	Advantages	Limitations

Studies	Typesof AI technologies	Characteristics	Advantages	Limitations
Komly et al. (2012); Ghnimi et al. (2018),	Multilayer perceptron neural network(MLP NN)	- Supervised learning - Back-propagation algorithm is widely used as training algorithm	- Easy to implement - High accuracy and consistent estimations when changes occur	- Slow speed of convergence - Numbers of hidden neurons always based on trial and error
Komly et al. (2016); [8]	Radial basis function neural network (RBFNN)	- Basis function used can be Gaussian or wavelets -Universal approximation	- High tolerance of noise - Fast training - Good capability in generalization	- Large number of hidden neurons needed
[1][7]	Support vector machine(S VM)	- Use quadratic programming to solve support vector	- Small data required - Global searching ability - High robustness against noise	- Large memory requirement and CPU time when trained in batch mode
[2][22]	Geneticalgori thm(GA)mec hanismsofbiol ogical	- Selection and genetic evolutionary theory - Universal approximation - Heuristic algorithm	- Good global optimization capability -Generate feasible solutions and allow researchers to choose from several approaches for obtaining best results	- Computational heaviness - Challenges in choosing network topology
[7] [15]	Fuzzy neural network (FNN)	- Human-like reasoning - Suitable for advanced control systems	- If-then rules easy to interpret -Implementation can be either from input to output or output to input -Able to accurately describe imprecise values of parameters	- Long computational time - Low robustness against noisy data
[11][21]	Adaptive network-based fuzzy inference system (ANFIS)	- Consist of antecedent and conclusion - Integrate gradient descent method and least square method to train parameters	- Efficient nonlinear approximation - Short learning time - Fast in reaching optimum results	- Optimum structures are based on trial anderror
[17][24]	Artificial neural network coupled with genetic algorithm (GA-ANN)	- Predicted output value of neural network can be used as the fitness function of GA	- Prevent local minimum - Fast convergence - High accuracy	- Computational heaviness - Unable to determine numbers of hidden neurons

All the algorithms of the presented studies (Table 1) act from the identification of the problem to the solution stage. However, most consulted articles point out that it is important that the researcher identifies the different data sources available and fully understands what types of data can be found in each source. Besides, the data should be thoroughly evaluated and proved that they are suited to the study proposal in the stage previous to the model development. The predictive model should be fitted to the input data to make accurate predictions, avoid ingunderfitting (when there is a lack of data) or overfitting (when there is an excess of data) that affects the quality of the results.

The AI techniques provide a high capacity to solve complex problems with uncertain, interactive and dynamic characteristics in an economical and efficient manner. The AI methods combined with LCA were applied in studies of agriculture, climate and engineering as presented by the works of [2][11][15].80% of the selected articles combined AI and LCA in industrial activities (Table 1), while 10% of the studies were in agricultural activities regarding the energy and water performances. AI techniques were used to automate the LCA models and improve the quality of the inventory (Table 1), maintaining low uncertainty levels.

4. Future gaps and opportunities

LCA studies rely on generic databases in the absence of specific ones due to time and cost constraints, which increases the uncertainty of the results, while AI techniques can provide missing and unavailable information. The inventory analysis presented by the articles in table 01 was computerized and AI techniques were applied to store, track missing data, estimate and calculate potential impacts using the developed models. Zidonien et al [24] and Nabavi-Pelesaraei et al. [17] point out that the integration between AI and LCA enables the construction of predictive models of machine learning to improve the assertiveness of decision making within organizational processes [17][24]. In addition, the integrated assessment using AI and LCA methods can increase the robustness of the model and efficacy in decision making. Nevertheless, AI algorithms should be used at scale and tested in a wide variety of techniques and industrial sectors to increase its reliability.

5. Conclusion

This study analyzed the integration of Artificial Intelligence (AI) and Life Cycle Assessment (LCA) methods from literature review using data mining. AI techniques support data collection in LCA studies, however there is a lack in the literature regarding the combination of these methods in an integrated model. Thus, this study identified some AI techniques, considering key advantages and limitations, in order to support its use in LCA studies.

References

[1] Azapagic A, Millington A, Collett A 2006. A methodology for integrating sustainability considerations into process design. *Chemical Engineering Research and Design.* **84** (A6): 439–452.

[2] Froemelt A, Dürrenmatt D J, Hellweg S 2018. Using Data Mining To Assess Environmental Impacts of Household Consumption Behaviors. *Environmental Science & Technology.* **52**(15) 8467-8478.

[3] Hosseinzadeh-Bandbafha H, Nabavi-Pelesaraei A, Majid K, Ghahderijani M, Chau K 2018. Application of data envelopment analysis approach for optimization of energy use and reduction of greenhouse gas emission in peanut production of Iran. *Journal of Cleaner Production.* **172**. 1327–1335.

[4] Hosseinzadeh-Bandbafha H, Safarzadeh D, Ahmadi E 2019. Combined life cycle assessment and artificial intelligence for prediction of output energy and environmental impacts of sugarcane production. *Science of the Total Environment.* **664**.1005-1019.

[5] Hosseinzadeh-Bandbafha H, Chau K W 2018. Integration of artificial intelligence methods and life cycle assessment to predict energy output and environmental impacts of paddy production. *Science of the Total Environment.* 631–632; 1279–1294.

[6] Hosseinzadeh-Bandbafha H, Safarzadeh D, Ahmadi E Nabavi-Pelesaraei A, Hosseinzadeh-Bandbafha E 2016. Applying data envelopment analysis to evaluation of energy efficiency and decreasing of greenhouse gas emissions of fattening farms. *Energy.*1-11.

[7] Hou P, Jolliet O, Zhu J, Xu M 2020. Estimate ecotoxicity characterization factors for chemicals in life cycle assessment using machine learning model. *Environment International.* 135.

[8] Huntingford C, Jeffers E S, Bonsall M B, MChristensen H, Lees T, Paramesh V, Arunachalam V, Nikkhah A, Das B, Yang H 2019. Machine learning and artificial intelligence to aid climate change research and preparedness. *Environ. Res. Lett.* **14**.

[9] ISO INTERNATIONAL ORGANIZATION FOR STANDARDIZATION 2006. 14044: Environmental managements – lifecycle assessments – requirements and guidelines. International Organization for Standardization.

[10] Kaab A, Sharifi M, Mobli H, Nabavi-Pelesaraei A, Chau K W 2019. Use of optimization techniques for energy use efficiency and environmental life cycle assessment modification in sugarcane production. *Energy.***181**. 1298-1320.

[11] Kaab A, Sharifi M, Mobli H, Nabavi-Pelesaraei A, Chauc K, Khoshnevisan B, Bolandnazar E, Shamshirband S, Shariati H M, Anuar N B, Mat Kiah M L 2014. Decreasing environmental impacts of cropping systems using life cycle assessment (LCA) and multi-objective genetic algorithm. *Journal of Cleaner Production.* 1-11.

[12] Khoshnevisan B, Rafiee S, Omid M, Mousazadeh H 2013. Applying data envelopment analysis approach to improve energy efficiency and reduce GHG (greenhouse gas) emission of wheat production. *Energy.* **58**. 588

[13] Khoshnevisan B, Rafiee S, Omid M, Mousazadeh H, Clark S 2013. Environmental impact assessment of tomato and cucumber cultivation in greenhouses using life cycle assessment and adaptive neuro-fuzzy inference system. *Journal of Cleaner Production.* 1-10.

[14] Khoshnevisan B, Rajaeifar M A, Clarkb S, Shamahirband S, Anuar N B, Shuib N L M, Gani A 2014. Evaluation of traditional and consolidated rice farms in GuilanProvince,Iran, using life cycle assessment and fuzzy modeling. *Science of the Total Environment.* **481** 242-251. data. Journal of Cleaner Production. 137. 1109-1117.

[15] Lee E K, Zhang W J, Zhang X, Adler P R, Lin S, Feingold B J, Khwaja H A, Romeiko X X 2020. Projecting life-cycle environmental impacts of corn production in the U.S. Midwest under future climate scenarios using a machine. *Science of The Total Environment.* 714.

[16] Liu K F R, Yu C W 2009. Integrating case-based and fuzzy reasoning to qualitatively predict risk in an environmental impact assessment review. *Environmental Modelling & Software.* **24**.1241-1251

[17] Nabavi-Pelesaraei A, Rafiee S, Mohtasebi S S, Hosseinzadeh-Bandbafha H, Chau K 2017. Energy consumption enhancement and environmental life cycle assessment in paddy production using optimization techniques. *Journal of Cleaner Production.*

[18] Nabavi-Pelesaraei A, Rafiee S, Mohtasebi S S, Pishgar-Komleh S H, Akram A, Keyhani A, Sefeedpari P, Shine P, Brandao M 2020. Integration of life cycle assessment, artificial neural networks, and metaheuristic optimization algorithms for optimization of tomato-based cropping systems in Iran. *The International Journal of Life Cycle Assessment.* **25**. 620-632.

[19] Park J-H, Seo K-K 2006. A knowledge-based approximate life cycle assessment system for evaluating environmental impacts of product design alternatives in a collaborative design environment. *Advanced Engineering Informatics.* **20.** 147-154.

[20] Ramanujan D, Bernstein W Z, Chandrasegaran S K, Ramani K 2017. Visual Analytics Tools for Sustainable Lifecycle Design: Current Status, Challenges, and Future Opportunities. *Journal of Mechanical Design.* **139**-111415-1n.

[21] Slapnik M, Istenic D, Pintar M, Udovc A 2015. Extending life cycle assessment normalization factors and use of machine learning – A Slovenian case study. *Ecological Indicators.* **50** 161-172.

[22] Sun A Y, Scanlon B R 2019. How can Big Data and machine learning benefit environment and water management: a survey of methods, applications, and future directions. *Environ. Res. Lett.* **14**(7).

[23] Sundaravaradan N, Marwah M, Shah A, Ramakrishnan N 2011. Data mining approaches for life cycle assessement. *In: Proceedings of the IEEE International Symposium on Sustainable Systems and Technology.* 1-6.

[24] Zidonien S, Kruopien J 2014. Life Cycle Assessment in environmental impact assessments of industrial projects: towards the improvement. *Journal of Cleaner Production.* 1-8.

[25] Venkatesh P et al 2018. Optimization of energy consumption and environmental impacts of arecanut production through coupled data envelopment analysis and life cycle assessment. *Journal of Cleaner Production.* **203**(1). 674-684.

Acknowledgments

The authors are grateful to the Coordination for the Improvement of Higher Education Personnel (CAPES)- Finance Code 001 and Industrial Engineering Graduate Program (PEI) at the Federal University of Bahia (UFBA)for supporting this research.

Exploring Secondary Markets to Improve Circularity: A comparative case study of photovoltaics and hard-disk drives

Liam Watts*, Julien Walzberg, Alberta Carpenter, Garvin A. Heath

National Renewable Energy Laboratory 15013 Denver West Parkway Golden, CO 80401, United States of America

*Email: liwatts@syr.edu

Abstract. Each year renewable energy generation increases notably with solar panel installations, but these panels have a limited lifespan and will produce between 2 and 4 million metric tons of waste by 2040. Similarly, there are currently between 20 to 70 million hard-disk drives (HDDs) reaching end-of-life (EOL) annually. The circular economy (CE) strives to recycle and reuse materials that are rare and expensive to obtain, minimizing waste. However, studying the potential circularity of photovoltaics (PV) and HDDs requires various data, for instance, on the maturity of the secondhand markets. In this context, the objective of the present study is to identify the current state of secondhand PV and HDD markets. After conducting a literature review, an automated data collection process was set up for that purpose. The analysis of the literature and collected data assess the maturity of the secondhand PV modules and HDDs markets and highlight differences between them.

Keywords: Circular economy, agent-based modelling, photovoltaics, hard-disk drives, secondhand market.

1. Introduction

The commercial market consists of countless product innovations as companies push the boundaries of existing technologies. The U.S. economy largely consists of linear flows of goods (i.e., so called "take make waste" approach) where products are manufactured, sold, and then landfilled [3]. A linear economy produces a large amount of waste as products reach their end-of-life (EOL) and no longer function as intended. As the solar panel market grows, the photovoltaics (PV) industry is expected to generate between 2-4 million tons of waste in the United States by 2040 [11]. With a growing data-storage demand [6], around 20 to 70 million EOL hard-disk drives (HDDs) are generated per year. Besides waste management issues, the growing demand of critical and precious metals used in these two technologies means they can be threatened by sudden supply restrictions [20]. These problems can be solved by changing to a circular economy which reinforces the concepts of reduce, reuse, and recycle (3R) [15]. A circular economy improves on the 3R methodology by adding refusing (e.g. material replacement) repairing, and repurposing to help narrow, close, and slow the waste generation loops [15]. In addition, this CE approach can lead to

IOP Conf. Series: Materials Science and Engineering 1196 (2021) 012029 doi:10.1088/1757-899X/1196/1/012029

other waste reduction methods such as the *sharing economy* and *product service systems* that extend the life of both products and their components [15].

The circular economy requires strong communication and collaboration between the various actors of the supply-chain (extraction, manufacturing, distribution, and use) so that pathways exist to recover EOL products and their components. In practice this has led to the establishment of industrial eco-parks, where waste and co-products from one industry are utilized by another [2][7]. Recent research, for instance, proposes to apply this concept to the PV industry [13]. In another research project, the CE concept of closed loops was applied to recover the rare earth elements (i.e., neodymium and dysprosium) from used HDDs and reuse them in new drives [10].

However, the current available data pertaining to the development of the CE is severely lacking. For example, strides in CE for PV have been made in the EU through work with Circusol [23] yet few data were made easily accessible. The issue is not only restricted to energy technologies: recent research highlights that the lack of data or ambiguity of current data is preventing the progression of the CE model [19]. The lack of data issue has been mentioned in a few CE studies, for instance, in a New Zealand study on electronic waste [1], an Italian study on construction waste [5], and a Trinidad and Tobago study on plastics [14]. However, if we get past this issue, the circular economy has the potential to greatly reduce waste and environmental impacts [3] and decrease economic costs, and increase employment opportunities for the economy at large [16].

To successfully transition to a circular economy, the current market and material flows needs to be understood. Several types of modeling may help study the CE transition, the choice of the model ultimately depending on the research question [24]. For instance, methods from industrial ecology such as material flow analysis or life cycle assessment can provide information on the sustainability of the CE transition [24]. Agent-based modeling (ABM), by contrast, can help study the CE transition in a different way by providing information on the effect of the intertwined decisions of multiple actors [24]. While developing an analytical model is an alternative solution, market behaviors are often erratic and do not align with other stakeholders' behaviors, thus documenting an ABM through observation may prove more reliable [25]. This approach, however, requires a large quantity of various data, such as recycling and manufacturing costs, prices of secondhand products, and the number and type of actors involved in the CE transition [24]. However, as already mentioned, information surrounding product repair and secondary markets is often non-existent. Despite the effort required to understand these unknown markets, they can be easily broken up into case studies and thoroughly examined.

The PV and the HDD markets are facing a growing need for material recovery and recycling. For instance, the developing PV market is growing about 9% annually worldwide, while the HDD companies are transitioning to a more circular model [10][11][21]. Currently about 50 million HDDs reach their end-of-life annually and need to be shredded or recycled [10]. Due to the recent development of the PV market and the long product life (about 30 years), only about 1.5 million panels are reaching their EOL, but around 32.4 million new panels are being installed each year, which brings an increasing focus on PV EOL management. While still small, the secondary HDD market is far more robust than the PV market and sells 1.5-3 million drives annually that do not include HDDs for internal reuse by large cloud storage companies like Google, Facebook, and Microsoft [21]. In addition to being larger than the PV market, the HDD market is far more developed, containing third party hard drive wiping and restoration companies and some secondary market regulations to hold a standard. Both PVs and HDDs contain precious metals or rare earth elements (REEs) that can be extracted for reuse if the module or disk drive cannot be reused as a unit [17]. For example, silver accounts for about half the cost of a PV module, yet is less than 1% of the module's weight, dissuading companies from pursuing recycling efforts for other PV components due to high costs and low benefit [11].

Several studies have acknowledged the existence of a secondhand PV market, yet lack detail to properly characterize the market. Tao and colleagues state that refurbished modules may be sold at 50% the price of new modules [22]. In addition, the IRENA-IEA 2016 report estimates that the secondhand module ratio might be closer to 70% of the original sales price, but does not provide much detail on how this number was inferred [11]. Regarding to HDDs, one of the International Electronics Manufacturing Initiative (iNEMI) projects reported prices for used drives [10] but this study was interested in the possibility of finding more data-points. Thus, this study aims to provide both a quantitative and qualitive analysis of both the secondhand PV and HDD markets. One of the main findings is the characterization of the price ratio between secondhand and firsthand PV modules and HDDs. In understanding the relationship between these markets, it is possible to document the underdeveloped state of the secondhand PV market (as compared to used electronics such as HDDs) and work to improve its traction within a sustainable economy.

2. Materials and Methods

This section introduces the approach of our project. Figure 1 summarizes the different steps undertaken to reach our research objectives which are to:

- Contribute to the CE research by collecting data on the primary and secondary PV and HDD markets;
- Identify the shortcomings of the secondhand PV market to improve the available knowledge about secondhand modules.

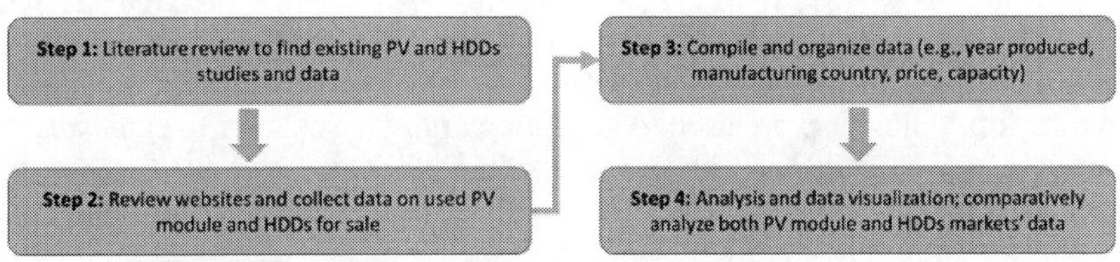

Figure 1: Flow chart of data collection and analysis method.

There are various ways of collecting market data. A literature review or interview can be conducted, and more recently, web scraping has become a common practice [12]. In this study, several methods were used due to wide range of markets and availability of data. For both used PV modules and used HDDs, the types of data collected were the year of production of the product, its origin (manufacturing location), the model name, the product's capacity (Watt-peak (Wp) for PV modules and Terabytes (TB) for HDDs), and the unit price on the secondary market. For new PV modules and HDDs, only the average yearly price was collected as the primary focus of the study was on secondary markets.

The primary source for new PV prices were collected from the Wood Mackenzie 2019 report (Figure 1, step 1). Secondhand PV market prices were collected using the Beautiful Soup Python plugin, that automatically collected PV prices and other descriptors daily from Energybin.com during a period of 10 months (Figure 1, step 2 and Figure 2). Other secondhand data from Kinectsolar.com was scraped prior to this project and has been updated for this analysis. Finally, a third source of used PV modules was manually collected from GreatSolarPanels.com since the website design and limited inventory made writing a scraping script unnecessary. In total, 20 websites were looked at for collecting secondhand PV data, but only three contained secondhand modules for sale. The lack of websites selling secondhand modules was a challenge, since even some websites that had inventory were duplicates of websites we had already used. Moreover, the websites frequently changed the layouts of their webpages, which required updates of the web scraping code and may have caused some data points to be missed. When all collected data was compiled into a data frame, there were around 2,000 data points, but when duplicates were removed our number dropped down to 117 data points (Figure 1, step 3). This is due in part to how our web scrapers collected data daily and the lack of module sales on the websites. Over the collection period, it was clear that most modules take a long time to sell (often a month or two). Finally, modules were often sold in batches, meaning that one data points represent between 2 to 2992 modules being sold (for a total of 13,367 modules). For example, in March 2021, 624 SUN-250 PV modules were for sale on the Energybin website, representing a single data point. Once organized, data were plotted in various way to analyze and find trends in the PV secondary market and to enable its comparison with the HDDs secondary market (Figure 1, step 4).

A similar method was used to collect data on HDDs secondary market. The New HDD market price-per-terabyte (PPT) information was collected from Coughlin (2020) (Figure 1, step 1).The used HDD market prices were automatically collected from diskprices.com, a website which scrapes secondhand HDDs for sale on all amazon platforms internationally for the same duration as PV modules (Figure 1, step 2). Similarly to the PV data, duplicates were removed (Figure 1, step 3), leading to 17,781 data points. Because the year of the HDD model was not indicated, a sample of 127 HDDs were then selected to collect that missing information. In the HDD sub-sample, one data point is corresponding to one used HDD. In addition to collecting used HDD prices, third party EOL HDD companies were contacted. Re-Tek, a data-wiping and electronic refurbishment company provided an internal report, which was used in this study (Figure 1, step 1). Once all data were collected, plots were drawn to analyze trends and enable comparison with the PV modules secondary market (Figure 1, step 4).

For the analysis (Figure 1, step 4), once the data collection period had ended, the data was organized and cleaned up using the pandas Python plugin. The data was then converted into an identical format from each website and duplicates were removed in the data frame. Once each row of data was unique and homogenous, the seaborn and matplotlib plugins were used to generate various visuals for the analysis. Scatter plots and box plots were used as well as some basic linear regression plots. The violin plot function from seaborn was also applied to get a sense of the underlying distribution of the data. Results are presented in the next section.

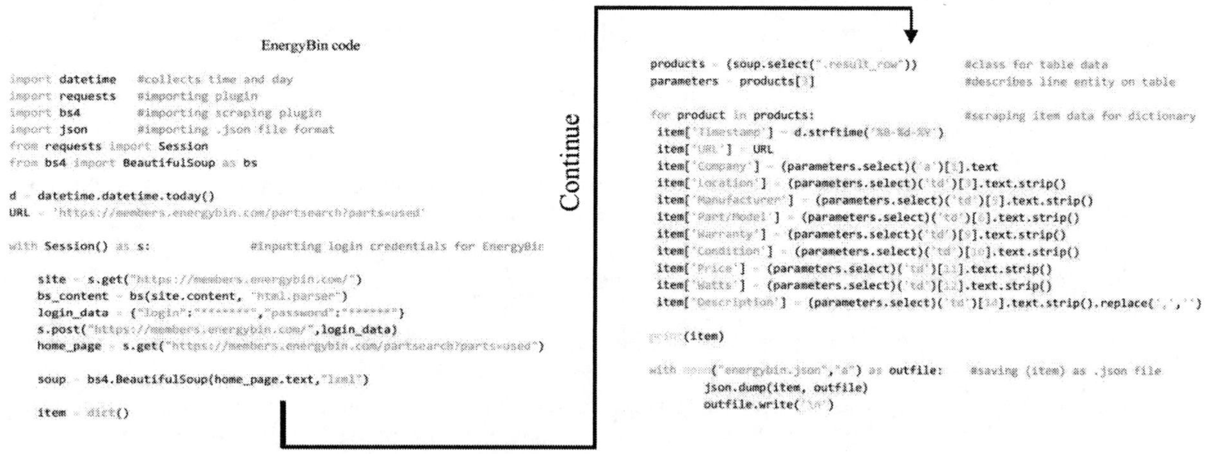

Figure 2. Python Beautiful Soup scraping code.

3. Results

Despite having primary data for PV and HDD prices and ratios, additional market data gives a more holistic view of the PV market. One of the biggest differences between the PV and HDD market is the average useful lifetime of the products (Table 1). PVs have a useful life that is five times longer than HDDs. This

IOP Conf. Series: Materials Science and Engineering 1196 (2021) 012029 doi:10.1088/1757-899X/1196/1/012029

fact combined with the recent commercial adoption of PVs in the early 2000's shows why the annual product end-of-life for the PVs is much smaller than the annual installed capacity. On the other hand, HDDs have been used in computers since the 1960's and thus the market is more developed (Table 1).

Table 1: Product market metrics by type

	PV	HDDs
Average lifetime	30 years (IRENA & IEA, 2016)	6 years (iNEMI, 2019)
Annual installed capacity (units)	32.4 million (IRENA & IEA, 2016)	70 million (Shehabi et al., 2016)
Annual EOL product (units)	1.5 million (IRENA & IEA, 2016)	50 million (iNEMI, 2017)
Annual used sales (estimated units)	10-25 thousand	1.5-3 million

The secondhand PV market represents 0.05% of total annual sales, which is two factors of magnitude smaller than the secondhand HDD market (4.2% of total sales). In addition, in looking at the data from a waste perspective, the secondhand PV market absorbs about 1.0% of the waste, while the secondhand HDD market absorbs 6% of the waste generated annually. So, while there is room for waste reduction efforts in both markets, the secondhand HDD market absorbs six times more waste than the secondhand PV market. The annual used sales were estimated from our primary data collection and the literature [8][18].

Next, we present an analysis of the primary data collected. First, Table 2 presents the summary statistics for both the PV and HDD secondary markets. The table shows that most of the secondhand modules were manufactured in 2013, while the oldest module for sale was from 2000. Given their shorter lifetime, used HDDs were, in average, manufactured in 2017.

Table 2: Primary data summary statistics by product type (Std = standard deviation).

	PV				HDDs			
	Mean	**Std**	**Min**	**Max**	**Mean**	**Std**	**Min**	**Max**
Capacity (Wp/TB)	264	48	75	370	4.4	3.6	0.32	16
Year produced	2013	3.77	2000	2019	2017	2	2012	2020
Used PPW/PPT ($)	0.23	0.13	0.1	0.6	33.75	16.25	13.75	81.47
New PPW/PPT ($)	0.72	0.17	0.4	1.26	38.74	13.55	30	100
Used/new price ratio	0.31	0.21	0.11	1.14	0.92	0.47	0.19	2.34

Figure 3-a shows that there is a small trend of increasing used/new price ratios as the years progress, but overall, the secondhand PV module prices tend to vary more in recent years (Figure 3-a). Figure 3-a also shows the points by website, revealing that no similarities exist between each website's pricing structure. In Figure 3-b we compared the average used price-per-watt (PPW) with the new PPW obtained from the Wood Mackenzie Report. The blue horizontal line was used to represent the average used PPW across the years since projected linear regression model did not fit the data set well. There is no relationship between the average secondhand PPW and the average PPW for new modules, showing a lack of price structure in the secondhand PV market.

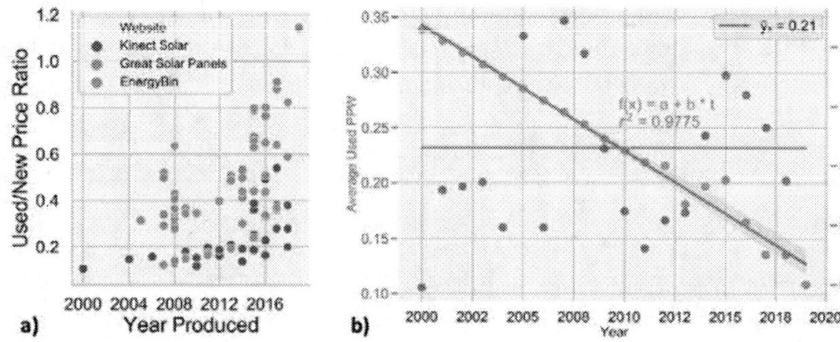

Figure 3: a) scatterplot of PV modules used/new price ratios with color by website; 117 data points, b) Average used PV module price in comparison to new module price by year (a=8.59, b=-4.23, t=year).

Figure 4 shows the average used/new price ratios for PV and HDDs as a function of time. While both products show higher ratios in the more recent years, the figure further demonstrates that the PV secondary market seems less mature with a more erratic relationship.

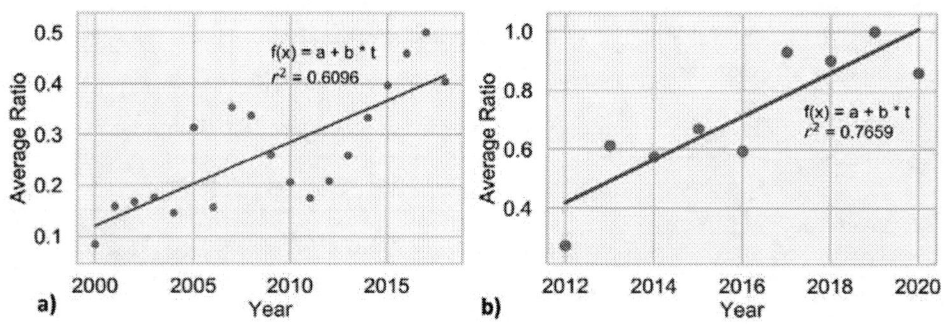

Figure 4: a) PV average used/new price ratios over the years b) HDD average used/new price ratios over the years.

Secondhand PV modules are typically sold at less than the new module price (one outlier at 114% of new PV price) (Table 2); HDDs have greater variability (Figure 5-a). This can be explained since websites list secondhand HDDs with product specifications (e.g., Xbox, Mac, PC etc.) but HDDs are not unique in function and are cross compatible in most cases. This type of market manipulation or price inflation also exists in the PV market (when a specific model is highly demanded due to its compatibility with existing PV projects but rare) but did not appear in our dataset. Finally, Figures 5-b) and 5-c) show the PPW and PPT for all PV modules and HDDs.

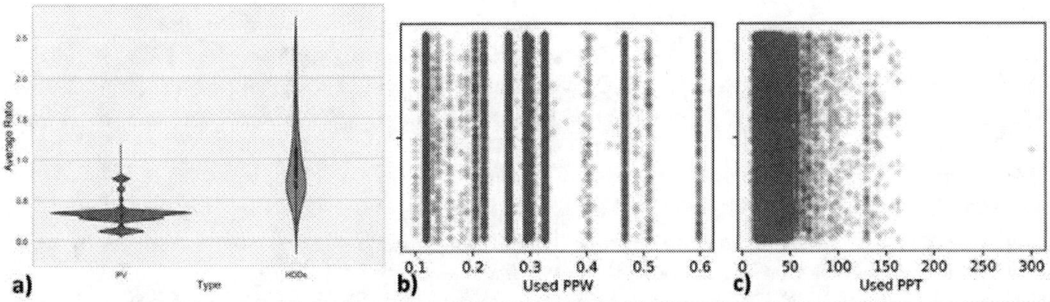

Figure 5: a) Violin plot showing used/new PV and HDD price ratios (13,367 datapoints and 127 data points respectively), b) the 13,367 PV modules PPW and c) the 17,781 HDDs PPT.

4. Conclusion

In understanding the relationship of both the new and used solar panels and HDD markets, this study documented the underdeveloped state of the used solar panel market. This can improve the traction of secondary markets in the sustainable economy. This joint study with the HDD market provides valuable insights on how the PV market can evolve with more regulations, refurbishment companies, and general engagement of stakeholders. Overall, the used PV market is underdeveloped and needs further continuous research to understand what factors (e.g., improved warranties/regulations for used modules) could boost its traction in a circular economy. More research would also lead to better estimations regarding the secondary markets' sizes as new data are available (Table 1). Future research could compare other emerging secondary markets to the mature secondary market for used cars to provide insights in how such emerging secondary markets could be best developed to contribute to a more circular economy.

References

[1] Blake V, Farrelly T, Hannon J, 2019. Is Voluntary Product Stewardship for E-Waste Working in New Zealand? A Whangarei Case Study. *Sustainability* **11**(11), 3063.
[2] CIRAIG 2015. Circular economy: a critical literature review of concepts. Polytechnique Montréal.
[3] Circle Economy, 2021. Circularity Gap Report 2021.
[4] Coughlin T, 2020. HDD market history and projections. https://www.forbes.com/sites/tomcoughlin/2020/05/29/hdd-market-history-and-projections/?sh=495998ec6682. (Accessed 07/2020 2020).
[5] Cristiano S, Ghisellini P, D'Ambrosio G, Xue J, Nesticò A, Gonella F, Ulgiati S 2021. Construction and demolition waste in the Metropolitan City of Naples, Italy: State of the art, circular design, and sustainable planning opportunities. *Journal of Cleaner Production* **293**, 125856.
[6] Gantz J F, Reinsel D, Rydning J 2019. The U.S. Datasphere: Consumers Flocking to Cloud. https://www.seagate.com/files/www-content/our-story/trends/files/data-age-us-idc.pdf (Accessed 12/31/2020 2020).
[7] Ghali M R, Frayret J-M, Ahabchane C 2017. Agent-based model of self-organized industrial symbiosis. *Journal of Cleaner Production* **161**, 452-465.
[8] Google 2019. Environmental Report.
[9] iNEMI 2017. Value Recovery from Used Electronics.
[10] iNEMI 2019. Value Recovery Project, Phase 2.
[11] IRENA and IEA 2016. End-of-Life Management: Solar Photovoltaic Panels.
[12] Krotov V, Silva L, 2018. Legality and ethics of web scraping.

[13] Mathur N, Singh S, Sutherland J W, 2020. Promoting a circular economy in the solar photovoltaic industry using life cycle symbiosis. *Resources, Conservation and Recycling* **155**, 104649.
[14] Millette S, Williams E, Hull C E 2019. Materials flow analysis in support of circular economy development: Plastics in Trinidad and Tobago. *Resources, Conservation and Recycling* **150**, 104436.
[15] Morseletto P, 2020. Targets for a circular economy. *Resources, Conservation and Recycling* **153**, 104553.
[16] Padilla-Rivera A, Russo-Garrido S, Merveille N 2020. Addressing the Social Aspects of a Circular Economy: A Systematic Literature Review. *Sustainability* **12**(19), 7912.
[17] Peiro L, Ardente F 2015. Environmental Footprint and Material Efficiency Support for product policy - Analysis of material efficiency requirements of enterprise servers.
[18] Schmid M 2020. Used Solar Panels Change the Game in the 2020s as Used Cars Did in the 1930s. https://resources.energybin.com. (Accessed 09/20/2020 2020).
[19] Schöggl J-P, Stumpf L, Baumgartner R J 2020. The narrative of sustainability and circular economy - A longitudinal review of two decades of research. *Resources, Conservation and Recycling* **163**, 105073.
[20] Schrijvers D *et al.* 2020. A review of methods and data to determine raw material criticality. Resources, Conservation and Recycling 155, 104617.
[21] Shehabi A, Smith S, Sartor D, Brown R, Herrlin M, Koomey J, Masanet E, Horner N, Azevedo I, Lintner W 2016. United States Data Center Energy Usage Report. ; Lawrence Berkeley National Lab. (LBNL), Berkeley, CA (United States), p. Medium: ED; Size: 66 p.
[22] Tao M, Fthenakis V, Ebin B, Steenari B-M, Butler E, Sinha P, Corkish R, Wambach K, Simon E S, 2020. Major challenges and opportunities in silicon solar module recycling. *Progress in Photovoltaics: Research and Applications* **28**(10), 1077-1088.
[23] Tsanakas J A, van der Heide A, Radavičius T, Denafas J, Lemaire E, Wang K, Poortmans J, Voroshazi E 2020. Towards a circular supply chain for PV modules: Review of today's challenges in PV recycling, refurbishment and re-certification. *Progress in Photovoltaics: Research and Applications* **28**(6), 454-464.
[24] Walzberg J, Lonca G, Hanes R J, Eberle A L, Carpenter A, Heath G A 2021. Do We Need a New Sustainability Assessment Method for the Circular Economy? A Critical Literature Review. *Frontiers in Sustainability* **1**(12).
[25] Wilensky U, Rand W 2015. An introduction to agent-based modeling: modeling natural, social, and engineered complex systems with NetLogo. MIT Press.

Acknowledgment

This work was authored in part by the National Renewable Energy Laboratory, operated by Alliance for Sustainable Energy, LLC, for the U.S. Department of Energy (DOE) under Contract No. DE-AC36-08GO28308. Funding provided by the Advanced Manufacturing Office and the Office of Strategic Programs. The views expressed in the article do not necessarily represent the views of the DOE or the U.S. Government. The U.S. Government retains and the publisher, by accepting the article for publication, acknowledges that the U.S. Government retains a nonexclusive, paid-up, irrevocable, worldwide license to publish or reproduce the published form of this work, or allow others to do so, for U.S. Government purposes.

Towards Artificial Intelligence in Urban Waste Management: an early prospect for Latin America

J C B F Bijos[1,*], L M Queiroz[2], V M Zanta[2], K P Oliveira-Esquerre[1]

[1] Industrial Engineering Department, Federal University of Bahia, Salvador, BA 40210-630, BR

[2] Sanitary and Environmental Engineering Department, Federal University of Bahia, Salvador, BA 40210-630, BR

* **E-mail:** juliabijos@outlook.com

Abstract. The expectations about the increase of solid waste generation are a global concern in order to mitigate the negative impacts caused by this scenario. Recent experiences show advantages in applying Artificial Intelligence (AI) in Urban Solid Waste (USW) management. Although, developing countries face many structural and governance barriers that limits the real potential of applying these technologies. This study presents a worldwide outlook about the application of AI in USW management and identifies the reality of Latin America countries in this new context. It is especially important to improve monitoring and to create data management platforms. Also, this study collaborates on the minimization of possible technological boundaries posed to Latin America region and besides, it can be considered a reference study for countries in similar conditions.

Keywords: urban solid waste, artificial intelligence, machine learning, waste management.

1. Introduction

The mitigation of negative impacts of solid waste generation is an important challenge for developing countries, since they faced a fast population growth, non-controlled urbanization and a lack of financial resources to keep a capable management system (Guerrero et al., 2013; Caruso; Gattone, 2019).
Latin America countries deal with this reality, since they face difficulties to set adequate final disposition, struggling to eradicate open dumps and to disseminate the use of sanitary landfills (Margallo et al., 2019). The countries of Latin America and Caribbean (LAC) region count 8.42% of world population. In addition to that, the rate of urban solid waste generation is 1.09 kg/capita/day. When compared to the world average of 0.74 kg/capita/day it is considered a high value. About 50% of the generated waste is disposed in sanitary landfills. Even though, that is a significant volume being inadequately discarded. (Kaza et al., 2018)
The improvement of urban waste management is urgent since the high volume and vast composition generated in this region allows the use of different techniques for treatment. Technology has a significant potential to support the improvements and advances in the waste sector. Nowadays, Artificial Intelligence (AI) is considered a vast field of knowledge with a huge potential to establish methods and tools for promoting assertive decisions by analysing data, forecasting quantities, automating process to obtain sustainable and efficient practices. These benefits made AI to play a significant role to promote social good and the usage of AI have been emphasized for minimizing inequalities in LAC region.

The LAC countries are in an early stage of defining the basis and policies to use AI and the mainly fields of interests are health, education, agriculture and finance sectors (Panduro & Roman, 2020, Carrillo-Larco et al., 2020). The emerging expectation of using AI promoted an increase in published studies in the field. Cechinel et al. (2020) investigated the increase of publications in Learning Analytics in Latin America from 2011 to 2019. The authors analysed 282 papers based on applied techniques, goal, ethical issues, to cite some. It was demonstrated an enhanced interest to research and to apply AI to set solutions in Latin America region leaded by Brazil and Ecuador in number of publications.

Despite the efforts and advances to constitute AI as a tool for decision-making in many sectors, there are no considerable discussions on literature about the improvement specifically in the sanitation field through AI in Lati America context. The vast majority of the published cases and reviews of AI applications in waste management refers to developed countries (Abdallah et al., 2020; ISWA, 2019).

Thus, to investigate the impacts of applying AI on the waste management of Latin American countries is a relevant task, since these developing countries present limitations related to financial and structural aspects (Leal Filho et al., 2016) that limits an efficient management. To overcome that and consolidate technology as a useful tool, it's necessary to adapt AI based technologies to local reality, as cited by De-Arteaga et al. (2018).

This paper aims to present an initiatory discussion about the perspectives of Latin America to establish AI based solutions in USW management while it indicates a concise outlook of the worldwide applications in the waste sector.

2. Materials and Methods

The methodology of this study comprised a search on bibliography databases as Web of Science and Google Scholar. The key-words investigated was "solid waste management", "artificial intelligence", "waste management", "machine learning", "urban waste management", "urban waste" for the period corresponding of years 2000 to 2021. In addition, many articles were found by consulting cited references.

3. An overview about recent trends of AI usage in USW management

Artificial Intelligence (AI) can be understood as a set of techniques and tools that enable a system to accomplish tasks that initially were exclusive for the human being capacity. AI is composed by the representation of knowledge allied to data manipulation, creativity, capacity of auto correction, ability to perceive and respond to external stimulus (Harkut et al., 2016; Garcia et al., 2018).

Garcia et al. (2018) cite that the intelligence of machines can be classified in levels or kinds. Thus, the classes of AI are comprised in computer vision, robotics, automated reasoning, natural language processing (NLP) and machine learning. This last class is especially important because it deals with the proposition of learning algorithms that guide the decision-making process through analysing inferences, complex associations, predictions and data trends. Machine Learning enables a system to learn and adapting to new conditions by combining rules of computing, statistics and optimization (Garcia et al., 2018).

The several benefits provided by AI poses it as an essential tool to attain the reduction of environmental impacts, development of tools and actions to improve lifespan of equipment, recycle of components and improve the circularity of materials and energy in USW management. Then, this field of sanitation is a potential target to improvements by AI.

3.1. Machine Learning for prediction of USW generation

The increase on solid waste generation is an environmental challenge that affects later steps of the process: storage, collection, transportation, treatment, recycle and final disposition in adequate form (Khan & Samadder, 2014). In this sense, the use of Machine Learning techniques is an efficient way to extract patterns and understanding factors that influences waste generation and the expected composition. It enables to predict the volume generated or the composition expected, developing a numerical or categorical based approach. Based on bibliographic searches, these applications major apply the methods Cluster Analysis, Artificial Neural Networks, Support Vector Machine (SVM), Tree Based Methods and K- Nearest Neighbours (KNN).

IOP Conf. Series: Materials Science and Engineering 1196 (2021) 012030 doi:10.1088/1757-899X/1196/1/012030

The most cited features to investigate the USW generation profile are climactic variables (average temperature, precipitation rate, wind velocity), demographic variables (population density, age range, family size) and socioeconomic variables as income level, educational level, employment rate (Dias et al., 2012).
In addition, variables related to the waste collection (collection volume, weight, composition) are essential to analysis and georeferenced data can improve the prediction based on locations (Johnson et al., 2017, Kontokosta et al., 2018, Liu et al., 2019, Kannangara et al., 2018).

3.2. Collection improvements: estimation of bin level and route optimization
Some challenges faced in this stage of management are the high spend of fuel, atmospheric emissions, need for estimation the ideal time for collection and the estimation of the ideal level of waste bins. These difficulties justify the collection stage to concentrate about 70% of operational costs as cited by Ferrer e Alba (2019). In Table 01 are described some applications for the development of smart bins.
Another approaches (Khan e Samadder, 2016; Kim et al., 2006) proves advantages on the route optimization through the use of GIS tools (*geographic information system)*. Although, the limitations to optimize many observations (Sulemana et al., 2018) is a reason for associate GIS tools with Machine Learning (ML). This association allows improvements on spatial analysis by applying image processing techniques and also, allows the perception of urban area in a social and structure perspective. Besides, integration of the two tools, reduces analysis time (Tohidi and Rustamov, 2020).
Optimizing routes allows to reduce distance travelled and quantity of vehicles in circulation, which causes improvements on traffic and reduction on emissions of pollutants (Kontokosta et al., 2018). This corroborates that modelling is a powerful tool for the decision-making and the generation of insights for planning and implementing sustainable and efficient actions.

3.3. Automatic sorting
The automatic sorting is considered a technology to approximates the USW management to sustainable practices and optimization of recycling process (Costa et al., 2019).
In this stage of the process, the use of image detection enables the automatic classification of the waste. This is advantageous since it enhances the potential to an adequate segregation of components for reuse or recycle.
Costa et al. (2019), Arayakandy et al. (2019) and Bansal et al. (2019) successful applied Machine Learning to automatic classification of waste composition by images, as described in Table 01.
These applications confirm that the automatic sorting achieved by AI contributes also for developing equipment, which enhances the benefits. The cited approaches demonstrate the potential of AI to reduce errors on sorting which increases the efficiency of the valorisation process. In addition, AI plays a significant role for reducing the health risks for employees by minimizing the manual contact between them and the solid waste fractions.

Table 1. AI based applications for waste collection and waste sorting

	Author	Application	Method
Estimation of bin level	Zhao et al. (2017)	Identifies different fill levels of bins by changes in motor vibrations.	Cluster Analysis (K-Means)
	Aziz et al. (2018)	Classification of the fill level of bins and prediction of reminiscent days for waste collection	SVM for classification and Hidden Markov Model (HMM) for prediction
	Costa et al. (2019)	Classification of waste images in paper, glass, plastic and metal	Best performance using CNN (Convolutional Neural Network),
Imag e	Arayakandy et al. (2019)	Classification of waste images in plastic, glass, paper, metal, cardboard and non-recyclable	SVM

Bansal et al. (2019)	Detection of valuable waste in real time video and development of robotic arm to allocate the waste in adequate compartments	CNN

4. Perspectives and future challenges

The lack of adequate waste segregation associated to inefficiencies to execute alternatives treatments as composting or anaerobic treatment, causes loss in (I) the reuse and recycling potential, and (II) energy and economic recovery of components. This reflects a remedial vision of the process, which limits the perception about the economic value of the waste and poses the final disposition in sanitary landfills as a treatment stage (Leal Filho et al., 2016).

The difficulties faced to manage the sector are barriers to the modernization, especially related to the application of smart technologies in less economically favoured countries. Cultural and economic aspects, legal and infra structural and technological availability (Leal Filho et al., 2016; Guerrero et al., 2013) poses differences among developing and developed countries. Because of that, when prospecting solutions based on AI, primarily using Machine Learning, it is important to consider the conditions of each country, as mentioned for De-Arteaga et al. (2018).

In the face of this, the following discussion is based on three categories of improvements based on the local conditions.

4.1. Data Management

On a report of Mont et al. (2020) it was indicated that every 12 countries investigated (Argentina, Brazil, Chile, Colombia, Costa Rica, Dominican Republic, Ecuador, Mexico, Paraguay, Peru, Trinidad and Tobago, Uruguay) present some kind of policy to consolidate AI as a digital strategy for many sectors. A notable topic of this expectation is the consolidation of data management systems.

The bibliography investigation highlighted the importance of data management to improve the monitoring of collection, treatment and final disposition. In addition, the aggregation of georeferenced data from GIS tools generates a data description level that enables the creation of precise models. In this sense, a data management system plays a leading role in the potential to use AI based systems. The data management and share are considerable challenges, that needs integrated actions on public and private sectors.

Although, LAC countries struggle to establish a data management system and it is common to note low data availability, or little detailed data, outdated records or without standardized labels, also contradictions in records (Deus et al., 2016). It is important to note that these are topics of interest for investigations, given the prominence of the usage of Machine Learning techniques based on scarce data (De-Arteaga et al., 2018) or operations with heterogeneous data (Caruso & Gattone, 2019). Besides, resampling methods and data imputation places a fundamental role to increase the quantity of observations on databases and reduction of bias in models. The integration of data bases from different sources is a relevant topic, particularly when dealing with restrictions to collect new data (Guerrero et al., 2013; Meza et al., 2019).

Relevant efforts to establish the USW management improvement by data systems are verified in public sector of Brazil, Ecuador, Colombia and Argentina with the main aim to adequate monitor waste generation, provide reports for citizens, enhance the reuse or recycle of waste and manage the financial resources of the waste sector.

4.2. Waste collection

The use of GIS tools is a great way of gathering georeferenced data, and poses advantages to the route optimization by the integration of data. Bueno-Delgado et al. (2019), Khan & Samadder (2016) worked on route optimization and ideal allocation of waste bins, achieving success on reducing costs and collection planning.

The access to internet in urban areas allows the utilization of platforms or applications that inform the users the route and time to collection, give instructions about waste segregation and guarantee transparency on information. The application "I got Garbage" was successfully applied in India, to inform citizens the collection planning and promote instructions to adequate segregation (Kaza et al., 2018). Initiatives based on the interactions with citizens, as mobile applications, are potentially feasible. Although, those kinds of initiatives are not well established in the region and are mainly developed by private sector. It is not common the public sector disseminates that.

Latin American countries has a vast potential to use internet as a tool for the communication with citizens and establishment of connectivity-based resources. The penetration rate of the internet on region is 72%, Brazil is the country that presents the majority of internet users. (Statista, 2021)

In this sense, to inform citizens the route and schedule of collection trucks, or to allow citizens to use a mobile application to report infractions on waste collection is an option to approximate them of the decision-making process. Although, this effort needs to be complemented by sensibilization to avoid biased information.

4.3. Manufacturing and treatment

The use of technologies on manufacturing process allows the increase of materials reuse, considering the life cycle of components. This can be achieved by the use of software that guides the choose of materials (Kaza et al, 2018).

As cited before, manual waste sorting is susceptible to errors and health risks to employees. Because of that, the automatic sorting is an interesting way for the optimization of the sorting process, increase of components recovery and protection of employees.

In addition, other interesting application is the monitoring of sanitary landfills by drones. They can be used to verify the capacity of the landfill, to detect gas leak or verifying spaces to allocate equipment, successfully applied in Brazil (Mello et al. 2017; da Silva et al. 2020; da Silva et al., 2018). Figure 01 points the mainly applications and potential impacts in every stage cited.

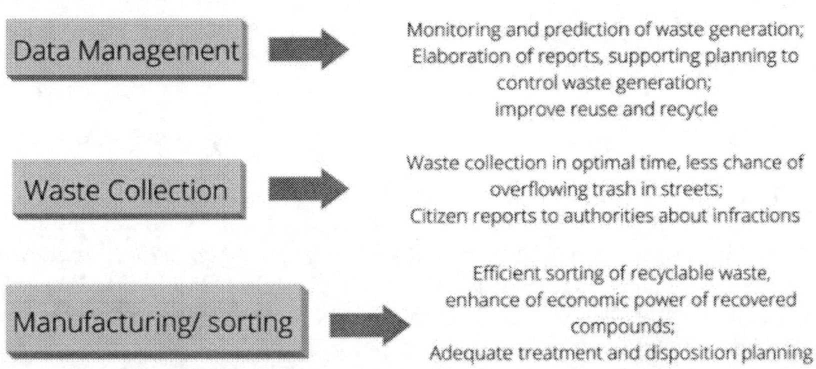

Figure 1. Potential impacts of AI based solutions

5. Conclusion

This study exposed the conceptual basis of AI, evidencing its importance the establish decision-making tools. It is presented an initial overview about the usage of AI in USW management in three stages: generation, collection and automatic sorting of the solid waste.

In view of that, the main challenges and opportunities of applying AI in the UWS management in Latin America is discussed. Despite the limitations on data quality, Latin America countries can benefit on the investigation of Machine Learning topics of interest as data imputation, database integration, among others. Also, it is observed a great potential of the region to benefit of internet-based tools since the spread of internet access are vast and present an increase trend.

Thus, data management, waste collection and manufacturing or monitoring are the main opportunities to apply AI and Machine Learning based solutions. It is interesting to note that the consolidation of AI in developing countries is gradual in order to guarantee the sustainability of actions in front of the technological and human resources available. Also, it is a worldwide interest to know how AI based tools are adapted to less favoured nations. This study briefly shares the potentialities of AI field in USW management and evidences progresses that can be achieved by that.

6. References

[1] Guerrero L A, Maas G, Hogland W 2013 Solid waste management challenges for cities in developing countries *Waste Management* **33** 220–32

[2] Caruso G, Gattone, S A 2019 Waste Management Analysis in Developing Countries through Unsupervised Classification of Mixed Data *Soc. Sci.* **8** 6 186.

[3] Margallo M, Ziegler-Rodriguez K, Vázquez-Rowe I, Aldaco R, Irabien A, Kahhat R 2019 Enhancing waste management strategies in Latin America under a holistic environmental assessment perspective: A review for policy support *Science of The Total Environment* **689** 1255–75

[4] Kaza S, Yao L, Bhada-Tata P, Van Woerden, F 2018 *What a Waste: A global snapshot of solid waste management to 2050* (World Bank Group: Washington)

[5] Panduro A, Roman S 2020 Personalized medicine in Latin America *Personalized medicine* **17** 5 339–43

[6] Carrillo- Larco R M, Castillo- Cara M, Anza- Ramirez C, Bernabé-Ortiz A 2021 Clusters of people with type 2 diabetes in the general population: unsupervised machine learning approach using national surveys in Latin America and the Caribbean *BMJ Open Diabetes Research and Care* **9** 1 1–8

[7] Cechinel C, Ochoa X, Lemos H S, Carvalho J B N, Rodés V and Queiroga E M 2020 Mapping Learning Analytics initiatives in Latin America *Br J Educ Technol.* **51** 4 892–914.

[8] Abdallah M, Talib M A, Feroz S, Nasir Q, Abdalla H, Mahfood B 2020 Artificial intelligence applications in solid waste management: A systematic research review *Waste Management* **109** 231–46.

[9] International Solid Waste Association (ISWA) 2019 How Industry 4.0 transforms the waste sector

[10] Leal Filho W, Brandli L, Moora H, Kruopien J, Stenmarck A 2016 *Journal of Cleaner Production* **112** 4377–86

[11] De-Arteaga M, Herlands W, Neill D B, Dubraswki A 2018 Machine Learning for the Developing World *ACM Trans. Manage. Inf. Syst* **9** 2

[12] Harkut D G, Kasat K, Harkut V D 2016 Introductory Chapter: Artificial Intelligence – Challenges and Applications In.: *Artificial Intelligence - Scope and Limitations* IntechOpen

[13] García C G, Núñez-Valdez E R, García-Díaz V, G- Bustelo C P, Lovelle J M C 2018 A Review of Artificial Intelligence in the Internet of Things *International Journal of Interactive Multimedia and Artificial Intelligence* **5** 4 1–13

[14] Khan D and Samadder S R 2014 Municipal solid waste management using Geographical Information System aided methods: A mini review *Waste Management & Research* **32** 11 1049–62

[15] Dias D M, Martinez C B, Barros R T V, Libânio M Modelo para estimativa da geração de resíduos sólidos domiciliares em centros urbanos a partir de variáveis socioeconômicas conjunturais *Engenharia Sanitária e Ambiental* **17** 3 325–32.

[16] Johnson N E, Ianiuk O, Cazap D, Liu L, Starobin D, Dobler G, Ghandehari M 2017 Patterns of waste generation: A gradient boosting model for short-term waste prediction in New York City *Waste Management* **62** 3–11

[17] Kontokosta C E, Hong B, Johnson N E, Starobin D 2018 Using machine learning and small area estimation to predict building-level municipal solid waste generation in cities *Computers Environment and Urban Systems* **70** 151–162

[18] Liu J, Li Q, Gu W, Wang C The impact of consumption patterns on the generation of municipal solid waste in China: Evidences from provincial data *Int. J. Environ. Res. Public Health* **16** 10 1–19

[19] Kannangara M, Dua R, Ahmadi L, Bensebaa F 2018 Modeling and prediction of regional municipal solid waste generation and diversion in Canada using machine learning

approaches *Waste Management* **74** 3–15
[20] Ferrer J, Alba E 2019 BIN-CT: Urban Waste Collection based on Predicting the Container Fill Level *Biosystems* **181**
[21] Khan D, Samadder S R 2016 Allocation of solid waste collection bins and route optimisation using geographical information system: A case study of Dhanbad City, India *Waste Management & Research* **34** 7 666–76.
[22] Kim B, Kim S and Sahoo S 2006 Waste collection vehicle routing problem with time windows *Computers & Operations Research* 33 12 3624–42
[23] Sulemana A, Donkor E A, Forkuo E K and Oduro-Kwarteng S 2018 **2018** 1–12
[24] Tohidi N and Rustamov B R 2020 A Review of the Machine Learning in GIS for Megacities Applications In IntechOpen. *Geographic Information Systems in Geospatial Intelligence* **32** 137–44
[25] Costa B, Bernardes A, Pereira J, Zampa V, Pereira V, Matos G, Soares E, Soares C and Silva A 2018 Artificial Intelligence in Automated Sorting in Trash Recycling *Proc. Anais do XV Encontro Nacional de Inteligência Artificial e Computacional* Porto Alegre: SBC pp 198–205
[26] Arayakandy A, Gupta A and Thakur R 2019 Design and Development of Classification Model for Recyclability Status of Trash Using SVM *International Journal for Research in Applied Science & Engineering Technology (IJRASET)* **7** 3, 2146–50
[27] Bansal S, Patel S, Shah I, Patel A, Makwana J and Thakker R 2019 AGDC: Automatic Garbage Detection and Collection *Robotics*
[28] Aziz F, Arof H, Mokhtar N, Shah N M, Khairuddin A S M, Hanafi E, et al. 2018 Waste level detection and HMM based collection scheduling of multiple bins *PLoS ONE* **13** 8
[29] Zhao Y, Yao S, Li S, Hu S, Shao H and Abdelzaher T F 2017 VibeBin: A Vibration-Based Waste Bin Level Detection System *Mobile, Wearable and Ubiquitous Technologies* v. 1 n. 3 1–23
[30] Mont C G, Del Pozo C M, Pinto C M and Alcocer A V M C 2020 Artificial Intelligence for social good in Latin America and the Caribbean: The Regional Landscape and 12 Country Snapshots Banco Interamericano de Desarrollo p 142
[31] Deus R M, Battistelle R A G, Silva G H R 2015 Solid waste in Brazil: context, gaps and trends Engenharia Sanitaria e Ambiental **20** 4 685–98
[32] Meza J K S, Yepes D O, Rodrigo-Ilarri J and Cassigara E 2019 Predictive analysis of urban waste generation for the city of Bogotá, Colombia, through the implementation of decision trees-based machine learning, support vector machines and artificial neural networks *Heliyon* **5** 1–11
[33] Bueno-Delgado M V, Romero-Gázquez J L, Jiménez P and Pavón-Mariño P 2019 Optimal Path Planning for Selective Waste Collection in Smart Cities *Sensors* (Switzerland) **19** 9 1–14
[34] Statista [homepage] 2021 Internet use in Latin America - Statistics & Facts [accessed at 20 mar 2021] Available at: https://www.statista.com/topics/2432/internet-usage-in-latin-america/#dossierSummary.
[35] Mello C C S, Salim D H C and Teixeira L P T 2017 Uso de VANT como ferramenta auxiliar na gestão integrada de aterros sanitários *Proc. IX Simpósio Brasileiro de Engenharia Ambiental, XV Encontro Nacional de Estudantes de Engenharia Ambiental e III Fórum Latino Americano de Engenharia e Sustentabilidade* 2292–2302.
[36] Silva P R da, Ribeiro E A W, Serpe B M, Tobon G M, Meira E, Mattedi M A, Gaia C J, Spierss M R, Ludwing L, Castro C A, Jacks P G and Martins A O 2020 Transferência de Tecnologia para Monitoramento Ambiental por Drone em Aterros Sanitários Regulados e Recursos Hídricos no Vale do Itajaí – SC *Metodologias e Aprendizado* **2** 52–7.
[37] Silva J R 2019 Automação do processo de compostagem: uso de sensores para monitoramento e controle de parâmetros de um processo sustentável Master Thesis in Production Engineering Universidade Federal de Goiás – UFG 75 p.

Acknowledgments

The Authors gratefully acknowledge the Foundation to Support Research in the State of Bahia (FAPESB), the Coordination of Improvement of Higher Education Personal (CAPES), the National Council of Scientific and Tecnological Development (CNPq) and the Federal University of Bahia (UFBA).

Environmental Product Declaration (EPD) For Clay Roof Tiles-Case Study: Production Plant of Clay Roof Tiles in Republic of Serbia

Aleksandar Drpić[1,*], Tea Spasojević-Šantić[2], Zagorka Radojević[3]

[1]Institute for testing materials-IMS Institute, Bulevar vojvode Mišića43, 11000 Belgrade, Republic of Serbia
[2] Institute for testing materials-IMS Institute, Bulevar vojvode Mišića43, 11000 Belgrade, Republic of Serbia,
tea.spasojevic@institutims.rs
[3]Institute for testing materials-IMS Institute, Bulevar vojvode Mišića43, 11000 Belgrade, Republic of Serbia, zagorka.radojevic@institutims.rs

Email corresponding author: aleksandar.drpic@institutims.rs

Abstract. Environmental Product Declaration (EPD) is a standardized way of quantifying the impact of a product or system on the environment. This study evaluates the environmental impact of 1 tone of clay roof tiles, produced in production plant in Serbia, defined and grouped by use of Product Category Rules (PCR). The purpose of this study is to determine: (1) life cycle stages of the product which have the major impact on the environment expressed as environmental impact categories; (2) the processes of clay roof tile production which have the most important effects on the environment also expressed as environmental impact categories. The LCA analysis has been conducted with the One Click LCA software, developed by Bionova Ltd, Finland. All processes have been modelled based on the inventory data given in the Ecoinvent database (v3.6). According to the results in this study and observed from the aspect of the product life cycle, the production process has the major impact on the environment, and from the aspect of the resources used, the major impact on the environment has the consumption of energy and the use of raw materials.

Keywords: LCA, EPD, PCR, roof tiles.

1. Introduction

In Europe, the building sector consumes 40% of the total energy and raw materials, emits 36% of the total greenhouse gas emissions (GHG) and produces 15% of the total industrial waste. This sector consumes more raw materials than any other industrial sector and it also involves high energy consumption [1]. [2] concluded that the most commonly used construction materials are steel, cement and ceramic materials and are also the major energy consumers among construction materials. Those conclusions showing us that is important to establish a more efficient construction industry from environmental aspect, but also from the aspect of energy efficiency. Building products like clay roof tiles generates environmental impacts during its entire life cycle stages, such as consumption of resources, water, energy, waste, transport of raw materials, vehicle used for transportation, packaging materials, etc. [3]. Growing awareness of the environmental performance of construction products and buildings brings about the need for a suitable method to assess their environmental performance and to provide proper assistance for applying this information for further use in the field of building certification [4]. Environmental Product Declarations (EPDs) belong to the type III environmental declarations. Environmental product declarations (EPD) for construction products are based on reliable and verifiable LCA informations. The EPD contains verified and precise information about the environmental impact of products and the potential for improvement based on the scientifically proven facts [5]. Two important mechanisms for EPD development are: Product category rules (PCR) and Life Cycle Assessment (LCA). Life Cycle Assessment (LCA) is a recognized approach to assess the environmental impacts of life cycle of the product or a service starting from the extraction of raw materials until the end-of-life (recycling, reusing and landfilling) [6,7], helping with the identification of environmental impact of the product and determination of influences for unit process from the environmental performance aspect. LCA is a tool to assess and quantify the potential environmental impacts by collecting and characterizing the inputs (product, material or energy flow that enters a unit process) and outputs (product, material or energy flow that leaves a unit process) in different processes of the product's life cycle. [3]. It has also been used as a tool for decision-making when it's necessary to determine environmental performance of products and the environmental comparison of different building elements [8]. This approach is named cradle-to-grave LCA. LCA results can be used to prioritize and make strategic planning decisions regarding the design or redesign of products or processes [3]. The objective of this study is to provide the European manufacturers of clay construction products, i.e. clay bricks, blocks, pavers and roof tiles, guidance for the development of Environmental Product Declarations (EPDs) for clay construction products based on a LCA approach (EPD from cradle to grave) [3]. Life Cycle Assessment (LCA) includes product stage (A1-A3), construction process stage (A4-A5), use stage (B1-B7), end of life stage (C1-C4+D). For this purpose, LCA results can provide information that serves as a basis for implementing strategies to reduce pollution, preserving resources, and reduce waste within a company, industry, or government sector. [9]. The methodology used to develop the EPD is based on standards EN ISO 14025[10]., ISO 21930 [11] and EN 15804 [5]. In accordance with the requirements of the standard EN 15804 [5], each EPD needs to cover all life cycle stages of the product (module A-D), as well as informations about the product, reference service life (RSL), declared unit, functional unit, system boundary, allocation, period under review, all input and output flows, and environmental impact categories assessment with results interpretation and recommendations. A comparison or an evaluation of EPD data is only possible if all the data sets to be compared were created according to EN 15804 and the building context, respectively the product-specific characteristics of performance, are considered [5].

2. Material and methods

For the purposes of this study, the CML methodology is used. CML is the methodology of the Institute of Environmental Sciences Faculty of Science University of Leiden, Netherlands [5]. The main purpose is to group environmental impact categories and expressed them as emissions into the environment. The environmental impact categories are following:

- global warming potential (GWP),
- ozone depletion potential (ODP),
- acidification potential (AP),
- eutrophication potential (EP),
- formation of ozone of lower atmosphere (POCP)
- abiotic depletion potential (ADP-elements) for non-fossil resources (ADPE),
- abiotic depletion potential (ADP-fossil fuels) for fossil resources (ADPF).

The LCA analysis has been conducted with One Click LCA software developed by Bionova Ltd, Finland [12]. All inputs and outputs are taken from Ecoinvent database (v3.6) [13]. According to the PCR for clay construction products, in case of insufficient input data, the maximum cut-off of input flows for a module is 5% for energy use and mass, while it is maximum 1% for unit processes [3]. The functional unit was 1 t of clay roof tile ready to be used for a period of 50 years. The declared unit was 1 t of clay roof tile. The system boundary includes: raw material extraction and processing (A1), transport to the manufacturer (A2), manufacturing (A3), transport to the building site (A4), installation into the building (A5), waste transportation (C2), waste processing (C3) waste disposal (C4) and external impacts (excluded from totals) (D). The input and output flows for production of 1t of clay roof tiles were considered in this study: the mass of raw materials (clay), transportation vehicle, truck engine type, distances to a different locations, wooden pallets, thermofoil, water consumption, energy consumption for natural gas and electricity, hazardous and non-hazardous waste. These input and output flows are provided from production plant in Republic of Serbia and refer to 2019. The production stage of clay roof tiles includes the following processes: clay extraction, primary processing of clay, clay preparation and shaping, drying, firing, quality management control, packaging and shipping. Clay roof tiles are produced by molding (extruding and / or pressing), drying and firing of clay mass without addition of additives. In the production of clay roof tiles, only natural resources are used. Clay is exploited from a clay pit located near the plant and water is added to clay to achieve the required moisture content, and additionally, electricity and natural gas are used as energy resources. The red color of the tile is achieved during firing at high temperatures (over 1020°) and it is resistant to UV radiation. Clay roof tiles produced in this plant have a properly defined microstructure that provides excellent mechanical properties and exceptional resistance to high and low temperatures. This study does not exclude any hazardous materials or substances.

3. Results and discussion

The results of product life cycle assessment are presented in three different categories: environmental impacts, resource use and waste generation.

Table 1 represents environmental impact assessment results obtained in the One Click LCA software for clay roof tiles production in production plant in Republic of Serbia for entire life cycle (EPD type: cradle-to-grave).

Observing the energy use in the production stage (module A1-A3), there is a significant impact for the following categories: abiotic depletion potential for fossil resources (ADPF), ozone depletion potential (ODP), acidification potential (AP), and global warming potential (GWP). Consumption of raw materials in the production stage (A1) has a significant impact for the following categories: abiotic depletion potential for non-fossil resources (ADPE) and eutrophication potential (EP).

Transport to the manufacturer (A2) has the lowest impact on the environment for the following categories: global warming potential (GWP), abiotic depletion potential for fossil resources (ADPF), ozone depletion potential (ODP) and acidification potential (AP). Also, waste disposal (C4) has low impact on the environment for the following categories: global warming potential (GWP), ozone depletion potential (ODP) and acidification potential (AP).

Resource use per 1 tone of clay roof tiles are shown in the table 2 and waste categories and output flows per 1 tone of clay roof tiles are shown in table 3. There is an environmental benefit at the end-of-life stage (module D) due to reusing of clay roof tiles (reuse rate of 90%) after the deconstruction stage (C1) for another application-as fillings in roads construction.

Table 1. Environmental impact results of 1 tone of clay roof tiles [14].

Parameter	Unit	A1	A2	A3	A4	A5	B1-C1	C2	C3	C4	D
GWP	[kg CO_2-eq.]	1,084E1	3,096E0	1,482E2	6,668E1	2,15E−2	-	4,267E1	9,748E0	5,168E−1	−1,356E1
ODP	[kg CFC11-eq.]	1,154E−6	5,514E−7	3,291E−5	1,125E−5	7,149E−9	-	7,572E−6	1,914E−6	1,718E−7	−1,491E−6
AP	[kg SO_2-eq.]	5,822E-2	6,474E−3	1,129E0	1,577E−1	8,668E−5	-	8,886E−2	4,697E−2	2,084E−3	−6,124E−2
EP	[kg $(PO_4)^{3-}$-eq.]	2,266E−2	1,422E−3	9,432E−2	3,591E−2	1,677E−5	-	1,951E−2	8,295E−3	4,032E−4	−2,414E−2

POCP	[kg Ethene-eq.]	4,034E−3	3,845E−4	4,897E−2	8,336E−3	6,356E−6	-	5,807E−3	1,818E−3	1,528E−4	−3,989E−3
ADPE	[kg Sb-eq.]	2,328E−3	1,125E−4	3,114E−4	2,336E−3	2,002E−7	-	1,543E−3	1,202E−4	4,812E−6	−9,512E−4
ADPF	[MJ]	1,455E2	4,609E1	5,1E3	9,917E2	6,099E−1	-	6,329E2	1,704E2	1,466E1	−2,022E2
Caption	GWP=Global warming potential; ODP=Ozone depletion potential; AP=Acidification potential; EP=Eutrophication potential; POCP= Formation of ozone of lower atmosphere; ADPE=Abiotic depletion potential for non-fossil resources; ADPF= Abiotic depletion potential for fossil resources										

Table 2. Resource use per 1 tone of clay roof tiles [14].

Parameter	Unit	A1-A3	A4	A5	B1-C1	C2	C3	C4	D
PERE	[MJ]	6,312E2	1,271E1	4,952E−3	-	1,089E1	2,551E0	1,19E−1	−1,178E1
PERM	[MJ]	0	0	0	-	0	0	0	0
PERT	[MJ]	6,312E2	1,271E1	4,952E−3	-	1,089E1	2,551E0	1,19E−1	−1,178E1
PENRE	[MJ]	5,323E3	1,006E3	6,163E−1	-	6,485E2	1,734E2	1,481E1	−2,304E2
PENRM	[MJ]	0	0	0	-	0	0	0	0
PENRT	[MJ]	5,323E3	1,006E3	6,163E−1	-	6,485E2	1,734E2	1,481E1	−2,304E2
SM	[kg]	1,4E0	4,677E−1	1,66E−4	-	3,144E−1	7,904E−2	3,991E−3	−2,309E−1
RSF	[MJ]	0	0	0	-	0	0	0	0
NRSF	[MJ]	0	0	0	-	0	0	0	0
FW	[m^2]	5,16E−1	1,724E−1	6,7E−4	-	1,113E−1	8,453E−2	1,611E−2	−1,085E0
Caption	PERE = Use of renewable primary energy excluding renewable primary energy resources used as raw materials; PERM = Use of renewable primary energy resources used as raw materials; PERT = Total use of renewable primary energy resources; PENRE = Use of non-renewable primary energy excluding non-renewable primary energy resources used as raw materials; PENRM = Use of non- renewable primary energy resources used as raw materials; PENRT = Total use of non-renewable primary energy resources; SM = Use of secondary material; RSF = Use of renewable secondary fuels; NRSF = Use of non-renewable secondary fuels; FW = Use of net fresh water								

Table 3. Waste categories and output flows per 1 tone of clay roof tiles [14].

Parameter	Unit	A1-A3	A4	A5	B1-C1	C2	C3	C4	D
HWD	[kg]	2,925E0	1,473E0	5,732E−4	-	7,539E−1	2,631E−1	1,378E−2	−8,687E−1
NHWD	[kg]	6,383E1	6,714E1	4,173E0	-	4,184E1	4,022E2	1,003E2	−3,644E1
RWD	[kg]	1,215E−2	6,351E−3	4,052E−6	-	4,341E−3	1,089E−3	9,74E−5	−1,011E−3
CRU	[kg]	0	0	0	-	0	-	0	0
MFR	[kg]	0	0	0	-	0	0	0	0
MER	[kg]	0	0	0	-	0	0	0	0
EEE	[MJ]	0	0	0	-	0	0	0	0
EET	[MJ]	0	0	0	-	0	0	0	0
Caption	HWD = Hazardous waste disposed; NHWD = Non-hazardous waste disposed; RWD = Radioactive waste disposed; CRU = Components for re-use; MFR = Materials for recycling; MER = Materials for energy recovery; EEE = Exported electrical energy; EET = Exported thermal energy								

It was also considered which inputs and outputs have significant impact on the environment for all impact categories. According to the table 4, for GWP (global warming potential), most contributing inputs and/or outputs are: energy for electricity (36,08%), transported mass (34,19%). For ODP (ozone depletion potential), most contributing inputs and/or outputs are: transported mass (30,27%), energy for electricity (29,29%), energy for natural gas (21,55%). For AP (acidification potential), most contributing input and/or output is energy for electricity (61,37%). For EP (eutrophication potential), most contributing inputs and/or outputs are: energy for electricity (33,69%), transported mass (27,79%). For POCP (formation of ozone of lower atmosphere), most contributing input and/or output is energy for electricity (48,8%). For ADPE (abiotic depletion potential for non-fossil resources), most contributing inputs and/or outputs are: transported mass (47,44%), operation with clay (28,68%). For ADPF (abiotic depletion potential for fossil resources), most contributing inputs and/or outputs are: energy for natural gas (44,69%), transported mass (21,31%).

Table 4. Most contributing environmental impact categories for 1 tone of clay roof tiles [14].

Impact category	Energy for electricity	Transported mass	Energy for natural gas	Operation with clay	Waste from clay roof tiles	Wooden pallets
GWP (%)	36,08	34,19	6,38	3,78	3,0	-
ODP (%)	29,29	30,27	21,55	2,23	-	-
AP (%)	61,37	15,84	7,27	3,88	2,98	-
EP (%)	33,69	27,79	4,82	11,54	-	5,85
POCP (%)	48,8	19,04	10,48	5,59	-	4,42
ADPE (%)	1,4	47,44	-	28,68	1,45	1,54
ADPF (%)	18,2	21,31	44,69	2,16	2,2	-

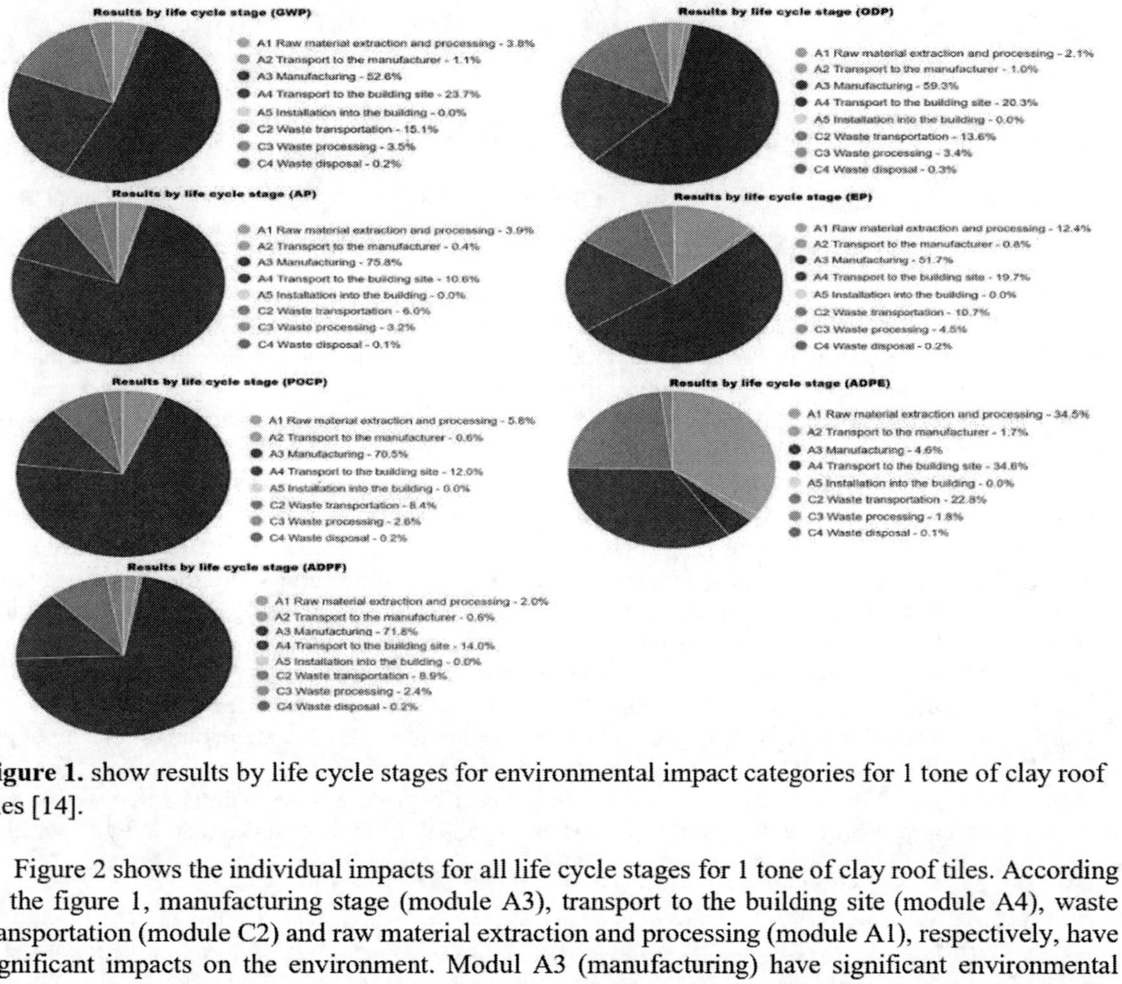

Figure 1. show results by life cycle stages for environmental impact categories for 1 tone of clay roof tiles [14].

Figure 2 shows the individual impacts for all life cycle stages for 1 tone of clay roof tiles. According to the figure 1, manufacturing stage (module A3), transport to the building site (module A4), waste transportation (module C2) and raw material extraction and processing (module A1), respectively, have significant impacts on the environment. Modul A3 (manufacturing) have significant environmental impact for the following categories: AP (71,53%), ADPF (65,94%), and POCP (65,02%). Modul A4 (transport to the building site) have significant environmental impact for the following categories: ADPE (28,15%), GWP (20,55%), and ODP (17,82%). Modul C2 (waste transportation) have significant environmental impact for the following categories: ADPE (18,59%), GWP (13,15%), and ODP (12%). Modul A1 (raw material extraction and processing) have significant environmental impact for the following categories: ADPE (28,05%), EP (11,22%), and POCP (5,36%). Regarding the lowest environmental impact, it was noticed that following modules, C3 (waste processing), A2 (transport to the manufacturer), C4 (waste disposal) and A5 (installation into the building), respectively, with their

IOP Conf. Series: Materials Science and Engineering 1196 (2021) 012031 doi:10.1088/1757-899X/1196/1/012031

individual values for all 7 impact categories do not exceed 5%.

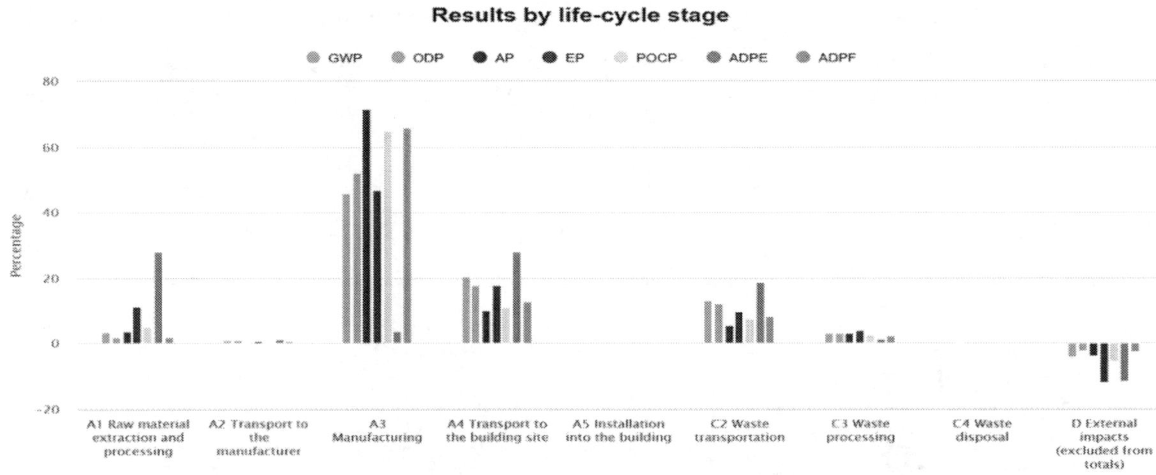

Figure 2. Environmental impact results for the life cycle stages of 1 tone of clay roof tiles [14].

4. Conclusions

Environmental Product Declarations (EPD) for construction products provides important information about products and their use. „Green products" and „cleaner production process" are priority nowadays, and EPD is used to obtain all the necessary information about construction products which have an impact on the environment during their entire life cycle.

Based on the EPD, it is possible to determine which production processes save resources and energy, and which are more financially payable, as well as which are "eco-friendly" in order to reduce the impacts on the environment and human health. The most significant inputs and/or outputs on the environment according to the LCA results are: energy for electricity, the mass of tiles that are transported to the location, energy for natural gas, operation with clay. It is concluded that due to high energy consumption during the roof tile production, as well as due to the transportation of the finished products to long distances, the primary energy and global warming potentials increase. Reducing the temperature during firing process, minimizing transportation of raw materials, controlling emitters during primary processing in order to reduce emissions of hazardous substances into the atmosphere, are some measures that must be implemented according to Serbian law in order to reduce environmental impacts. However, according to the results, there is an environmental benefit at the end-of-life stage due to reusing of clay roof tiles after the deconstruction stage for another application, as fillings in roads construction. This reduces the overall impact on the environment. In the nearest feature, wen manufacturers want to put any eco-label on their product, they will be obligated to publish EPD for that product on ECO platform.

5. References

[1] Almeida MI, Dias AC, Demertzi M, Arroja L, (2015). Contribution to the development of product category rules for ceramic bricks. In: *Journal of Cleaner Production* **92**, 206-215. https://doi.org/10.1016/j.jclepro.2014.12.073

[2] Bribián I Z, Capilla A V, Usón A A (2011). Life cycle assessment of building materials: comparative analysis of energy and environmental impacts and evaluation of the eco-efficiency improvement potential. *Build. Environ.* **46**, 1133-1140

[3] TBE PCR 2012-TBE PCR for clay construction products, Guidance document for developing an EPD, Tiles & Bricks Europe aisbl, www.tiles-bricks.eu

[4] Passer A, Lasvaux S, Allacker K, Lathauwer D, Spirinckx C, Wittstock B, Kellenberger D, Gschösser F, Wall J, Wallbaum H, (2015). Environmental product declarations entering the building sector: critical reflections based on 5 to 10 years experience in different European

countries. *Int J Life Cycle Assess* **20**:1199–1212. doi: 10.1007/s11367-015-0926-3
[5] CEN (2013), EN 15804:2012+A1:2013: Sustainability of construction works - Environmental product declarations - Core rules for the product category of construction products, European Committee for Standardization, Brussels
[6] Curran M, de Baan L, De Schryver A M, van Zelm R, Hellweg S, Koellner T, Sonnemann G, Huijbrejts M A J (2011). Toward meaningful end points of biodiversity in Life Cycle Assessment. *Environ. Sci. Technol.* **45**, 70-79
[7] EEA European Environmental Agency, (1997). Life Cycle Assessment: a Guide to Approaches Experiences and Information Sources. European Environmental Agency, Brussels, p.119. available at: www.eea.europa.eu/publications/GH-07-97-421 595-EN-C
[8] Souza D M, Lafontaine M, Charron-Doucet F, Bengoa X, Chappert B, Duarte F, Lima L, (2015). Comparative Life Cycle Assessment of ceramic versus concrete roof tiles in the Brazilian context. *Journal of Cleaner Production* **89**. 165-173. https://doi.org/10.1016/j.jclepro.2014.11.029
[9] Hunkeler D, Rebitzer G (2005). The Future of Life Cycle Assessment. *Int. J. LCA*, Vol. **10**, No. 5, Ecomed publishers, Landsberg, Germany, pp. 305 – 308
[10] CEN (2010), EN ISO 14025: Environmental labels and declarations - Type III environmental declarations - Principles and procedures, European Committee for Standardization, Brussels
[11] ISO (2017), ISO 21930: Sustainability in buildings and civil engineering works — Core rules for environmental product declarations of construction products and services, International Organization for Standardization, Geneva
[12] OneClickLCA software, developed by Bionova Ltd, Finland, all informations available at: https://www.oneclicklca.com/
[13] Ecoinvent database, version 3.6, Developed by Ecoinvent Association, Switzerland, all information available at: https://www.ecoinvent.org/
[14] Environmental Product Declaration according to EN ISO 14025, ISO 21930 and EN 15804 for clay roof tiles, (2020). EPD-IMS-2020-02-EN, Institute for testing materials-IMS Institute, Republic of Serbia

Ecosystem Services – An Approach for Sediment Management

Ross O'Sullivan*, **Joseph Harrington, Brano Batel**
School of Building & Civil Engineering and Sustainable Infrastructure Research & Innovation Group, Munster Technological University, Ireland

*Corresponding Author Email: ross.osullivan@mycit.ie

Abstract. The ecosystem services approach involves classifying, describing and assessing the monetary value of natural resources in terms of the benefits that humans can derive from nature. An ecosystem services assessment has been developed and applied to the sediment management aspect of a harbour development for a scale typical for Ireland. Two sediment management techniques have been analysed for the sediment generated from dredging for the project; wetland creation and the widely practised disposal at sea. The wetland creation approach is consistent with the eco-friendly concept of 'Building with Nature' and also the circular economy. The ecosystem services assessment applied to the disposal at sea option yielded a negative overall monetary value (-€191,878/year) and the wetland creation option resulted in a positive overall monetary value (€1,403,275/year). These assessment results indicate that a 'Building with Nature' solution beneficially using dredged sediment can provide significant monetary benefit in the form of ecosystem services, for a sediment management project of this scale. This ecosystem services approach can contribute significantly to assessment of a sediment management projects and can inform stakeholders and policy makers of the benefits of 'Building with Nature' solutions.

Keywords: Dredging, ecosystem services, sustainable, sediment management, resource, circular economy.

1. Introduction

Dredging is the removal of sediment from the bottom of harbours, lakes and other water bodies, such as navigational channels, berthing areas, and marinas. It is essential for the maintenance and development of ports and is a key feature of national and international infrastructure and of vital economic importance to worldwide trade and commerce. It is estimated that between 200 and 250 million tonnes of marine sediment are dredged annually in the EU [2] with the Netherlands, Belgium, France and Germany accounting for approximately 70% of this [35]. The quantities of dredge sediment generated internationally is also significant with the United States dredging 230 million tonnes annually [43]. In Ireland approximately 1.8 million tonnes is dredged annually [23][36], which is significantly lower than for other countries, but the importance of dredging Ireland's navigational channels is critical as Ireland's main ports and harbours account for 99% of Ireland's imports and exports by volume and 95% by value [38].

In Ireland, dredged sediment is classified as a waste under the Waste Management Act and requires a license or a permit (depending on the quantity), if the sediment is treated or stored on land [10]. The

Waste Framework Directive, indicates that dredged sediment can be used as a resource and not as a waste, if certain criteria can be met [17].

Sediment management is a key aspect of a dredging project and in general dredge sediment may be managed by disposal, treatment and/or beneficial use. The disposal approach may be: unconfined disposal at sea, confined disposal at sea or land disposal. There are many techniques used for the treatment of sediment and depend on the sediment characteristics and the intended use of the sediment. Beneficial use options are many and one general approach to categorization is as follows [22]: Engineering uses (substitute for land based resources e.g. sand, aggregate, rock), Environmental Enhancement (wetland creation, sediment cell maintenance and fill for abandoned mines) and Product Uses (sediment used to form useful products e.g. manufactured topsoil, road sub-base construction and landfill liner).

Numerous studies provide greater detail on the various sediment management beneficial use options including United States Army Corp of Engineer (1987, 2007, 2015), PIANC (1992, 2009a, 2009b) and the OSPAR Commission (2009). Recent research in a European context includes the PRISMA Project (2014), the CEAMaS Project (2015), the USAR Project (2016) and the SURICATES Project (2018), all funded by the EU Interreg Programme.

Although Ireland generates a low sediment volume relative to many other countries, it encounters many of the same challenges experienced by larger countries. The main sediment management challenge is seeking an alternative to disposal at sea for primarily clean uncontaminated sediment, in particular the finer grained fraction of the sediment. The management method(s) suitable for dredge sediment depend on a number of factors including physical characteristics, contamination levels, volume of sediment, how and where the sediment is dredged, available beneficial use opportunities, engineering acceptability and local practice. These factors are considered when undertaking a feasibility study of the sediment management options including completion of technical, economic, environmental and ecosystem services analyses in addition to legislative and societal aspects. This paper primarily focuses on the ecosystem services aspect of dredge sediment management.

2. Ecosystem Services Approach to Sediment Management

2.1. Background to Ecosystem Services

Ecosystems are dynamic environments and can include a wide variety of species. This biodiversity is critical to the health of an ecosystem. Ecosystems can come under pressure in two forms, natural and human [19]. From a dredging perspective, human pressures on ecosystems include the construction of marine infrastructure projects (e.g. port developments) and the act of dredging itself can be destructive to habitats, if not managed correctly. It may be noted that the negative changes that ecosystems have experienced due to human pressures have resulted in significant gains to human well-being and economic development [26].

Westman (1977) first investigated the complex link between ecosystems and economic systems [47]. Ehrlich & Ehrlich (1981) coined the term 'ecosystem services' and in the following decade, ecologists and scientists further examined the concept that ecosystems are life-support systems, providers of ecosystem services and economic benefits [8][12][18][28]. In the late 1990's the concept of valuing the world's ecosystem services and natural capital gained widespread attention [6][7]. Annan (2000) called for a Millennium Ecosystem Assessment (MEA) by submitting a report to the UN General Assembly entitled "We the Peoples: The Role of the United Nations in the 21st Century" [1]. In 2001, the MEA was initiated under the auspices of the United Nations. The MEA purposely did not focus on valuing ecosystems, but rather on the complex relationships between ecosystems and human well-being (Figure 1).

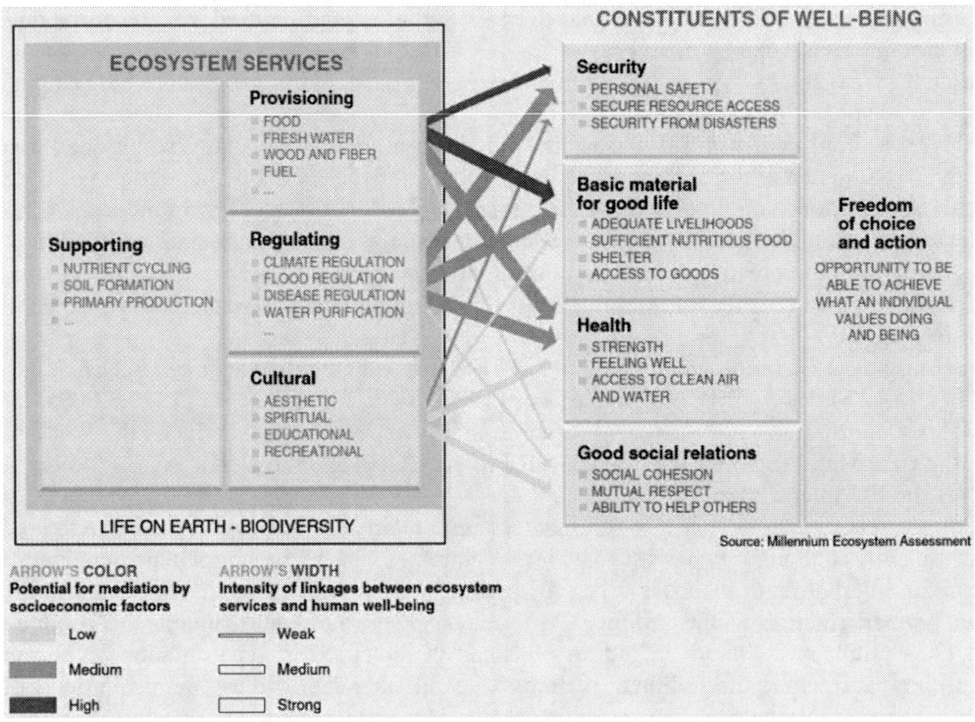

Figure 1. Linkages between ecosystem services and human well-being.
(Millenium Ecosystem Assessment, 2005)

MEA (2005) found, that human pressures have changed ecosystems rapidly over the past 50 years. These changes to ecosystems have provided gains to human well-being and economic development, but at the expense of the ecosystems and that reversing the degradation of ecosystems is possible but involves significant changes in policies and practices that are not currently being undertaken . These changes can be achieved by recognising the importance of ecosystem services. MEA (2005) categories ecosystem services as follows: supporting, provisioning, regulatory and cultural.

The Economics of Ecosystems and Biodiversity Group (2010) emphasized the global economic value of biodiversity; this can assist decision makers in determining the monetary value of ecosystems.

2.2. Ecosystem Services and Sediment Management

Historically, the construction of water infrastructure projects involving dredging has been an essential activity in a country's development and prosperity. This is even more prevalent today with the continuous growth of trade, resulting from increased prosperity and world population growth. Marine infrastructure projects can be disruptive to the environment. As a result, society's response has been to evaluate the impacts of dredging on the environment. This has led to international and national legislation being developed such as the Habitats Directive [16] and the Water Framework Directive [17] in Europe. In general legislation is becoming more stringent for environmental protection, which can negatively impact on the implementation of innovative solutions for sediment management projects.

The dredging industry has recognised this worldwide shift towards sustainability with PIANC (2008) leading the way for the dredging industry demonstrating the importance of understanding the natural system and using natural processes to achieve a sustainable project. De Vriend & Van Koningsveld (2012) illustrated real examples of 'Building with Nature' solutions to sediment management projects [9]. Bridges et al. (2014) illustrated examples of successful 'Engineering with Nature' sediment management projects with the United States Army Corps Engineers presenting further case studies [3][44]. Further guidance on achieving sustainability in dredging projects while adding value to the natural and socio-economic system has recently been presented [5].

Maintenance of a port's navigable areas is key to facilitating the international trend to larger ships and an increasing import-export shipping trade. In a market economy, the assets of a port are appraised in terms of job creation, GDP and import-export trade [24]. Most ports operate at the mouth of an estuary where there is typically an abundance of ecosystems. It may be stated that these ecosystems were previously classified, in general, as without 'economic good'. The challenge that the dredging industry faces is assessing the economic value of ports with the economic value that ecosystems provide. This is where the concept of ecosystem services becomes an important contributor, where the benefits that humans derive from ecosystems can be quantified in monetary terms. By taking an ecosystem services approach towards sediment management projects, stakeholders have the opportunity to develop innovative and sustainable sediment management strategies such as 'Building with Nature' solutions.

The concept of taking an ecosystem services approach to a sediment management project is developed in this paper. An ecosystem services assessment is an important contributor in undertaking a full analysis of different sediment management options. Ecosystem services can contribute to ensuring that the analysis for a dredging project is sustainable and that the project is implemented satisfactorily in an environmentally sensitive location. The objective of an ecosystem services approach is to classify, outline and assess the monetary benefits that society derives from ecosystems. The benefits that humans derive from ecosystems are delivered in sediment management projects, but these benefits may not be considered as providing an economic good by the project stakeholders. An ecosystem services assessment provides quantifiable economic results which can be included as a key part of an overall assessment of a sediment management project.

3. Application of an Ecosystem Services Approach

3.1. The General Framework

In order to determine the positive and negative monetary values that humans derive from ecosystems, previously defined as 'ecosystem services', the unit cost (€/hectare/year) of the ecosystem is combined with the change in the land use (hectare). The change in land use refers to the positive/negative change in an ecosystem area that results from the activities associated with the development. The calculation of the ecosystem services in monetary terms, can then contribute to a full analysis for the different sediment management options for a dredging project. This paper focuses, in particular, on the ecosystem services aspect of a sediment management project.

It should be noted that the most suitable sediment management option for a dredging project is generally based on a cost-benefit analysis of the proposed options and the ecosystem services assessment typically may not feature. The methodology and application developed in this paper addresses this by undertaking an ecosystem services assessment as shown in Figure 2.

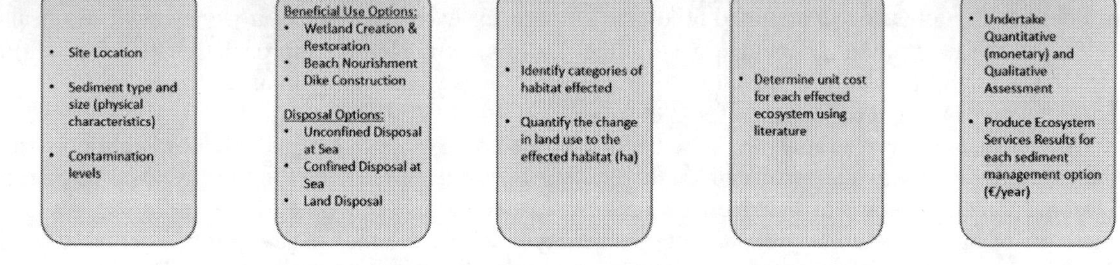

Figure 2. Ecosystem services Assessment – Methodology.

3.2. Assessment Methodology

This section provides greater detail on the methodology used for calculating the ecosystem services value for a sediment management project. There are five general steps in the ecosystem services assessment.

Step 1 involves inputting the site-specific details including the site location, physical characteristics of the sediment and contamination levels of the sediment. The physical characteristics and contamination levels dictate the potential amount of sediment available for beneficial use.

Step 2 involves selecting the sediment management options to be assessed. This typically involves selecting a number of beneficial use options and often including the traditional disposal at sea option.

Step 3 involves identifying the categories of habitats potentially impacted by the sediment management project and quantifying the change in land use for these impacted habitats. The category of habitat typically found in dredging projects include: offshore, at shore, estuary, rivers, artificial structures, hard substrate and terrestrial. In Ireland, for example, the category of habitats is publicly available as GIS files through the National Parks and Wildlife Service website [27]. The change in land use can be quantified (hectares) by comparing the current situation against the proposed sediment management option. This step is critical to the accuracy of the ecosystem services assessment as the land use change is combined with the unit cost (Step 4) to determine a monetary value (Step 5).

Step 4 involves sourcing the relevant unit costs associated with an ecosystem service, in terms of the value that these ecosystems provide to humans. Unit costs for this paper were primarily found from national [14] and international sources [25].

Step 5 involves undertaking a quantitative (monetary) and qualitative assessment of the ecosystem services for each sediment management option. The changes in land use (in hectares) multiplied by a unit cost (€/hectare/year) allows a monetary value to be assigned to the benefits that humans derive from ecosystems as €/year.

$$Ecosystem\ Services = Change\ in\ Land\ Use\ (Hectares)\ x\ Unit\ Cost\ (€/hectare/year) \qquad (1)$$

4. Ecosystem Services Assessment – An Application to a Harbour Development

An ecosystem service assessment is initially applied in a preliminary analysis for a specific 'Building with Nature' type sediment management option for a harbour development project, assumed to be undertaken in Ireland. The aim is to investigate the impact on ecosystem services when comparing a 'Building with Nature' sediment management option to the widely practiced approach of unconfined disposal at sea.

The sediment management project is assumed to involve dredging 140,000m^3 of dredge material, clean and uncontaminated. The project assumes extending an existing quay structure, deepening berthing pockets for ships and construction of an adjacent amenity area. 90,000m^3 of this is assumed to be fine-grained material with the remaining 50,000m^3 being rock. A suitable wetland creation site is assumed as a sediment management option and located 3km sail distance from the dredging site. The project beneficially uses an assumed 50,000m^3 of rock for harbor reclamation works, with disposal of 90,000m^3 of fine-grained sediment to sea. For the 'Building with Nature' approach the 90,000m^3 of fine-grained sediment is assumed suitable for placement at the wetland site with an average depth of fill of 1m. This gives an area of wetland of 90,000m^2 or 9ha.

An ecosystem services analysis is then applied based on the approach presented in Section 3. Table 1 and Table 2 present the results of the ecosystem services assessment for the Disposal at Sea and Wetland Creation sediment management options, respectively. The 'Category' heading refers to the category of habitat. 'Habitat' refers to the habitat type affected either negatively or positively by the dredging project. 'Change in land use' is the change (positive or negative) in terms of hectares that the dredging project has impacted on the specific habitat type. The 'Total' is the sum of all the ecosystem services affected by the change in land use for a specific habitat type. The 'Overall Effect' is a qualitative summary of the affect that the dredging project has on a specific habitat type.

The Disposal at Sea option yields a negative overall value of -€191,878/year (Table 1). Table 1 shows that the overall result is mainly impacted by the Subtidal Deep habitat. A change in land use of -11.43 hectares occurs as a result of the existing quay structure seaward extension as part of the development and thus having a negative impact on this habitat type.

The Wetland Creation option results in a positive overall value of €1,403,275/year (Table 2). It can be seen from Table 2 that the +9 hectares of land use for the bare tidal flat habitat mainly impacts on the overall result. The +9 hectares is created using 90,000m^3 of sediment and with fill depth placement of 1m at the relocation site to create the wetlands.

To put the results from the ecosystem services assessment in context, a direct cost analysis of such a harbour development project was carried out using an economic model developed by Munster Technological University [20][21]. The direct cost of the sediment management project is approximately €6.3m and €7.6m for the wetland creation and unconfined disposal at sea options, respectively.

Table 1. Indicative Ecosystem services Results - Disposal at Sea Option.

Category	Habitat	Change in land use	Total	Overall Effect
		ha	*(€/y)*	*POS/NEG*
Shore	Beach	-0.86	-1,197	NEG
Estuary	Subtidal deep	-11.43	-193,478	NEG
	Bare Tidal Flat			
Artificial Structures	Artificial reefs	1.74	1,570	POS
Hard Substrate	Rock Rubble	-0.71	-641	NEG
Terrestrial	Dunes, Planted Area	0.16	424	POS
	Grassland	0.5	1,444	POS
Summary			-191,878	POS/NEG

Table 2. Indicative Ecosystem services Results – Wetland Creation Option.

Category	Habitat	Change in land use	Total	Overall Effect
		ha	*(€/y)*	*POS/NEG*
Shore	Beach	-0.86	-1,197	NEG
Estuary	Subtidal deep	-11.43	-193,478	NEG
	Bare Tidal Flat	9	1,595,153	POS
Artificial Structures	Artificial reefs	1.74	1,570	POS
Hard Substrate	Rock Rubble	-0.71	-641	NEG
Terrestrial	Dunes, Planted Area	0.16	424	POS
	Grassland	0.5	1,444	POS
Summary			1,403,275	POS/NEG

5. Discussion and conclusions

This paper presents an ecosystem services approach that may be applied to assess sediment management scenarios for dredging projects providing an additional important contribution to informing the decision-making process in selecting the most sustainable sediment management option. This ecosystem services approach demonstrates the potential to quantify the monetary effects (both positive and negative) that ecosystem services provide for a sediment management project.

The application of an ecosystem services approach is applied in this paper to a harbour development project, assumed to be undertaken in Ireland, where two different sediment management scenarios are analyzed. The first scenario is the widely practiced unconfined disposal at sea approach and the second scenario is a 'Building with Nature' solution in the form of wetland creation. This approach is consistent with the European Union's (EU) adoption of a 'Circular Economy Action Plan', which aims to help close the loop of product lifecycles through greater re-use, bringing benefits both for the environment and the economy [15].

The analysis and application of ecosystem services to 'Building with Nature' solutions contributes to potentially ensuring more sustainable use of a natural resource in the form of dredged material in the context of the principles of the circular economy. Indicative results from the analysis presented in this paper show the positive ecosystem services monetary value where the sediment is beneficially used.

This ecosystem services approach may also contribute to sediment management permitting policy and practice. For example the Irish regulator for assessing dredged material disposal at sea applications is the Environmental Protection Agency (EPA). Ireland and similar to many other countries requires an assessment of the sediment management alternatives to disposal at sea as part of the permitting process. This paper illustrates the value of undertaking an ecosystem services assessment where beneficial use of sediment is considered as a project option. The inclusion of such an assessment as part of the permitting process would enhance the potential for application of innovative 'Building with Nature' approaches, potentially reducing dredge sediment quantities disposed at sea.

This research work is continuing with a range of other 'Building with Nature' type sediment management techniques being assessed in the context of ecosystem services including, for example, beach nourishment and dike construction and with application to specific sites in Ireland.

References

[1] Annan K A (2000) *We The Peoples: The Role of the United Nations in the 21st Century*. New York: United Nations.

[2] Barceló D, Petrovic M, Bortone G, Heise S, Palumbo L and Owens P N (2007) Sediment Risk Management and Communication: Sustainable Management of Sediment Resources *(SEDNET)*. Elsevier.

[3] Bridges T S, Lillycrop J, Wilson J, Fredette T, Suedel B, Banks C and Russo E (2014) Engineering with nature promotes triple-win outcomes. *Terra et Aqua,* **135**(2), 17-23.

[4] CEAMaS (2015) *Civil engineering applications for marine sediments*. Retrieved March 14, 2021, from http://www.ceamas.eu

[5] CEDA / IADC (2018) *"Dredging for Sustainable Infarstructure"* (1st ed.). The Hague, the Netherlands: CEDA / IADC.

[6] Costanza R, d'Arge R, de Groot R, Farber S, Grasso M, Hannon B, . . . van den Belt, M. (1997) The value of the world's ecosystem services and natural capital. *Nature*, 253-260.

[7] Daily G, Postel S, Bawa K S and Kaufman L (1997) Nature's Services: Societal Dependence On Natural Ecosystems.

[8] De Groot R S (1992) Functions of Nature: Evaluation of Nature in Environmental Planning, Management and Decision Making. *Polar Record*, 345.

[9] De Vriend H and Van Koningsveld M (2012) *Building with Nature. Thinking, acting and interacting differently.*

[10] Department of the Environment Heritage and Local Government (1996) *Waste Management Act.* Dublin, Ireland: Dept. of the Environment, Hertitage and Local Government.

[11] Ecoshape (2008) *About Ecoshape*. Retrieved June 05, 2019, from https://www.ecoshape.org/en/about-ecoshape/

[12] Ehrlich P R and Mooney H A (1983) Extinction, Substitution, and Ecosystem Services. *BioScience,* **33**(4), 248-254.

[13] Ehrlich P and Ehrlich A (1981) *Extinction. The Causes and Consequences of the Disappearance of Species* (Vol. 1st edition). New York: Random House.

[14] Environmental Protection Agency (2014) *Valuing Ireland's Coastal, Marine and Estuarine Ecosystem Services.*

[15] European Commission (2020) *A new Circular Economy Action Plan For a cleaner and more competitive Europe.* Brussels.

[16] European Union (1992) Council Directive 92 /43 /EEC on the conservation of natural habitats and of wild fauna and flora. Official Journal of the European Union.

[17] European Union (2000) Directive 2008/98/EC Of the European Parliament and of the Council of 19 November 2008 on waste and repealing certain Directives. Official Journal of the European Union.

[18] Folke C, Hammer M and Jansson A-M (1991) Life-support value of ecosystems: a case study of the Baltic Sea Region. *Ecological Economics,* **3**(2), 123-137.
[19] Galaz V, Moberg F, Downing T E, Thomalla F and Warner K (2008) *Ecosystems Under Pressure.*
[20] Harrington J R, Murphy J, Coleman M, Jordan D and Szacsuri G (2016a) Financial modelling and analysis of the management of dredged marine sediments – development of a decision support tool. *Journal of Shipping and Trade.*
[21] Harrington J R, Murphy J, Coleman M, Jordan D, Debuigne T and Szacsuri G (2016b) Economic modelling of the management of dredged marine sediments. *Geology, Geophysics & Environment,* **42**(3), 311-324.
[22] Harrington J and Smith G (2013) *Guidance on the Beneficial Use of Dredge Material in Ireland.* Cork, Ireland: Environmental Protection Agency.
[23] Harrington J, Sutton S and Lewis A (2004) Dredging and dredge disposal and reuse in Ireland – a small island perspective. *Proceedings of World Dredging Congress XVII*, B**4-5**, pp. 1-14. Hamburg, Germany.
[24] IADC (2013) Ecosystem Services & Dredging.
[25] IADC (2016) Towards Integrated Marine Infrastructure Project Assessment. The Hague, the Netherlands.
[26] Millenium Ecosystem Assessment (2005). *Ecosystems and Human well-being: current state.* Washington, DC: Island Press.
[27] National Parks and Wildlife Service (2021) Maps & Data. Retrieved from https://www.npws.ie/maps-and-data
[28] Odum E P (1989) Ecology and our endangered life-support systems. *Journal of Tropical Ecology*, 283.
[29] OSPAR Commission (2009) Assessment of the environmental impact of dredging for navigational purposes. OSPAR Commission, New Court 48 Carey Street London WC2A 2JQ, United Kingdom
[30] PIANC (1992) Beneficial use of dredged material. EnviCom Working Group 19. Brussels.
[31] PIANC (2008) Working with Nature. EnviCom Working Group 176. Brussels.
[32] PIANC (2009a) Long term management of confined disposal facilities for dredged material. EnviCom Working Group 11. Brussels.
[33] PIANC (2009b) Dredged Material as a Resource. EnviCom Working Group 104. Brussels.
[34] PRISMA (2014) Promoting Integrated Sediment Management – Final Report. (http://www.prismaprojects.eu/-images/Final%20report%20PRISMA.pdf)
[35] SedNet (2004) An Overview of Sediment and Desdged Material Treatment.
[36] Sheehan C and Harrington J (2012) Management of dredge material in the Republic of Ireland – A review. *Waste Management,* **32**(5), 1031-1044.
[37] Sheehan C, Harrington J and Murphy J (2009) Dredging and Dredge Material. *Terra et Aqua* (Number 115), 3-14.
[38] Shields Y, O' Connor J and O' Leary J (2005) Ireland's Ocean Economy & Resources. *Marine Foresight Series,* No. **4**.
[39] SURICATES (2017) *Sediment Uses as Resources In Circular And Territorial EconomieS.* Retrieved April 29, 2021, from https://www.nweurope.eu/projects/project-search/suricates-sediment-uses-as-resources-in-circular-and-territorial-economies/
[40] TEEB (2010) Integrating the ecological and economic dimensions in biodiversity and ecosystem service valuation.
[41] United States Army Corps Engineers (1987) Beneficial uses of dredged material. EM 1110-2-5026, US Army Corps of Engineers, Washington, DC
[42] United States Army Corps Engineers & United States Environmental Protection Agency (2007) Identifying, planning, and financing beneficial use projects using dredged material:

beneficial use planning manual. U.S. Environmental Protection Agency & U.S. Army Corps of Engineers, Washington, DC
[43] United States Army Corps Engineers (2015) Dredging and Dredged Material Management - Engineering Manual. U.S. Army Corps of Engineers, Washington, DC
[44] United States Army Corps Engineers (2018) Engineering with Nature - An Atlas (1st ed.). Vicksburg, MS: U.S. Army Engineer Research and Development Center.
[45] United States Army Corps Engineers (2018) How can the dregding sector join the global shift towards sustainability. *Terra ET Aqua,* March(**150**), 19-26.
[46] USAR (2016) Using sediments as a resource. Retrieved March 19, 2021, from http://wrt.org.uk/project/usar/
[47] Westman W E (1977) How Much Are Nature's Services Worth? *Science*, **197**(4307), 960-964.

Acknowledgements
The authors wish to acknowledge the funding received from the EU Interreg NWE Programme for the SURICATES (Sediment Uses as Resources In Circular And Territorial EconomieS) Project.

Numerical Analysis of inlet vortex in the scaled engine

M Biglarian[1], **A Najafi**[2,*] **and S M Negharchi**[3]

[1] Department of Mechanical Engineering, Sharif University of Technology, Tehran, Iran

[2] Department of Wind Engineering, Flensburg University of Applied Sciences, Flensburg, Germany

[3] Department of Mechanical Engineering, Babol Noshirvani University of Technology, Babol, Iran

* **E-mail:** ahmadreza.najafi@stud.hs-flensburg.dei

Abstract. A jet engine that works near the ground and generates high thrust at low speed can experience the flow separation from the ground surface up to its inlet and, as a result, the formation of a famous vortex. This vortex extended from the ground surface to the fan inlet is known as the inlet vortex. In the present study, a jet engine model scaled down by 1/30 relative to the real jet engine with 3 million hexahedral elements is simulated using Computational Fluid Dynamics (CFD). The obtained results agree with empirical relationships, and the inlet vortex can be analyzed in scales much lower than the original prototype. Also, the inlet vortex gets weakened by increasing free stream velocity and completely degraded at higher velocities. By understanding how this phenomenon occurs and can be dealt with, the Foreign Object Damages (FOD) such as compressor surge, fan vibration, and particle ingestion into the engine core can be prevented.

Keywords: Computational Fluid Dynamics (CFD), Jet Engine, Inlet Vortex, Foreign Object Damages (FOD).

1. Introduction

Since the emergence of jet engines in airlines in 1950, the intake of foreign objects into the engine inlet has been one of the popular subjects of different studies [1]. When the engine works in static or pseudo-static modes (during taxiing and take-off), vortices may be formed between the engine and its inlet duct [2]. The inlet vortex has been well-known for six decades; however, its mechanism is not well understood, and different techniques are developing to prevent this phenomenon [1]. The collision of the inlet vortex and flow separation at the inlet lip significantly affects the engine performance [1]. In some cases, foreign object damage (F.O.D.) to the engine blades occur because vortex pitch debris into the intake section [3]. The stagnation pressure loss and flow deterioration at the fan inlet are among the effects of this phenomenon. So far, different tools have been invented to avoid or eliminate the inlet vortex. However, none of them are successful for all modes of external flow. The first system developed to prevent the inlet vortex formation was based on injecting excessive high-pressure air from the compressor to the ground surface near the stagnation point. By entering the stagnation point region near the ground, this airflow weakens and disappears the inlet vortex [4].

IOP Conf. Series: Materials Science and Engineering 1196 (2021) 012033 doi:10.1088/1757-899X/1196/1/012033

2. Geometry Specification

A 1/30 scaled engine model is simulated using CFD methods to recognize the inlet flow patterns in some regimes of free-stream flow in a real engine. This geometry is produced by CATIA software, and then it is imported to the ANSYS software environment. In this simulation, the inlet duct of the jet engine is considered two co-centric cylinders connected through an oval-shaped lip (Figure. 1). As shown in Figure 1, the inlet diameter (Di) and the height of the duct center above the ground (H) are 1/30 of the actual dimensions.

Since the effect of free-stream flow must be considered correctly; therefore, a cubic space around the jet engine is considered according to Figure 2 [5]. Dimensions of this cubic domain are 22 times as much as the characteristic length (inlet duct diameter, Di) in each direction. In Figure 2, the jet engine is located on the tailwind face, and other faces are named according to the flow passing through them; for example, the flow that comes from the engine upstream and goes downstream is named as follows: upwind and downwind. The inner and outer parts of the inlet duct of the jet engine and the ground, which are solid surfaces, are specified by black color in the Figure.

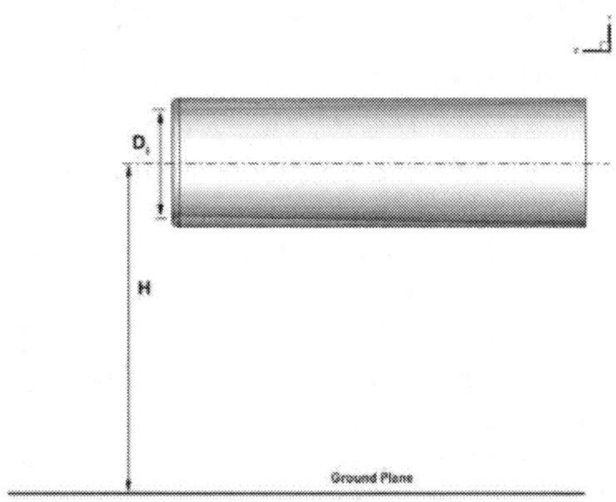

Figure. 1. Jet engine inlet modeled as a cylinder

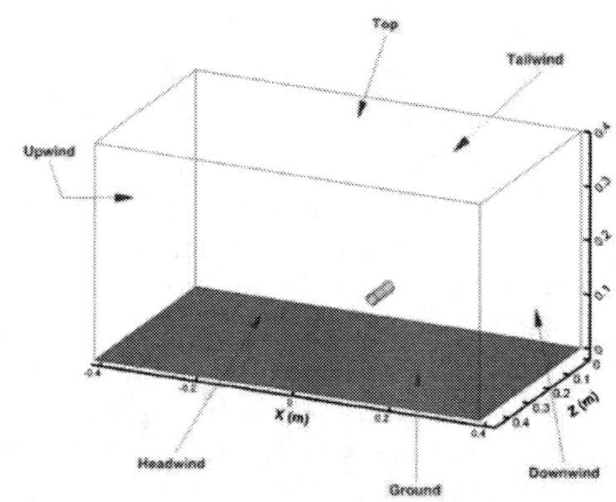

Figure 2. Applied computational domain and equivalent jet engine inlet like a cylinder

3. Jet engine scaling

All the dimensionless numbers must be constant for scaling the real model according to similitude and scaling laws. Two dimensionless groups which are obtained by similitude analysis are Reynolds number and Mach number. Reynolds number and Mach number are 5.6×10^6 and 0.388, respectively. If the real model is scaled down by 1/30, the flow velocity must be increased proportionally to keep the Reynolds number constant. In this case, the Mach number increases by 30 folds, and the flow regime transfers from the subsonic range to the supersonic one where Navier-Stokes equations are not valid anymore. Since Mach number is of greater importance than Reynolds number, Mach number is kept constant, and Reynolds number is multiplied by 1/30. Although Reynolds number changes, this simulation can be conducted because the flow regime is still turbulent.

4. Mesh generation and mesh independence study

The ICEM CFD software has been used to generate the structural mesh. In the present work, we first divided the geometry into smaller parts using the Multi-Block feature (Figure 3) and then, The mesh at the core of the inlet and around the lip was created using the O-GRID and C-GRID methods, respectively (Figure 4) [6]. To detecting better the effects of the inlet lip flow characteristics, the number of grids in this part is higher and gradually moving away from this area, the adjacent cells become larger (Figure 5). Also, the height of the genrated meshes has been reduced all around the inlet pipe and near the ground to capture the effects of the boundary layer precisely (Figure 6) [7]. The number of genrated mesh is about three million cells.

The location of minimum static pressure on the plane close to the ground must be examined for making solutions independent of the grid size. Figures 7 and 8 indicate the displacements in the x and z directions, respectively, which are non-dimensionalized relative to the inner diameter (D_i). As can be seen, two grids with 3 and 6 million have not significant changes relative to each other, and as a result, the grid of 6 million cells is selected as the simulation grid, and all analyses are conducted on this grid.

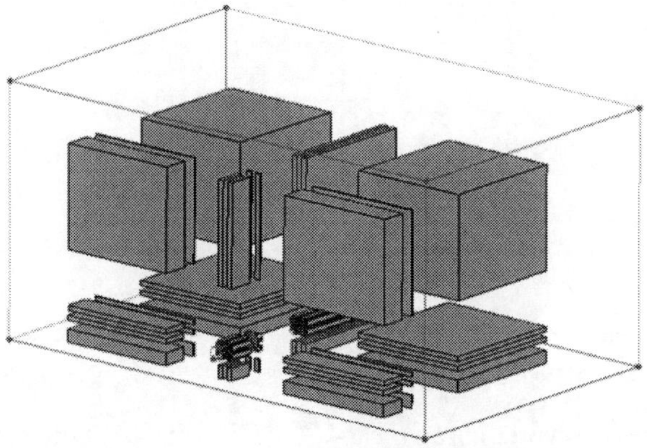

Figure 3: Using the Multi Block method

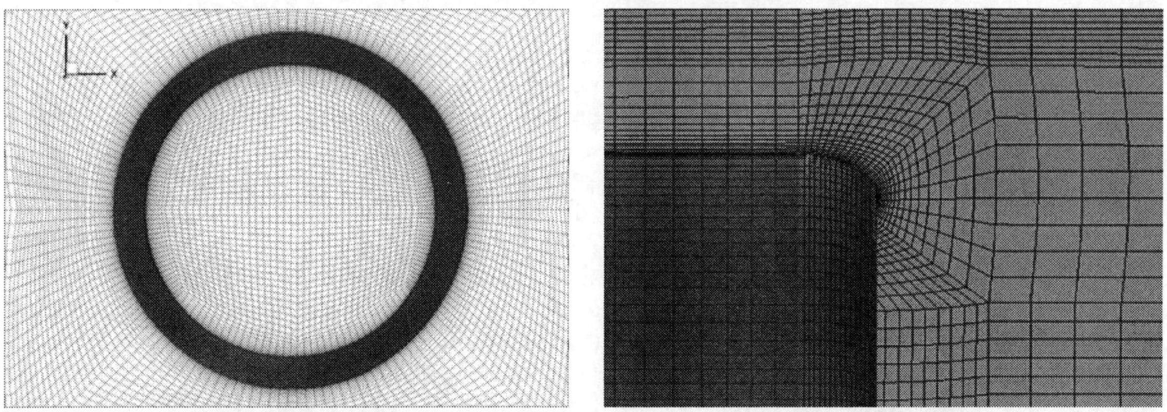

a) O-GRID mesh b) C-GRID mesh

Figure 4: generated mesh with O-GRID and C-GRID method around the inlet pipe

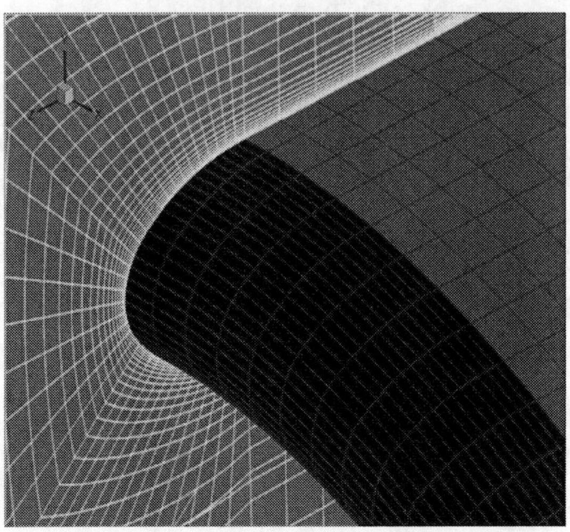

Figure 5: generated mesh around the edge of the

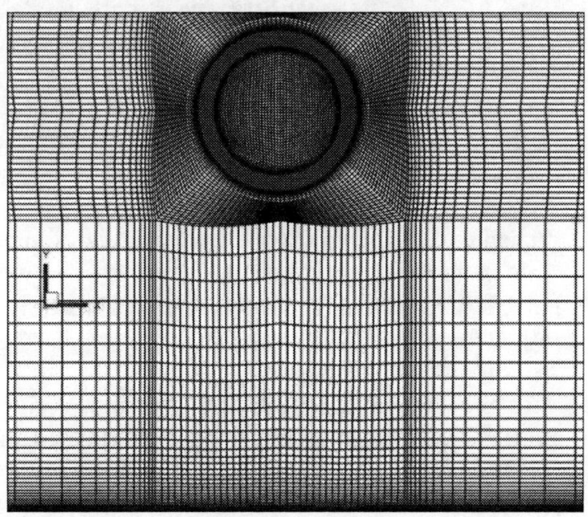

Figure 6: small grid near the ground and around

inlet pipe

the inlet pipe

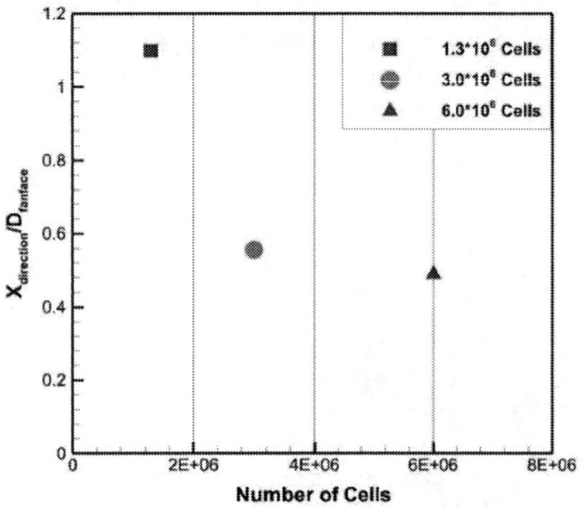

Figure 7. Static pressure variation in the x-direction against the cell number

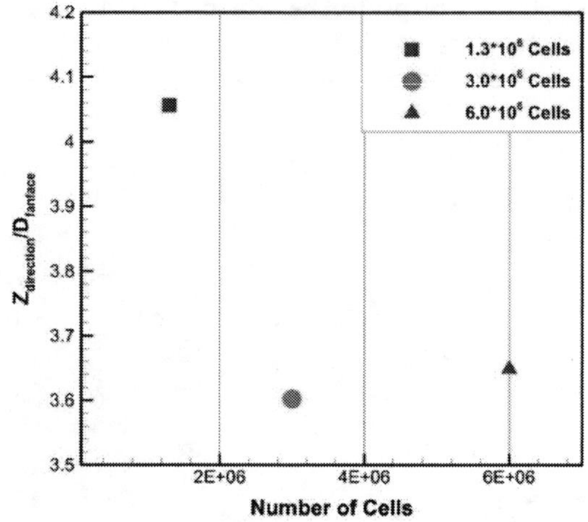

Figure 8. Static pressure variation in the z-direction against the cell number

5. Numerical simulation

ANSYS Fluent software is used to simulate the fluid flow based on the finite volume method. The pressure outlet boundary condition is applied to the duct outlet. In this section, the engine inlet fan is not considered for simplicity in calculations. In pressure outlet boundary conditions, the static pressure is calculated by equation (1) as follows.

$$\frac{P_0}{P_s} = (1 + 0.5 \times (K - 1) \times M^2)^{\frac{K}{K-1}} \tag{1}$$

Where P_0 is the stagnation pressure, P_S is the static pressure, K is the ratio of specific heat at constant pressure to that at constant volume, and M is the Mach number. Based on the given Mach number at engine inlet and total pressure, the static pressure at the engine inlet is obtained. Pressure far filed boundary condition is considered for five faces of the cubic domain, and the flow Mach number is calculated based on the free-stream velocity in each case. In this study, three values of free-stream velocity are assumed as 0, 4, and 6 m/s. The no-slip and adiabatic conditions are also considered for domain walls. Air is assumed as a compressible fluid, and its density is obtained from the ideal gas law. $k - \omega$ SST (Shear Stress Transport) model is used to simulate the turbulent flow[8,9]. This model applies the $k - \omega$ method for viscous sublayer and the $k - \varepsilon$ method for free-stream flow; for this reason, it is suitable for analyzing the flow separation and predicting the vortex formation. Also, this method is in good agreement with experimental data of vortex detection [10]. The available high-performance RAM allows to use of the coupled technique for the pressure-velocity correlation; because this coupling method occupies a large portion of the RAM due to simultaneously solving all equations. All scalars other than pressure are discretized by the second-order upwind discretization scheme to achieve higher accuracy. Also, the pressure is discretized by the PRESTO method because this method is very efficient for high-pressure gradients. The pseudo-transient method with time steps of 10^{-5} s is also applied to accelerate the convergence rate.

6. Results and discussions

In the case without outlet flow, the inlet vortex is formed, as shown in Figure 9. As can be seen, the vortex that separates from the ground surface and enters the inlet duct is observed clearly.

In order to specify the vortex center and the region where it is formed, a plane is created near the ground, and the curl characteristics of the flow field (fluid rotation) are examined in this place, as shown in Figure 10. This plane is parallel to the ZX plane (top view), and the inlet duct can be seen.

As it is observable, the generated vortex rotates around the y-axis (component perpendicular to the ground), the center of the vortex has the most significant rotational motion, and it is specified by black color. The tangent streamlines to this plane show that the flow taken from the surroundings separates from the ground right in the downside of the inlet lip and forms the inlet vortex.

Figures 11 and 12 indicate the local temperature variations from the top view (ZX plane) and side view (ZY plane). Based on Figure 11, the lowest temperature is attributed to the flow separation and inlet vortex formation; this reduction proceeds in Figure 12, and the temperature decreases at the inlet lip where the flow velocity reaches its maximum to Figure 9.

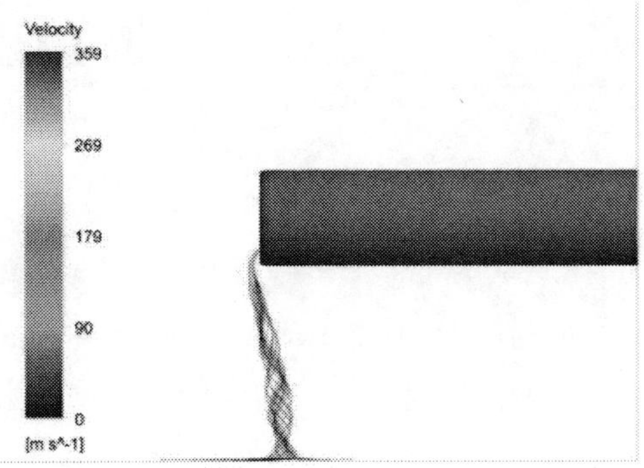

Figure 9. Side view of the inlet vortex formation in the case without external flow

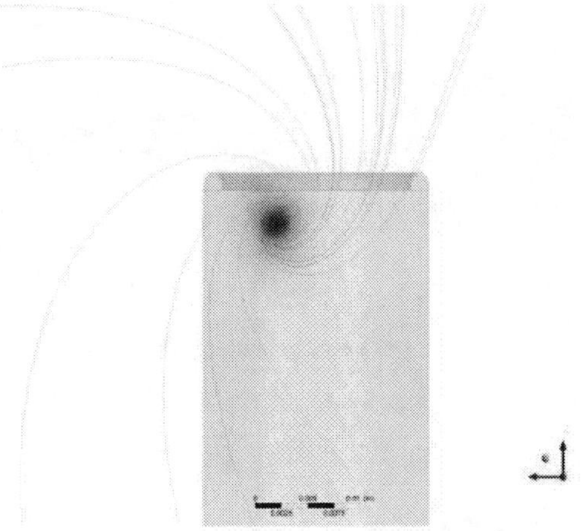

Figure 10. Y component of flow field velocity

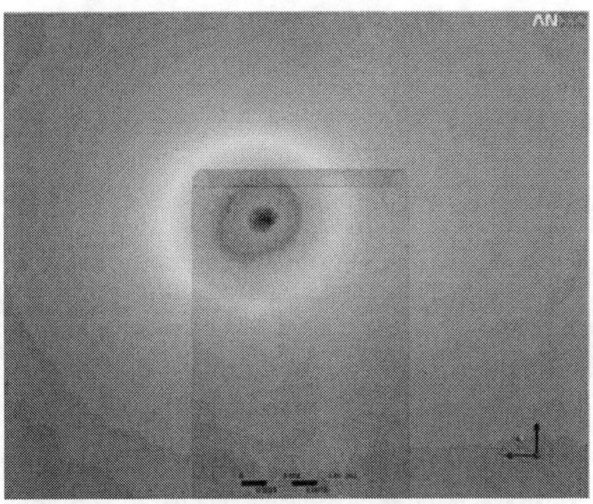

Figure 11. Top view of temperature contour parallel to ZX plane

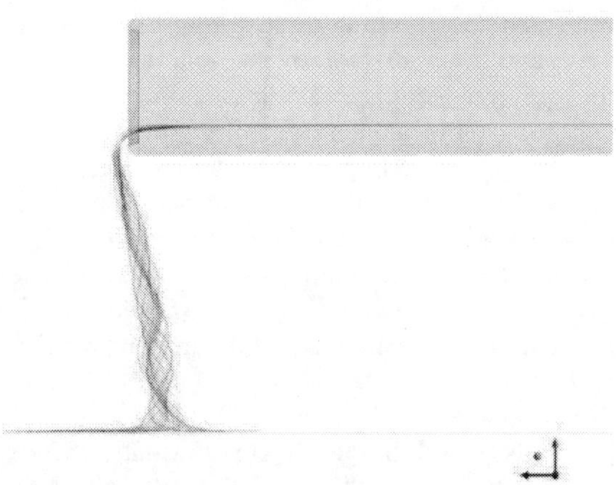

Figure 12. Inlet vortex and temperature variation on it

When the relative humidity of air is high (close to 100%), or the vapor particles are rich in the air, the temperature can be reduced to the dew point temperature and cause the water droplets to be formed [11]. The formation of these droplets can detect the inlet vortex.

The formation of this vortex leads to the flow asymmetry at the inlet, which can cause the fan vibration and corrosion of compressor blades.

According to equation (2) [1], the boundary of the inlet vortex formation is specified by dimensionless velocity and height. V_i and V_∞ are the inlet velocity and external flow velocity, respectively. Based on this equation and for the dimensionless height of 2.6 (in this engine) and the inlet velocity of 132 m/s, the minimum free-stream velocity is 3 m/s to avoid forming an inlet vortex.

$$\frac{V_i}{V_\infty} = 24 \times \frac{H}{D_i} - 17 \tag{2}$$

Figure 13 shows the flow pattern at the engine inlet for a headwind flow with a velocity of 4 m/s. As can be seen, the inlet vortex is not created based on empirical relationships. In the following, the flow is indicated for the free-stream velocity of 6 m/s. In this case, all streamlines enter the engine core without separation from the ground surface.

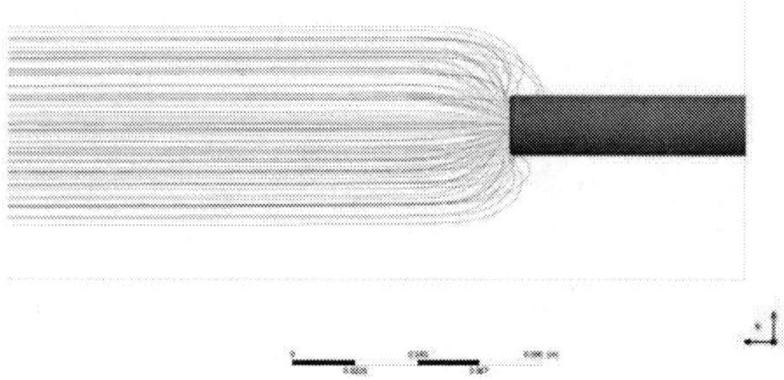

Figure 13. Inlet flow pattern for a free-stream velocity of 4 m/s

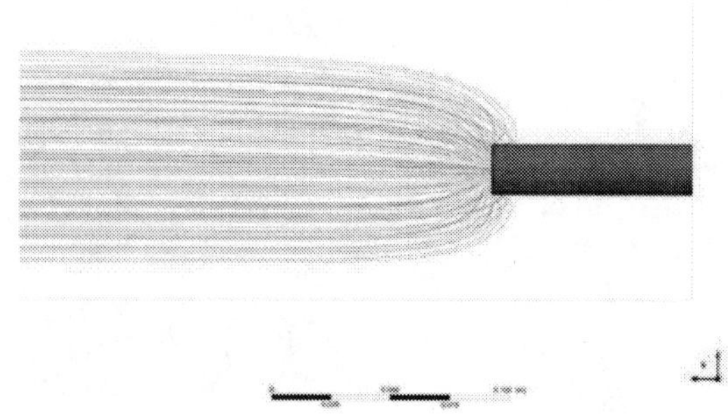

Figure 14. Inlet flow pattern when the free-stream velocity is 6 m/s

7. Conclusion

In this study, the inlet vortex was analyzed by ANSYS FLUENT software for a model scaled down by 1/30 relative to the real model of a jet engine. The obtained results showed that the inlet vortex is formed in the absence of free-stream flow and can be extended to the inlet duct of the engine. The flow separation occurs under the inlet lip on the ground. Also, the evaluation of temperature variation indicated that the fluid temperature could be reduced to the dew point temperature in the inlet vortex core, and the high value of the relative humidity led to the formation of water droplets. According to the presented empirical relationship, which determines the threshold of the inlet vortex formation using the dimensionless velocity and height, the free-stream velocity to avoid the formation of inlet vortex was equal to 3 m/s in this jet engine. The numerical results for flow velocity higher than this threshold show that the inlet vortex is not formed. It should be noted that the proposed model is scaled down by 1/30 relative to the real model; however, it can result in solutions that agree well with the results of empirical relationships. This scaling level can decrease the computation costs and accurately predict the formation of the inlet vortex in different types of engines. Therefore, different methods to prevent vortex formation and the inlet flow behaviour can be analyzed.

8. Reference

[1] N. Horvath, Inlet Vortex Formation Under Crosswind Conditions, 3 (2013).
[2] J.P. Murphy, D.G. MacManus, Inlet ground vortex aerodynamics under headwind conditions, Aerosp. Sci. Technol. 15 (2011) 207–215. https://doi.org/10.1016/j.ast.2010.12.005.
[3] W.H. Ho, A Consolidated Study Regarding the Formation of the Aero-Inlet Vortex, Stud. Comput. Intell. 416 (2012) 345–364. https://doi.org/10.1007/978-3-642-28888-3_14.
[4] L.G. Trapp, R. Da Motta Girardi, Crosswind effects on engine inlets: The inlet vortex, J. Aircr. 47 (2010) 577–590. https://doi.org/10.2514/1.45743.
[5] J. Chen, Y. Wu, O. Hua, A. Wang, Research on the ground vortex and inlet flow field under the ground crosswind condition, Aerosp. Sci. Technol. 115 (2021) 106772. https://doi.org/10.1016/J.AST.2021.106772.
[6] Z. Ali, P.G. Tucker, Multiblock Structured Mesh Generation for Turbomachinery Flows, (2014). https://doi.org/10.1007/978-3-319-02335-9.
[7] Z. Ali, J. Tyacke, P.G. Tucker, S. Shahpar, Block topology generation for structured multi-block meshing with hierarchical geometry handling, Procedia Eng. 163 (2016) 212–224.

https://doi.org/10.1016/j.proeng.2016.11.050.

[8] R.W. Winfree, The Application of Computational Fluid Dynamic Analysis to Jet Engine Inlet Flow Quality, J. Phys. A Math. Theor. 44 (2011) 1689–1699. https://doi.org/10.1088/1751-8113/44/8/085201.

[9] M. Biglarian, M. MomeniLarimi, B. Ganji, A. Ranjbar, Prediction of erosive wear locations in centrifugal compressor using CFD simulation and comparison with experimental model, J. Brazilian Soc. Mech. Sci. Eng. 41 (2019) 1–10. https://doi.org/10.1007/s40430-019-1610-5.

[10] D.E. Glenny, N.G.T.E. Pyestock, Ingestion of Debris into Intakes by Vortex Action, Aeronaut. Reasearch Counc. CP No. 111 (1970).

[11] C.J. Johns, The aircraft engine inlet vortex problem, AIAA's Aircr. Technol. Integr. Oper. 2002 Tech. Forum. (2002) 1–13. https://doi.org/10.2514/6.2002-5894.

Providing a targeted algorithm for replacing customers' Water meters aiming at reducing NRW: case study

E Ebadati[1], R shafaghat[2,*], MJ Raji Asadabadi[3], H Majidaei[4] and S M Negharchi[2]

[1] Department of Mechanical Engineering, Malek Ashtar University of Technology, Tehran, Iran

[2] Sea Based Energy Research Group, Babol Noshirvani University of Technology, Babol, Iran

[3] Department of Mechanical Engineering, Iran University of Science and Technology, Tehran, Iran

[4] Department of Mechanical Engineering, Shahid Beheshti University, Tehran, Iran

*** E-mail:** rshafaghat@nit.ac.ir

Abstract. Limited water resources, especially in Iran, which has an arid/semi-arid climate, highlight the need for accuracy in managing and controlling the consumption of this vital fluid. Although many approaches have been developed to reduce Non-Revenue Water (NRW) in water distribution systems, less attention has been paid to optimal prioritization for water meter replacement. In this research, an algorithm has been presented for the purposeful replacement of customers' water meters to increase the company's revenue. Gavankola village has been selected as a rural network in northern Iran with a high NRW rate (58%) as a case study. Examining the customers' water consumption in 20 periods, it was found that after replacing the water meter, the billed metered consumption (BMC) doubled and the revenue from it increased up to 5 times. Moreover, the study of two economic indicators and NRW revealed that class C water meters are less economical than class B water meters, even though they more effectively reduce apparent losses.

Keywords: Apparent losses, WM replacement algorithm, NRW, Consumption Schedules, water companies.

1. Introduction

Non-Revenue Water (NRW) is one of the most important water programs in the world. NRW represents the water that is delivered but not sold to customers. According to the WHO/UNICEF Joint Monitoring Program, the annual volume of NRW has been estimated to be 126 billion m^3, equivalent to $ 39 billion worldwide [1]. Apparent losses (AL) are among the first set of indicators in NRW that demonstrate a general management perspective on water efficiency and effectiveness, so-called basic indicators (level one) and have an effect on system management [2]. In general, AL includes water theft, billing errors, measuring instrument errors, human errors, management errors and operating errors. While these first factors are directly related to the management of water facilities and may be reduced by improving company procedures, WM errors are the most significant and difficult ones to determine the quantity. WM errors are amplified in networks vulnerable to water shortages, where users use private storage tanks to deal with intermittent water supply [3].

Measurement error is one of the most important components of AL that directly affects sales with revenue [4]. Water meters (WMs) are one of the most important tools to measure flow in the network [5]. Awareness of water companies about the amount of water produced and the total volume of water entering the network and BMC is crucial. In this regard, the accuracy of WMs is of utmost importance [6]. Since the major share of water companies' revenues is provided through customers' cost water, the existence of low-precision WMs can be an important factor in reducing revenue. Moreover, due to the inaccurate WMs, the correct estimation of BMC and planning to supply water resources is disrupted and causes a lot of damage to infrastructure assets [7]. Therefore, it is necessary to test the accuracy of installed and new WMs in accordance with a regular schedule and if necessary, they should be replaced. By replacing the customers' WMs and increasing the accuracy of the WM operation, the measured costs will become real, and as a result, the revenues of water companies will increase, and the loss will decrease [8]. A comprehensive review of research in this field has been presented in Table (1). Due to the similarity of consumption culture and economic conditions of water companies and the type of household WMs used, the most important studies in Iran and Islamic countries have been collected.

Table 1. Comprehensive review of studies on evaluating WM accuracy and its replacement.

Previous study (year)	Objective and parameters
London, UK (2006) [9]	Investigating AL and identifying the main causes of WM inaccuracy
Ferrara, Italy (2007) [4]	Presenting 5-step basic strategies to control AL
Palermo, Italy (2009) [10]	Identifying the effects of private storage tanks on AL and identifying the ascending relationship between the two parameters of WM life and WM error
Kerman, Iran (2011) [11]	Investigating the increase in BMC and consequently the increase in water price compared to the WM failure period
Patiala, India (2012) [12]	Conducting research and development on methods, calibration, and measurement facilities and regulating water flow in WMs based on statistical methods
Mohali, India (2013) [13]	Designing, developing, and presenting the benefits of automatic WMs and employing economic analysis to replace WMs in a variety of uses
Lithuania, Vilnius (2014) [14]	Analyzing accuracy of Class B and C WMs and the effect of each on AL reduction
Pretoria, South Africa (2015) [15]	Investigating AL due to measurement error, the relationship between WM life, WM error and BMC records and analyzing experimental data
Pretoria, South Africa (2015) [16]	Investigating the accuracy of smart WMs and their effect on reducing NRW and analyzing the cost of installation and operation
Borazjan, Iran (2016) [17]	Determining consumption schedules and conducting an economic analysis of WMR and determining the relationship between replacement time and WM life
Bushehr, Iran (2017) [18]	Analyzing the cost and revenue due to replacement of faulty WMs and reduction rate of NRW
Tehran, Iran (2018) [19]	Evaluation of testing household WMs with BMC less than ten m^3month^{-1} in reducing AL
Johannesburg, South Africa (2018) [20]	Improving the data collection process from customers' WMs and developing new methods in reducing AL of countries
Bialystok, Poland (2018) [21]	Optimizing WM measurements and the effect of WM diameter on the accuracy of WMs and NRW
Isfahan, Iran (2018) [22]	Cluster testing of WMs and investigating the effect of class, function, life, and diameter of WMs on WM accuracy
Antalya, Turkey (2020) [23]	Experimental analysis of the effect of different types of household WMs with different classes and the amount of pressure and water flow on the accuracy of the WMs
Samarang, Indonesia (2021) [24]	Determining the effect of faulty WMs, pipes leakage before installing WMs, and water quality on AL of low consumption customers
Bandang, Indonesia (2021) [25]	Identifying the types of data processing errors in the study area and the feasibility of setting up Internet remote reading system of WMs

IOP Conf. Series: Materials Science and Engineering 1196 (2021) 012034 doi:10.1088/1757-899X/1196/1/012034

In previous studies, most efforts have been made to improve WM accuracy and reduce NRW. However, few studies have evaluated the effects of replacement prioritization and class WM accuracy. This study proposes an algorithm to assess the accuracy and replacement of customers' WMs. One of the most important ways to increase the accuracy and become closer to realizing the results of this algorithm is to compare in a system. The region selected for the case study in this study is located in a rural climate with tourist attractions that have been less addressed in previous studies. Changes in the reduction of NRW and the increase in the revenue of the water company because of Water meter replacement (WMR) with different classes have been examined. In addition, changes in customers' consumption categories and replacement costs are considered. Assessing the probabilities of failure and relative error of the WM is another issue under consideration.

2. Case Study

The case study of this research is the network of Shirdarkola and Gavankola villages in Babol city in northern Iran. Mazandaran Province Water and Wastewater Company responsible for water supply and service of these villages. These villages are in good weather condition so that they have a tourist attraction and some of its customers are only among the water consumers on weekends. According to the tank outputs, the network is divided into four isolated zones (DMA), which can be seen in Figure (1). According to the measurements, the average inlet water entering the network is 17 Ls-1 and the selling flow rate is 7.2 Ls^{-1}. The NRW rate in this network is 58%, which is much higher than the World Bank standard (23%) [26]. Based on the data, the percentage of WMR rate in villages was about 3.5% from 2017 to 2020. However, the minimum percentage of WMR rate and its accuracy assessment in networks with high loss is estimated at 5% [1]. Therefore, it is necessary to take measures to reduce the amount of NRW caused by WMs errors and to reduce AL in these villages.
The number of village's customers in districts A, B, C and D were 423, 514, 240 and 413, respectively. These villages have 1590 subscribers, 94% of which are household customers.

Figure (2) shows the average BMC. As observed in the figure, about 60% of customers have an average BMC of less than 10 m^3month^{-1}. Since most of the customers in the villages are single units, BMC less than ten m^3month^{-1} can indicate the probability of WM failure as well as the high rate of NRW [27]. The status of customers' WMs based on WM life is shown in Figure (3). As can be seen from the figure, about 60% of the customers have a WM life of over 15 years. While the useful life of WMs is between 8 to 10 years depending on the operating and climatic conditions [28].

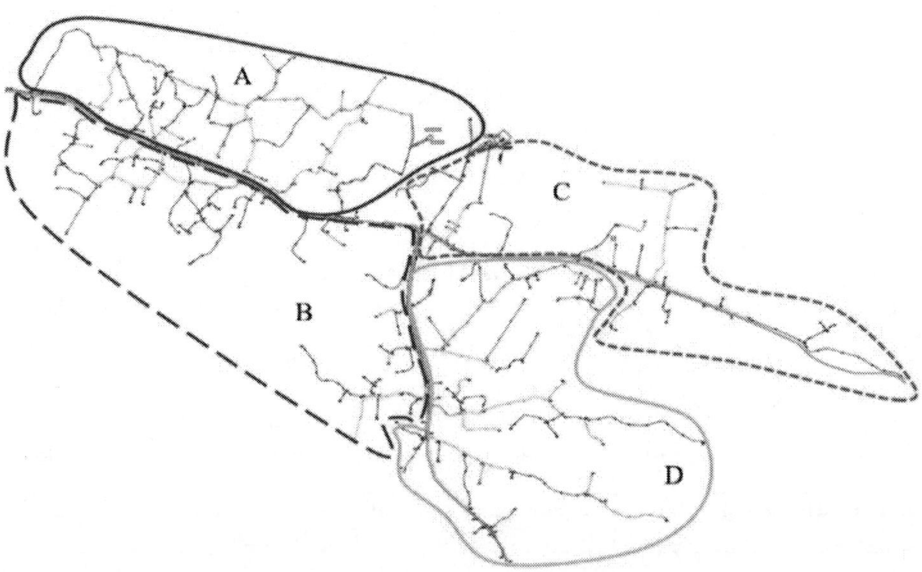

Figure 1. Districts of Shirdarkola and Gavankola village network.

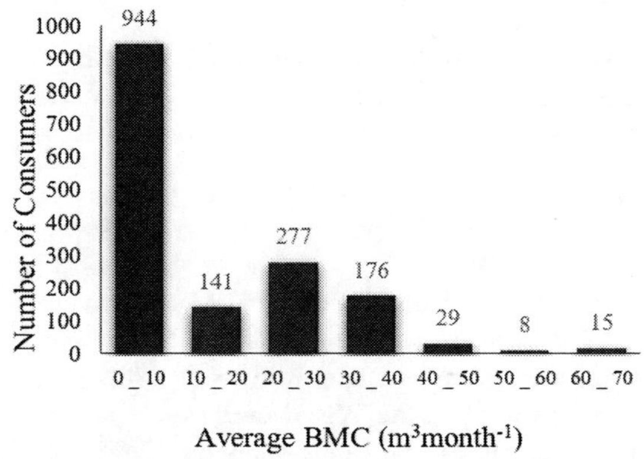

Figure 2. Average BMC of customers

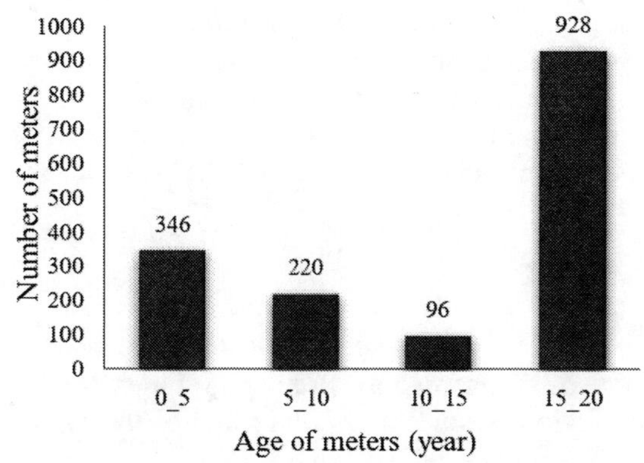

Figure 3. Status of customers' WMs

3. Research method

The primary purpose of this study is to determine the priorities for replacing faulty WMs in order to reduce NRW. This section has introduced the algorithm to evaluate the accuracy and replacement of faulty WMs (Figure 4). The basis for designing the introduced criteria is in previous studies (Table 1). The first step in replacing the WM is identifying the WM conditions based on BMC records, WM life and cluster testing [1]. The World Bank standard recommends performing cluster testing by selecting at least one percent of consumers' WMs [1]. Due to low manufacturing standards and the high average age of WMs in Iran, this value has been recommended to be 1 to 2.5 percent [29]. The WM status is detected first. Then, if the WM is defective, we act based on BMC records [30]. If the water WM is ok, we act in accordance with the life of the WMs. It should be noted that the life of the WM is not the only determining of failure and factors such as material, quality of consuming materials, climatic conditions, and accuracy will affect the manufacturing of the WM [9]. Prioritization for testing residential WMs is that residential WM testing with an average BMC between zero and 5 m^3month^{-1} is the top priority. BMC between 5 to 10 m^3month^{-1} is the second priority. Non- residential WMs (commercial, educational, public, etc.) will be evaluated as household WMs based on the type of use and under different defined tariffs and standards in water companies [23]. According to the Operational Regulations of Iran, the intended range for WMs errors in their replacement is 5% [29]. The data of this study were collected through customers' bills using software in Mazandaran Water Company. Primary information gathered includes the customer's first name and last name, file number, connection installation date, WMR date, WM body number, BMC average, WM size, number of units, username and reading status.

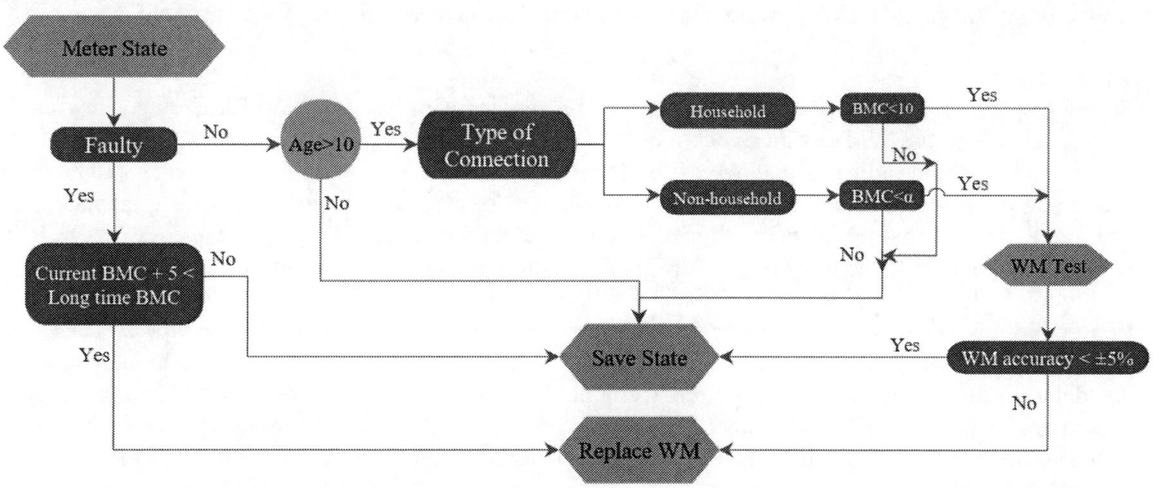

Figure 4. Algorithm for assessing the accuracy and priority of WM replacement

3.1. Identifying the risks of WM failure

Any device such as a WM can be defective. It is necessary to evaluate the probability and consequences of WM failure and defects to assess possible losses. Since the failure of a WM is a discrete event, it often occurs independently and continuously over time. The Poisson model (Equation 1) determines the probability of WM failure [31].

$$P_i = 1 - \exp\left(-BR_{ij} \times t_i\right) \tag{1}$$

In this formula, the BR parameter is the annualized failure rate (AFR) of WM i with class j. Moreover, the parameter P indicates the probability of WM failure i at time t (year). In this equation, t can also be regarded as a replacing index (representation) of the WM's operating year. According to the previous studies, the failure rate of the WM class and the life of the WM were assumed to be the constant values of 0.01 and 0.066, respectively [31]. In terms of the consequences of failure, the detection of more vulnerable WMs can lead to better management of the AL due to WM error. According to the rationale, the failure of the WMs will increase as the water count increases [31].

3.2. Relative WM error against WM life and total recorded water volume

To examine the relative error of the WM and determine its trend, the total volume of BMC during different years is recorded and classified based on the life of the WMs. The relative error of the WM is determined according to the following relation [32]:

$$\varepsilon_i = \frac{V_i - V_a}{V_a} \tag{2}$$

In this formula, the parameter ε_i represents the relative error of the WM, the parameter V_a represents the actual volume measured by the new WM, and the parameter V_a represents the amount of volume indicated by the old WM. Then, the relevant fraction can be converted into a decimal, where a negative value indicates that the WM gives a reading less than the actual consumption and a positive value means that the WM provides reading with more than the actual consumption [32].

4. Results and discussion

Over the last 15 years, 125 residential WMs have been replaced in the villages. In order to compare BMC and the revenue impacts of WMR with pre-replacement conditions, the performance of 48 replaced WMs in 2017 was reviewed. The minimum period for comparing and reviewing the costs consumption before and after WMR is three periods [33]. In this study, in order to increase the accuracy of BMC data, data were collected in 10 periods before and ten periods after WM replacement. The number of replaced meters in districts A, B, C and D were 14, 15, 8 and 11, respectively. All replaced WMs were Class B and one-half inches and were for household usage.

4.1. BMC analysis

Figure (5) shows the average BMC before and after replacement based on WM life. According to the figure, the life of the WMs of these customers in two categories was between 5 to 10 years and 15 to 20 years. This trend indicates that WM life alone is not the cause of WM failure because 10% of the first category WMs failed during their useful life. It is also observed that as the WM life increases, the average BMC decreases. However, after the replacement, the average BMC increases drastically. This is due to the increased accuracy of the WM, which increases the revenue of the water company.

Figure 6 shows the share of BMC classes before and after WM replacement. According to the figure, about 55% of customers have a BMC below ten m^3month^{-1} before WM replacement. Therefore, it can be concluded that the two parameters of WM life and reviewing BMC records alone cannot be the determining parameters for WM replacement. Consequently, it is necessary to identify eligible customers based on these two parameters to achieve the desired results. Then the cluster testing of the WM should be performed and if the WM accuracy is low, it should be replaced. It is worth noting that as the BMC classes change, not only is the volume of NRW reduced but also there is a significant increase in revenue of the water company. This is because the price of water changes exponentially as the BMC class changes and the cost of one m^3 of water in the class 0-5 m^3month^{-1} is not the same as the cost of one cubic meter of water in class 5-10 m^3month^{-1}. Figure 7 clearly shows the displacement of the BMC classes. In addition to the monthly increase in total BMC, the decrease in the share of the first BMC (0-5) and second BMC (5-10) classes and the increase in the share of the fourth (15-20), fifth (20-25), sixth (25-30), and the seventh (30-35) BMC classes is evident in this figure. Finally, based on the combination of the two diagrams in Figures (5) and (6), the average changes in BMC before and after WMs' replacement were calculated to be 9.63 and 19.28 m^3month^{-1}, respectively. Therefore, the results indicate a 100% increase in BMC after WMs replacement of these customers.

Furthermore, the effect of the accuracy of Class B and Class C WMs on BMC after the replacement has been evaluated. According to Figure. 7, due to the higher accuracy of the Class C WM than the Class B WMs, the NRW rate decreases more. By replacing the WMs of these customers with Class C WMs, BMC 2044 m^3 will increase annually. In other words, the amount of BMC per WM increases by one m^3month^{-1}. The results have been consistent with the global standard of accuracy of WM classes, OIML R49-1 [34], because of the higher accuracy of class C WMs. Therefore, it is logical to use Class C WMs to reduce NRW.

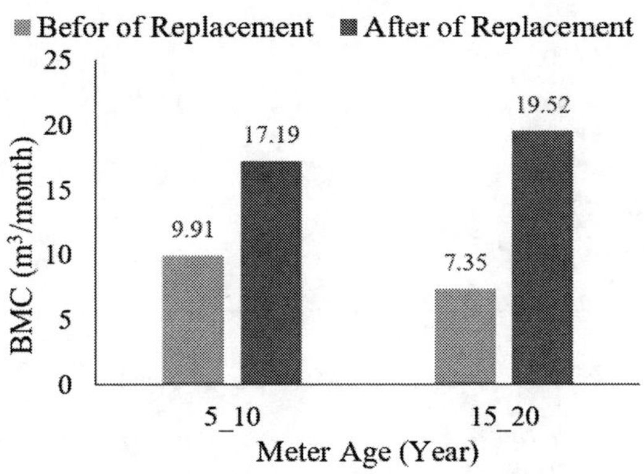

Figure 5. Average BMC before and after replacement based on WM life

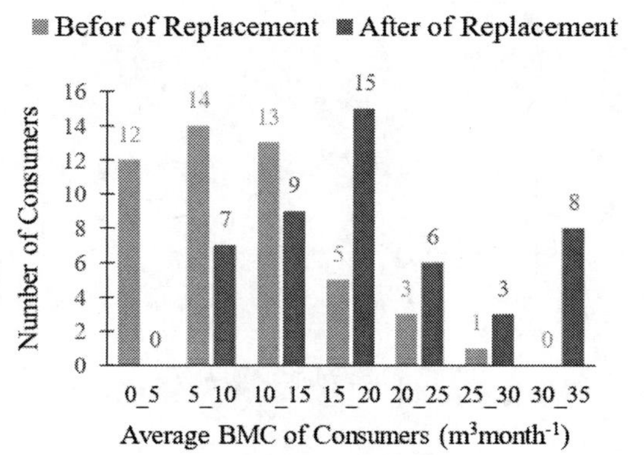

Figure 6. Share of BMC classes before and after WM replacement

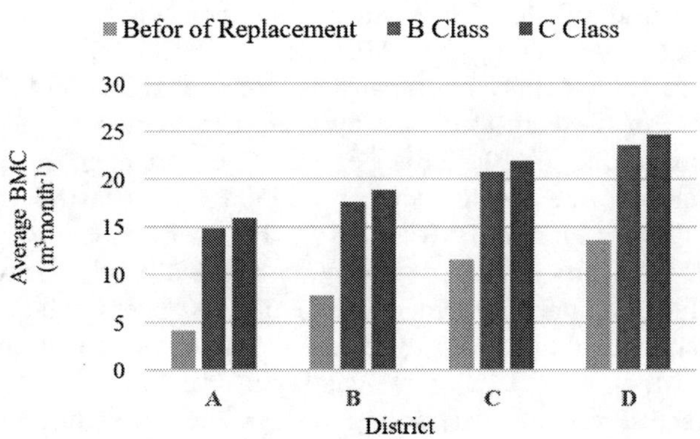

Figure 7. Effect of WM class on BMC of each region after replacement

4.2. Determining the probability of WMs' failure in Shirdarkola and Gavankola villages

In this section, the probability of WM failure is evaluated based on the WM life. Field studies show that all residential WMs used in the study region are class B and multi Jet dry WM. Based on the available records of the number of failures, the BR parameter was developed for these intended WMs and then was used to calculate the probability of the WMs' failure and fracture according to Equation (1). Figure.8 shows the failure probabilities of residential WMs of the village's customers in different years. The results reveal that the probability of WM failure highly depends on the BR value related to the class and type of WM and the WM life. In the present study, the minimum and maximum values of the probability of WM failure during the WM life of 1 to 20 years have been evaluated to be 0.0145 and 0.9398, respectively.

4.3. Estimation of relative WM error

Relative WM error was analyzed for all replaced WMs in villages. As shown in Figure. 9, the volume of BMC increases as the WM life increases. The pattern shows exponential growth. Growth in all cases of WM life is considered one year, which has a moderate and steady growth and reaches its peak in WMs with a life of 20 years. The minimum and maximum amounts of total water volume recorded are 1721 and 15594 m^3, respectively. If we analyze consumer information and add it as input, we expect this correlation diagram to improve.

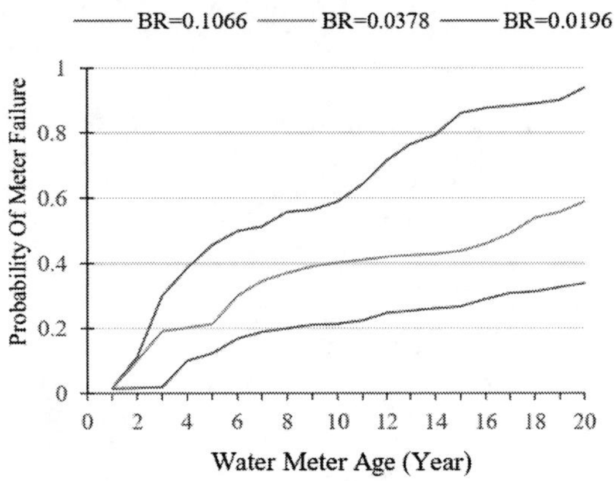

Figure 8. Diagram of the probability of failure and fracture of any type of WM over time

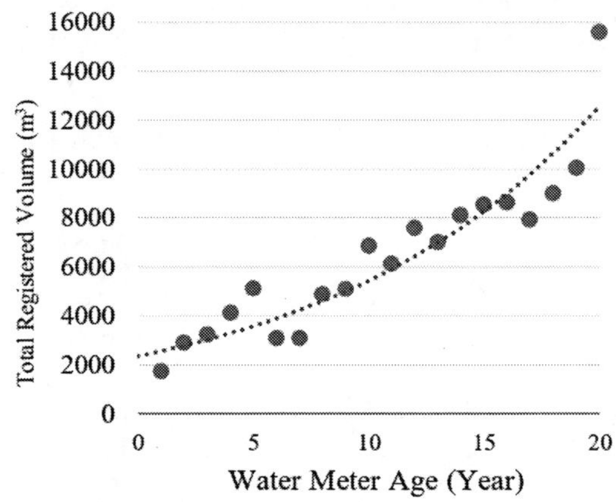

Figure 9. Diagram of the total volume of water recorded in relation to the life of the WM

4.4. NRW and Economic Analysis

As 48 WMs were replaced, the annual water sales increased from 5550 to 11107 m^3 (almost doubled). In addition to an increase in the BMC, the WMR has another benefit for the company, which is the increase in the price of water by changing the BMC classes. The amount of this increase in revenue for the villages is as shown in Table (2). The company had an annual income of approximately 86 Euros before replacing the WMs of these customers. However, after replacing these WMs, the water price received by the company is about 417 Euros per year, i.e., an increase of more than five times. However, the average selling price of water in Iran is 5500 Rials (0.0188 Euros) per cubic meter of water. Moreover, with the changes in BMC classes, the average selling price of water for these customers increased by 141%. It is predicted that by replacing all the priority, WMs of these villages (according to the algorithm), an annual income increase of 1000 Euros will be created.

Figure (10) shows a comparison between replacement and revenue with the WM class. The cost of WMR for WM Classes of B and C (based on the catalog of Iranensheab Company) is estimated at 1.8 and 2.8 million Rials, respectively [35]. According to this table, the return on investment (ROI) of the customers' WMR plan using Class B and Class C WMs is 8 and 13 months, respectively. Therefore, it can be concluded that if the company aims to reduce NRW, it is recommended to use class C WMs due to their high accuracy. It should be noted that Class C WMs are highly dependent on water quality and often have a shorter service life than Class B WMs. If the company's priority is in economic terms and return profit, it is better to use class B WMs because it has more economic benefits.

Table 2. Increase in water prices in accordance with the changes in BMC classes

Consumption Schedules (m^3month^{-1})	Old WMs			New WMs		
	Number of Customers	*BMC* (m^3year^{-1})	Annual income (Euro)	Number of Customers	*BMC* (m^3year^{-1})	Annual income (Euro)
0-5	12	236	2.3	0	0	0.0
5,10	14	1316	6.7	7	792	3.4
10,15	13	1906	18.8	9	1420	13.0
15-20	5	986	16.2	15	3067	48.7
20-25	3	793	23.7	6	1579	47.3
25-30	1	313	18.5	3	1014	55.5
30-35	0	0	0.0	8	3235	249.2
Total	48	5550	86.3	48	11107	417.1

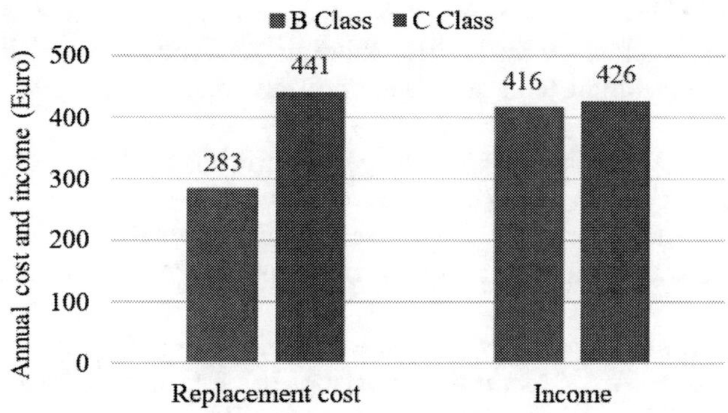

Figure 10. Comparison between replacement cost and annual revenue based on WM class

5. Conclusion

In this article, an algorithm was designed to replace and evaluate the accuracy of WMs and was implemented for the villages of Shirdarkola and Gavankola with 1590 customers as the study area. BMC customers whose WMs were replaced in 2017 were reviewed for 20 periods. The results were presented in three sections: BMC analysis, economic analysis and NRW analysis. The most significant achievements of this study are as follows:

- With the purposeful replacement of the customers' WMs, the BMC amount doubled and the NRW value decreased from 58% to 15%.
- It was specified that there is a direct relationship between WM life and BMC reduction.
- Replacing a Class C WM with a Class B WM will increase BMC and decrease NRW. However, if the goal is to achieve a lower return on investment, Class B WMs is more appropriate because of the lower purchase cost.
- The customers' WMR rates were analyzed in different years. It turned out that NRW caused by WM error in these villages should be a significant number. It was shown that the water company needs to take more effective measures to reduce AL.
- Although it was specified that testing and replacing customers' WMs with water consumption of more than ten m^3month^{-1} makes BMC closer to the actual consumption, it has no economic justification.
- Although older WMs are more likely to fail, this is highly dependent on the BR value of the WM classes and their types.

6. References

[1] Taha A W, Water Loss Assessment in Distribution Networks: Methods, Applications and Implications in Intermittent Supply, *CRC Press*, 2021.

[2] Thomas M, *The Impact of Non Revenue Water on Water Utilities-A Case Study of Mulonga Water Supply and Sanitation Company in Chingola*, in: Cavendish University, 2020.

[3] Mthuli N and Akpofure T, Assessment of apparent losses due to meter inaccuracy–a comparative approach, *Water SA*, **vol 45** (2019) pp 174-182.

[4] Alex R, Benoit V, Sandro G, John S T, Gilber M and Marcelo R P, Apparent water loss control: the way forward, *Water*, **vol 21** (2007) pp 45-47.

[5] Michael M H, Zvikomborero H, Shepherd M and Sinos M, Investigating the management of unaccounted for water for Lilongwe Water Board, Malawi, *Journal of Water, Sanitation and Hygiene for Development*, **vol 6** (2016) pp 362-376.

[6] Bulent S, Alev A, Özlem İ, Yakup K, Cumali K, Ayse M and Habib M, Management of water losses in water supply and distribution networks in Turkey, *Turkish Journal of Water Science and Management*, **vol 2** (2018) pp 58-75.

[7] Burak K and Eyup Ş, Exceedance probabilities of non-revenue water and performance analysis, *International Journal of Environmental Science and Technology*, (2021) pp 1-12.

[8] Jude C, Marcelo B B, Manoel D, Sofia P R, Administ G, Andres S and Joana R F, Taking water efficiency to the next level: digital tools to reduce non-revenue water, *Journal of Hydroinformatics*, **vol 23** (2021) pp 453-465.

[9] Francisco J A, Enrique C, Ricardo C and Jorge G-Serra G, Reducing apparent losses caused by meters inaccuracies, *Water Practice and Technology*, **vol 1** (2006).

[10] Antonio C, Chiara. M F, Gabriele F and Gafferdo L L, Evaluation of the apparent losses caused by water meter under-registration in intermittent water supply, *Water Science and Technology*, **vol 60** (2009) pp 2373-2382.

[11] Gholamreza Z. *Investigating the effectiveness of replacing broken meters of consumers for reducing Non-Revenue water in Kerman* (in Persian). In: International Conference on Water and Wastewater with Privatization Approach and Superior Models in Design, Construction and Operation.. Tehran, Iran, (2011) pp 1-19.

[12] Shiv K J, Sanjay Y, Ashis K B and Ravinder A, Global water flow measurement and calibration facilities: review of methods and instrumentations, *Mapan*, **vol 27** (2012) pp 63-76.

[13] Ria S, Manjit K and Hemant L, Design and development of automatic water flow meter, *International journal of computer science, engineering and applications*, **vol 3** (2013) p 49.

[14] Mindaugas R and Ramune A, *Analysis of Apparent Water Losses, Case Study, in: Environmental Engineering*. Proceedings of the International Conference on Environmental Engineering. ICEE, vol 9, Vilnius Gediminas Technical University, Department of Construction Economics, (2014), p 1.

[15] Frank A C and Jakobus E V, Apparent losses due to domestic water meter under-registration in South Africa, *Water SA*, **vol 41** (2015) pp 698-704.

[16] Deanroy M, Noble B, Nicholas K, Harrison M and Mohammed B, Determination of domestic water meter accuracy degradation rates in Uganda, *Journal of Water Supply: Research and Technology—AQUA*, **vol 64** (2015) pp 486-492.

[17] Babak D, Somaye K and Ali S, *Investigating the consumption pattern of consumers and economically analyzing the operation of replacing broken meters in Borazjan* (in Persian). In: First National Conference on Drinking Water Supply and Sanitation, Challenges and Solutions (ed J Abedi) , Isfahan, Iran, (2016) pp 75-92.

[18] Babak D, Somaye K and Ali S, *Investigating the role of replacing low performance worn meters on reduction of Non-Revenue water, Case Study: Genaveh Port* (in Persian). In: The first National Conference on Water Consumption Management and Waste Management. Shahid Beheshti University, Tehran, Iran, (2017) pp 23-36.

[19] Arash M, Hesam K H and Mina J. *The effect of replacing consumer's meters with residential and single-unit use in reducing apparent loss* (in Persian). In: Second Congress of Iranian Water and Wastewater Science and Engineering (ed M Heydarpoor), Isfahan University of Technology, Isfahan, Iran, (2018) pp 103-122

[20] Mthokozisi N and Akpofure E T, Assessment of apparent losses due to meter inaccuracy using an alternative, validated methodology, *Water Supply*, **vol 19** (2019) pp 1212-1220.
[21] Joanna G, Joanna M and Kamil Ś, *Analysis of the water meter management of the urban-rural water supply system*, in: E3S Web of Conferences, vol 44, EDP Sciences, 2018, p 00051.
[22] Hadi A and Forouzan B, *Comparison of two joint selection methods for meter replacement (Case study of Isfahan Water and Sewerage Company-Shahreza region)* (in Persian). In: Second Congress of Iranian Water and Wastewater Science and Engineering (ed M Heydarpoor), Isfahan University of Technology, Isfahan, Iran, (2018) pp 46-68
[23] Karadirek I E, An experimental analysis on accuracy of customer water meters under various flow rates and water pressures, *Journal of Water Supply: Research and Technology-Aqua*, **vol 69** (2020) pp 18-27.
[24] Putra R B, *Water meter for household customers with low water consumption in Perumda Surakarta*, in: IOP Conference Series: Earth and Environmental Science, vol 700, IOP Publishing, (2021) p 012061.
[25] Bramantyo R, Wildan A N, Mohamad R and Suripto D Y, *Utilization of water flow for powering the internet-based water metering system house old scale*, in: IOP Conference Series: Materials Science and Engineering, vol 1098, IOP Publishing, (2021), p 042044.
[26] Saeid M N and Rouzbeh S, Leakage estimation in water networks based on the BABE and MNF analyses: a case study in Gavankola village, Iran, *Water Supply*, **vol 20** (2020) pp 2296-2310.
[27] Ronald S M, Zinga S, Willem W, *The State of Non-Revenue Water in South Africa*, Water Research Commission, 2012.
[28] Francisco J A, Francisco J G, Javier Soriano and Laura P and Laura J, Performance analysis of ageing single-jet wate meters for measuring residential consumption, *Water*, **vol 10** (2018) p 612.
[29] Guideline 556 of Iran MOE, *Guideline for Determining Effective Parameters on Unaccounted for Water (UFW) and Water Losses Reduction Schemes; Code 556*, in: Ministry of Energy, Water Industry of Iran, Confirmation Group: Water and Wastewater Expert Committee for Preparation of Technical Standards and Standards, University of Tehran Water Institute, in Persian, (2017), pp 1-175.
[30] Tracy C B, Rodney A S and Kelvin R O, Smart metering: enabler for rapid and effective post meter leakage identification and water loss management, *Journal of Cleaner Production*, **vol 54** (2013) pp 166-176.
[31] Farhad Y and Ardalan I, An asset management approach to optimize water meter replacement, *Environmental Modelling & Software*, **vol 104** (2018) pp 270-281.
[32] Annlize M, Aluta M and Jan P, Domestic water meter optimal replacement period to minimize water revenue loss, *Water SA*, **vol 45** (2019) pp 165-173.
[33] Mthokozisi N and Akpofure T, *Meter accuracy degradation and failure probability based on meter tests and meter change data*, in: Proceedings of the 4th YWP-ZA Biennial Conference and 1st African YWP Conference, (2015), pp 1-10.
[34] Andres A M, Holger B and Holger M, Evaluation of domestic water measurement error: a case study, *AQUA—Water Infrastructure, Ecosystems and Society*, **vol 70** (2021) pp 217-225.
[35] I. ensheab.Co, See also https://www.iranensheab.com (accessed 01 March 2021), (2021).

The Effects of Airfoil Type and Wind Speed on the Aerodynamic Efficiency of Vertical Axis Wind Turbines

A Najafi[1], M Biglarian[2] and S M Negharchi[3,*]

[1] Department of Wind Engineering, Flensburg University of Applied Sciences, Flensburg, Germany

[2] Department of Mechanical Engineering, Sharif University of Technology, Tehran, Iran

[3] Department of Mechanical Engineering, Babol Noshirvani University of Technology, Babol, Iran

* **E-mail:** negharchi@nit.ac.ir

Abstract. This research focuses on improving the extraction of wind energy as one of the main renewable energy sources, in vertical axis wind turbines (VAWTs). In the current study, 5 turbine types with identical geometrical structures but with completely different blades have been analysed in various wind speeds. The numerical study was performed in two and three-dimensional (2D and 3D) modes using the finite volume method. The solution results were obtained in an unsteady state and assuming incompressible fluid flow. Also, k-ω SST turbulence model was employed for modelling the turbulent flow. To ensure that the numerical results are valid, the numerical results were compared with experimental data, and it was found that the numerical and experimental results are in good compatibility. Among the simulated turbines, the S1046 airfoil had the maximum coefficient of performance, a larger operating range, and a higher average efficiency compared to other turbines. For all of the airfoils, the results showed that the coefficient of performance reduces in a specific range of rotor blade tip speed ratio. Since the 2D results were in good agreement with the 3D results, 2D simulations can be used in the designing and optimizing processes of VAWTs.

Keywords: vertical axis wind turbines (VAWT), airfoil, blade tip speed ratio, numerical simulation.

1. Introduction

Among the non-fossil fuels, wind energy is one of the best sources of energy in the world, because not only it is accessible in the entire world, it is one of the renewable energy sources. The usage of wind as a clean and free source for electrical energy production is increasing [1,2]. Based on the estimations, in 2020 more than 10% of the world's total required energy will be produced using wind energy, and until 2040 this value rises between 10 and 30%. Wind energy has great importance because it is one of the cheapest and most accessible energy sources and can be significantly effective in the prevention of climate changes[3]. The primary form of this technology dates back to the 7th century BC in Afghan highlands, but this kind of turbine was first mentioned in Iranian historical texts [4].

Generally, wind turbines can be categorized into two main groups: horizontal axis wind turbines (HAWTs) and vertical axis wind turbines (VAWTs). Usually, the aerodynamic blades are used in HAWTs, but some VAWTs may use these blades. The efficiency of HAWTs can be significantly affected by the angle of the flow in which the turbine is placed in, however, VAWTs are not sensitive towards the wind direction [5]. This can be considered as an important advantage for this kind of turbine. VAWTs can be categorized into two groups, including the VAWTs that are driven by the lift force and the VAWTs that are driven by the drag force. Since the maximum efficiency in the first group of VAWTs is higher compared to the second group, the first group has attracted more attention [6,7]. In 2016, Sunny and Kumar simulated the aerodynamic model of a VAWT. They also performed some experimental tests in a wind tunnel, to evaluate the influence of the wind speed on the rotation speed of the turbine rotor and the obtainable torque from it. The experimental and numerical results both indicated that these turbines have an acceptable performance. Their research just focused on one design of VAWTs [4]. In 2020, He et al. studied different numerical methods for simulating VAWTs, and they compared the results of these methods with the experimental data. They aimed to find a faster and more accurate method that required a lower computational cost. These researchers recognized the limitations of the conventional methods by simulating a turbine, and they discussed these limitations. Finally, they recommended the usage of the hybrid meta-model. This research was also performed on one type of turbine [8]. In 2018, Liang and Li tried to study the effects of the maximum tangential force coefficient on enhancing the performance of VAWTs. The simulated turbine used NACA0015 airfoil, and the simulation was performed in 3D. Even though the simulations in this research were only done on one turbine type, this study was able to show the positive effect of design optimization on the turbine performance [9]. In the current work, the influence of changing the design of the turbine rotor through changing the airfoil has been examined on the aerodynamic characteristics of the turbine. Initially, the comparisons were done in 2D for the turbines with different airfoils, and then the 3D simulations for the turbine with the best 2D results were performed again. Based on the literature review, this can be considered as the novelty of the current work.

2. Geometry selection, meshing, and boundary conditions

In this research, the numerical simulations were done in 2D and 3D. At first, five airfoils that belong to different families were selected, and they were simulated in a domain. Initially, each of these airfoils was simulated and analyzed in the computational domain, then after evaluating the performance coefficient of these turbines separately, the best one in terms of the performance coefficient was simulated in 3D and its results were analyzed. The examined airfoils were a.NACA 008, b.NACA 63415, c.S1046, d.AH93W215, e.FLXV152. Figure 1 shows the design of these airfoils.

The simulated turbines in this study had three blades, a cord length of 12 cm, and a rotor radius of 30 cm. In the 3D simulation mode, these turbines had a height of 40 cm. These values are based on one of the references [9]. Figure 2 shows the 3D view of the designed turbine.

For meshing, triangular and tetrahedron grids were used in the 2D and 3D computational domains, respectively. Also, for observing the boundary layer and the separation phenomenon more precisely, boundary layer meshing was done in the vicinity of the airfoil [10]. Figure 3 shows the 2D and 3D mesh for the examined turbine and the computational domain.

2.1. Governing equations and dimensionless numbers

The general form of continuity and Navier-Stokes equations have been written as equations 1 and 2, respectively [11]. In these equations, $\bar{u}_i$ indicates the average speed, $\overline{u_i'}$ shows the fluctuation component of the speed, P is the mean pressure, ϑ represents the kinematic viscosity, and ρ is the fluid density [12].

$$\frac{\partial \bar{u}_i}{\partial x_i} = 0 \tag{1}$$

$$\frac{\partial \bar{u}_i}{\partial t} + \bar{u}_j \frac{\partial \bar{u}_i}{\partial x_j} = -\frac{1}{\rho}\frac{\overline{\partial p}}{\partial x_i} + \vartheta \frac{\partial^2 \bar{u}_i}{\partial x_j^2} - \overline{u_j' \frac{\partial u_i'}{\partial x_j}} \tag{2}$$

In this study, ANSYS Fluent software was employed for simulating the fluid flow and all of the equations were discretized using the second-order upwind discretization method. Since in the fluid flows with separation and adverse pressure gradient, k-ω SST turbulent modeling method presents accurate results and it has good performance in terms of stability and convergence, this model was

IOP Conf. Series: Materials Science and Engineering 1196 (2021) 012035 doi:10.1088/1757-899X/1196/1/012035

selected for the 2D and 3D simulations [15,16]. The k-ω SST model includes two equations: turbulent kinetic energy transport (Equation 3) and the turbulent kinetic energy dissipation rate transport (Equation 4):

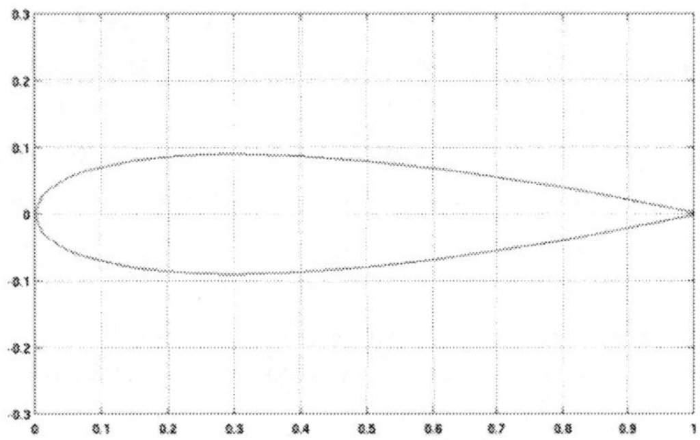
a. NACA 0018

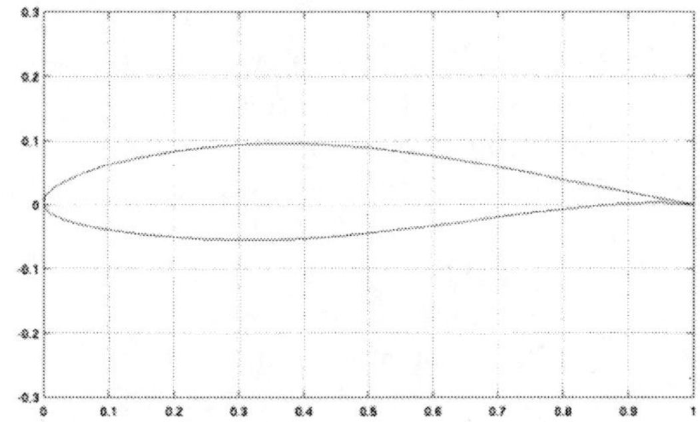
b. NACA 63415

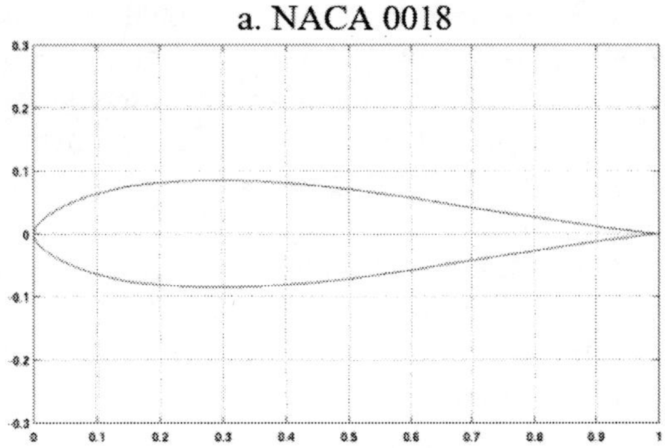
c. S1046

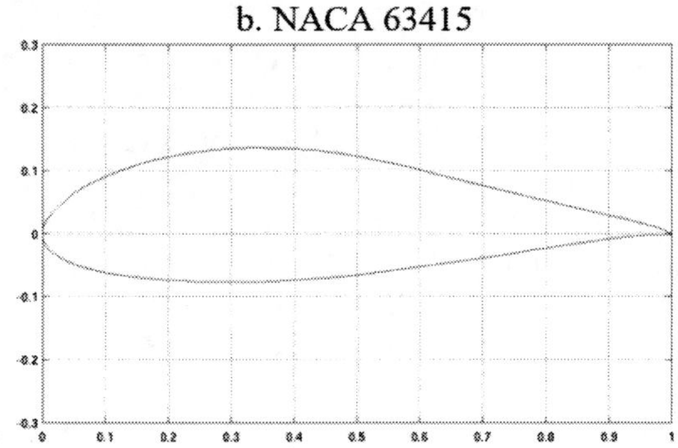
d. AH93W215

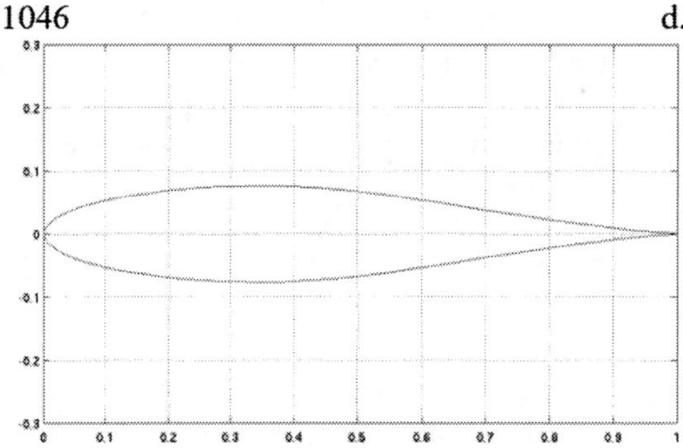
e. FLXV152

Figure 1. The examined airfoils [13].

Figure 2. The 3D view of the designed turbine for the simulation [14].

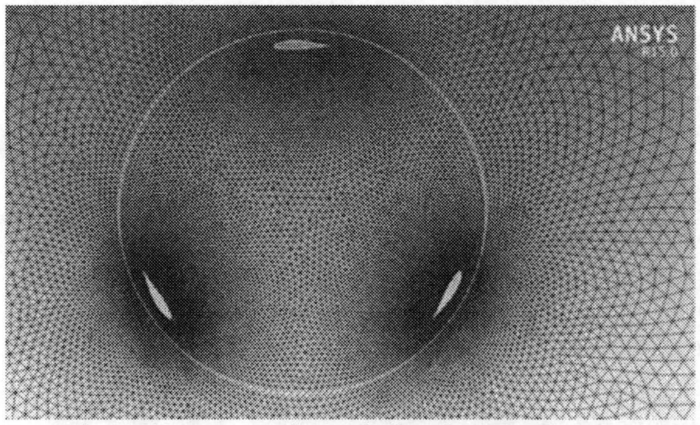

a. The 2D generated mesh for the NACA 63415 turbine

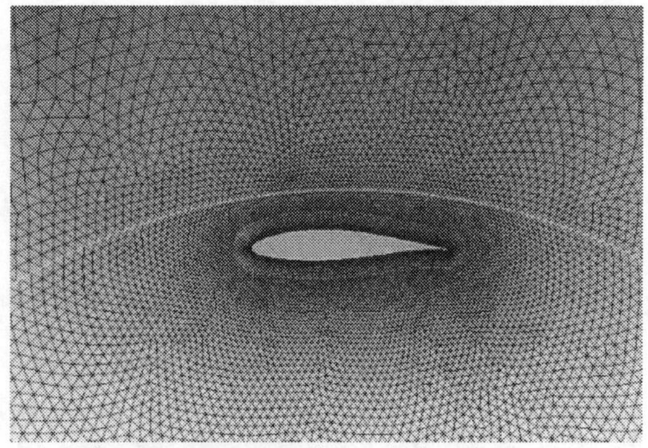

b. The 2D generated mesh for the rotor of NACA 63415 turbine

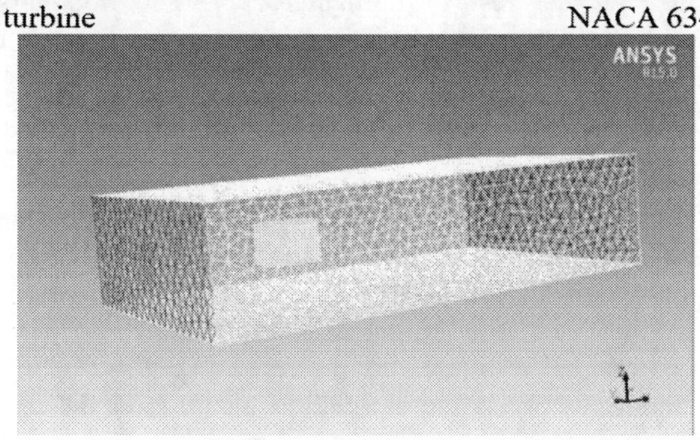

c. The 3D generated mesh for the S1046 turbine and its surrounding computational domain

Figure 3. The generated grid for the geometry in 2D and 3D simulation modes.

$$\frac{\partial}{\partial t}(\rho k) + \frac{\partial}{\partial x_i}(\rho k u_j) = \frac{\partial}{\partial x_j}\left[\Gamma_k \frac{\partial k}{\partial x_j}\right] + \tilde{G}_k - Y_k + S_k \tag{3}$$

$$\frac{\partial}{\partial t}(\rho \omega) + \frac{\partial}{\partial x_i}(\rho \omega u_j) = \frac{\partial}{\partial x_j}\left[\Gamma_\omega \frac{\partial \omega}{\partial x_j}\right] + G_\omega - Y_\omega + D_\omega + S_\omega \tag{4}$$

In which G, Γ, Y, S, and D indicate generation, diffusion, dissipation, user-defined source, and cross-diffusion items for the k and ω terms.

Also, in this study, sliding mesh as one of the most accurate methods for simulating flow in moving reference frames with several references has been employed [17]. For analysing the performance of wind turbines, the most important parameter is their coefficient of performance. This parameter has been presented in Equation 5. According to Equation 5, the coefficient of performance (C_p) is the ratio of the produced electrical energy from the turbine (P) to the power of the wind, in which the turbine is affected by (W).

$$C_p = \frac{P}{W} \tag{5}$$

The coefficient of performance can be significantly variable based on the flow velocity, turbine design, and energy conversion capability, but its maximum can be determined using the Betz coefficient. The value of the Betz coefficient shows the maximum accessible power for wind turbines. It shows that without any dissipation, and in ideal condition, the maximum obtainable energy in the wind turbines is 59.3% of the wind flow energy [7,18]. Another important parameter that is used for wind turbines is the Tip Speed Ratio or TSR. This parameter is defined as the ratio of the blade tip speed to the speed of the free flow (Equation 6) [19,20].

$$\lambda = \frac{\omega * r}{V} = \frac{v}{V} \tag{6}$$

$$\omega = 2\pi f \tag{7}$$

In these two equations, ω shows the rotational speed, v indicates the rotor tip speed, V is the wind speed, f represents the rotation frequency, and r is the rotor radius.

2.2. Boundary conditions

After discretizing the governing equations, some algebraic equations are obtained. To solve these equations, the flow conditions should be solved in the boundaries of the computational domain. The used boundary conditions in the 2D simulations was velocity inlet for inlet boundary, walls for blades and pressure outlet for the rest of domain[15,21]. The boundary conditions in the 3D simulations are similar, however, in the 3D simulations, the boundaries are surfaces and not lines [22].

2. 3. Validation and grid independence tests

3.1. Validation

For the validation of the obtained results for the simulations, both 2D and 3D results were validated by the wind tunnel test results that were presented in the reference [4]. Since the researchers in the reference [4] used NACA0021 airfoil, this airfoil was employed in the simulated turbine.

As it is illustrated in Figure 4, the numerical simulation results in 2D mode are in good compatibility with the experimental results of the reference [4]. After changing the simulation mode from 2D to 3D, the deviations between the simulation results and the experimental data becomes considerably smaller. The reason for this improvement is considering the effects of the 3D flow.

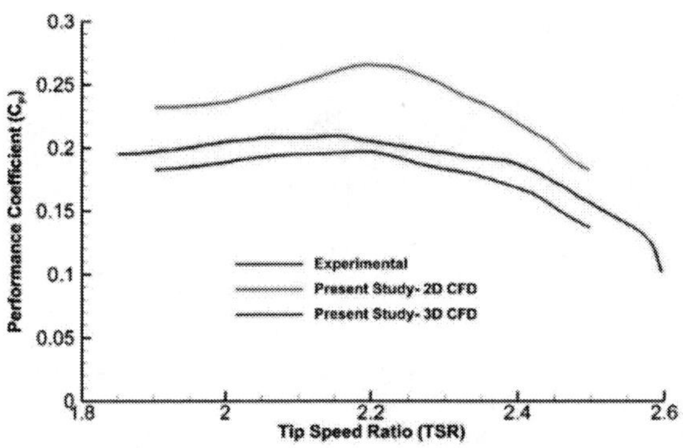

Figure 4. Validation of the simulations with the obtained results in reference [4].

3.2. Mesh independence test

Since the accuracy of the numerical solution can depend on the generated mesh, as shown in Figures 5 and 7, the mesh independence test was performed for both 2D and 3D simulation modes.

The results from NACA0018 airfoil in 2D simulation mode and for S1046 airfoil in 3D simulation mode shows that increasing the number of cells more than it is required, does not create a significant alternation in the obtainable results. For example, as it can be noticed in Figure 5, changing the number of cells from $6*10^4$ to $9*10^4$ changes the results less than 5%, which indicates that the results are independent of cell size. Furthermore, in the 3D simulation mode (Figure 6), the simulations were done with $7*10^5$, $7*10^5$, $8*10^5$, $10*10^5$, and $13*10^5$ cells. As it can be noticed from Figure 6, with one million of cells, the simulation reaches a desirable level of accuracy and increasing the cell numbers more than this, does not lead to a significant change in the results, therefore the generated grid with one million of cells is selected for the 3D simulations.

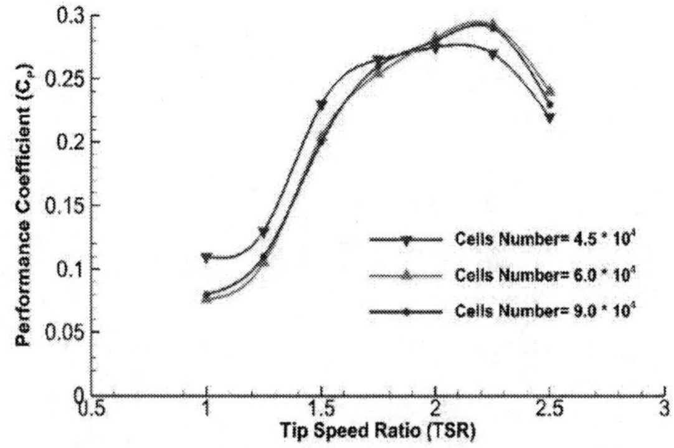

Figure 5. Mesh independence test in the 2D simulation mode for NACA 0018 airfoil.

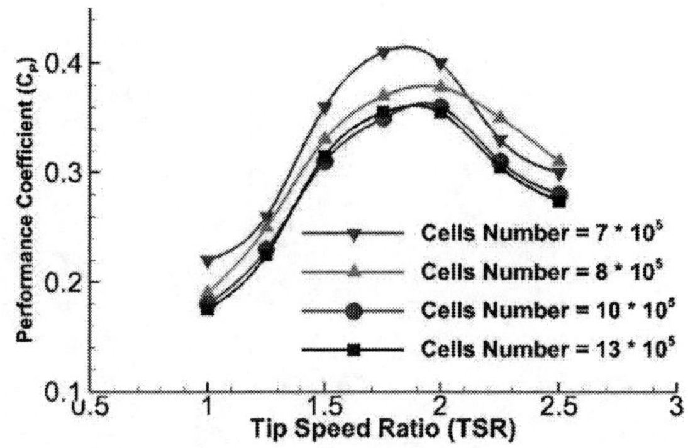

Figure 6. Mesh independence test in the 3D simulation mode for S1046 airfoil.

4. Results and discussions

4.1. Results of the 2D simulation

The coefficient of performance results for the turbine with different airfoils in 2D mode has been presented in Figure 7. As it can be observed in this figure, the lowest coefficient of performance belongs to NACA 63415 airfoil. The maximum coefficient of performance for this airfoil in the TSR of 2, reaches lower than 20%, which is significantly different from the Betz value. Unlike the other airfoils that have been shown in Figure 7, the AH93W215 airfoil reaches its highest performance in the TSR of 1.5-1.75. The AH93W215 airfoil presents a better performance compared to the NACA 63415 airfoil, in all of the TSRs. Furthermore, AH93W215 airfoil presents its best performance in the TSRs lower than 1.75, and overall, it has a better performance compared to NACA 0018 and FXLV152 airfoils, however, unlike these two airfoils the performance of AH93W215 decreases in higher TSRs. NACA 0018 airfoil exhibits similar behavior to the other airfoil in this family (NACA 63415), except that the overall performance of NACA 0018 airfoil is better.

By increasing the TSR, the coefficient of performance increases for the FXLV152 airfoil, and in the TSR of 2.25, the performance of this airfoil reaches its maximum, which in this TSR, is the highest performance coefficient compared to the other airfoils (except the S1046 airfoil). In almost all of the TSRs, the S1046 airfoil presents a better performance compared to other airfoils, and this difference is completely obvious in Figure 7. In the TSRs lower than 2, raising the TSR increases the performance coefficient of the S1046 airfoil, but after the TSR of 2, increasing the TSR leads to a reduction in the performance coefficient for this airfoil, so that in the TSR range of 2.25-2.5, the performance coefficient of this airfoil becomes similar to the performance coefficient of the FXLV152 airfoil. Overall, by reviewing Figure 7, it can be demonstrated that the S1046 airfoil has a better performance coefficient compared to other airfoils.

The highest coefficient of performance for each of the examined airfoils has been presented in Table 1. According to the table, NACA 63415 airfoil, at the maximum, reaches the performance coefficient of 0.18, which is half the maximum performance coefficient that can be obtained using the FXLV152 airfoil. The S1046 airfoil has a maximum performance coefficient of 0.41, which is significantly larger compared to other airfoils. Overall, the maximum coefficient of performance for all of the airfoils is far from the Betz limit, showing that there is dissipation in the fluid energy.

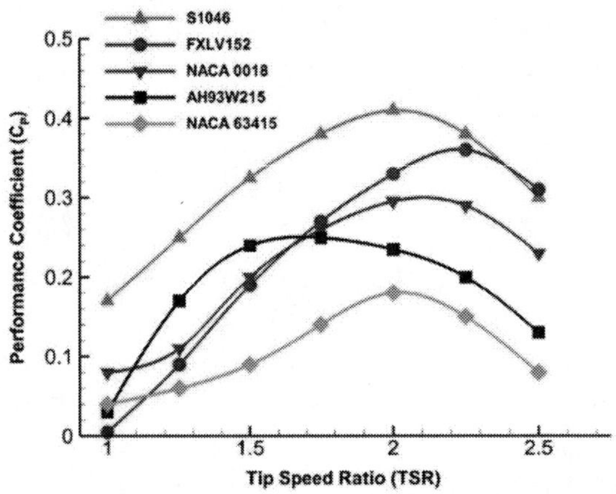

Figure 7. The changes in the performance coefficient (C_p) for the five simulated airfoils in 2D mode.

Table 1. The maximum performance coefficient (C_p) for each of the five turbines.

Airfoil name	NACA 0018	S1046	NACA 63415	AH93W215	FXLV152
The maximum performance coefficient (C_p)	0.29	0.41	0.18	0.25	0.36

Since the aerodynamic phenomena in all of the examined turbines are similar, the contours of velocity are just presented for the S1046 airfoil. Figure 8 shows these contours in four different TSRs. In the low TSRs (Figure 8-a) a larger amount of flow enters the turbine rotor. By increasing the TSR, the flow velocity in center decreases, so that in the TSR of 2.5, the turbine almost leads to blockage in the flow. Comparing Figures 8 and 7 reveals that by increasing the TSR, the amount of entered flow in the center of the turbine increases, and this raise leads to higher momentum transfer from the fluid to the turbine and consequently, increases the performance. However, in the TSR of 2.5 (Figure 8-d), the rotational speed of the rotor prevents the flow from entering, leading to the conversion of the turbine into a rotating cylinder and formation of Karman Vortex street behind the turbine.

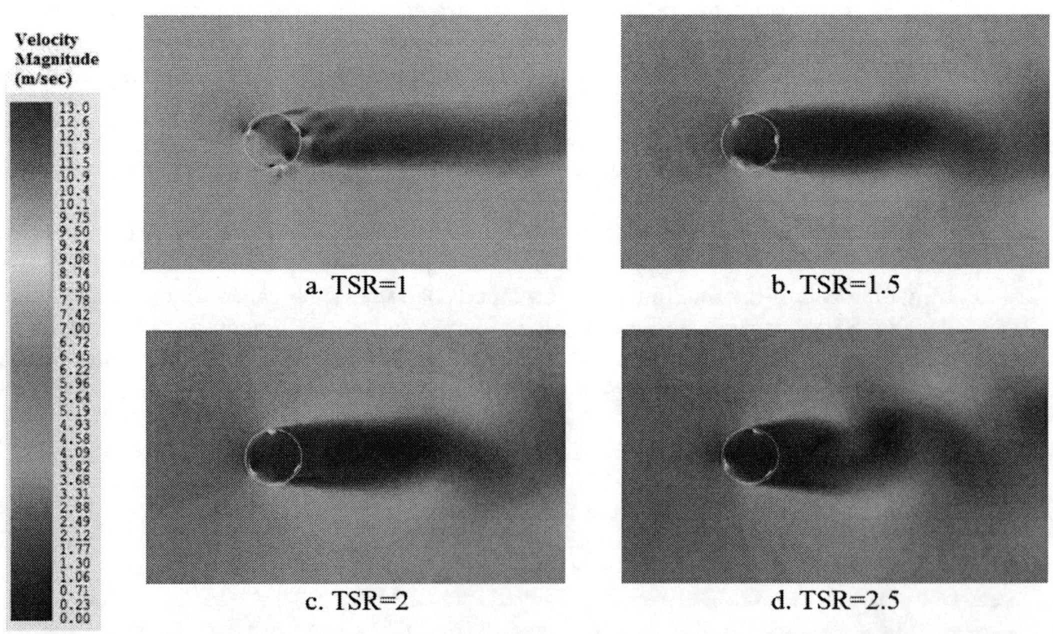

Figure 8. The changes of velocity magnitude around the S1046 airfoil in different TSRs.

By observing the presented velocity vectors in Figure 9, it can be concluded that the rotation of blades around their axis makes the airflow move in the blade's movement direction. Therefore, the movement direction of the flow changes from the horizontal to rotational, and the velocity magnitude of the flow around the blade, becomes close to the velocity magnitude of the blade. In lower distances from the blade, the direction and the velocity of the flow becomes more similar to those of the blade. When the blade passes from a region, the velocity changes, and after that, the horizontal flow that enters the turbine makes the rotational flow to move toward the downstream. All of the vortexes move from the upstream towards the downstream like a trail, and they cause energy dissipations in the flow. The most noticeable phenomena in the velocity vector figures are separation from the trailing edge, rotational flows, and vortexes around them. These phenomena are observed for all of the turbines, however, their intensity and sizes are different for each of the turbines. This difference is one of the factors that determines the aerodynamic performance for each of the turbines.

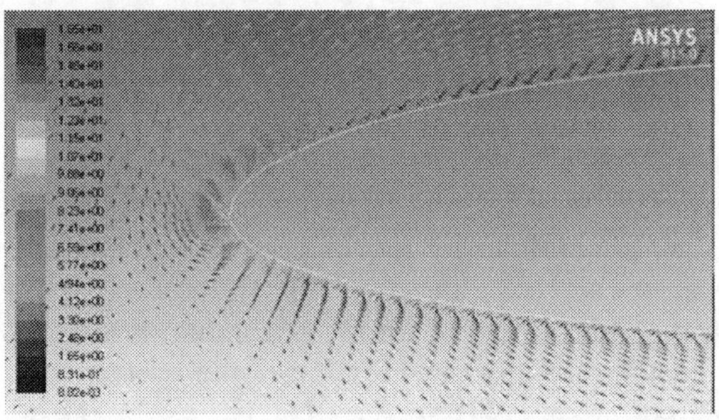

a. The velocity vectors in the leading edge of the blade (It is in the flow direction)

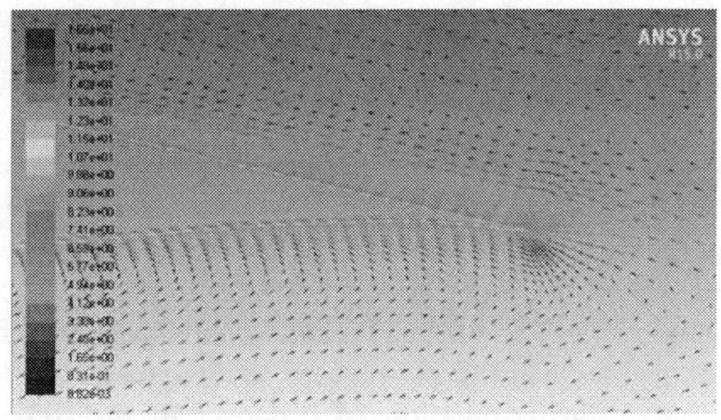

b. The velocity vectors in the trailing edge of the blade (It is in the flow direction)

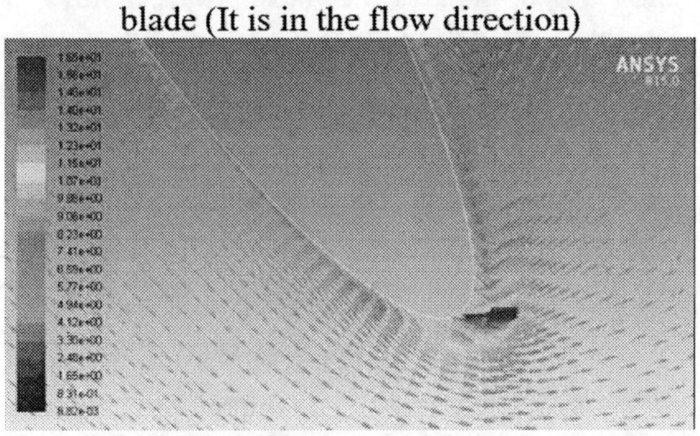

c. The velocity vectors in the leading edge of the blade (It is almost perpendicular to the flow)

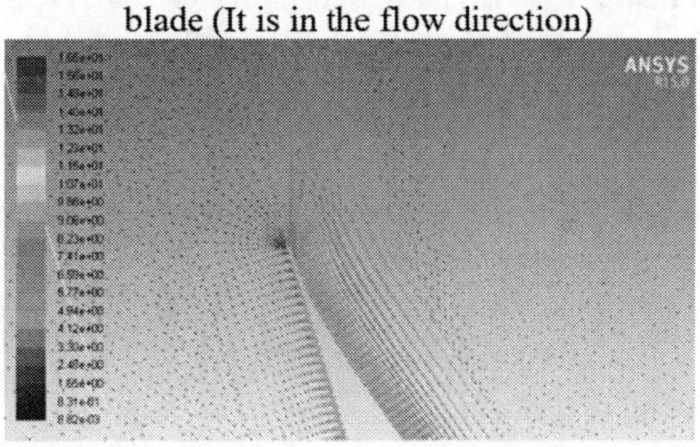

d. The velocity vectors in the trailing edge of the blade (It is almost perpendicular to the flow)

Figure 9. The velocity vectors around the airfoil.

The pressure distribution around the S1046 airfoil in the TSR of 2 has been presented in Figure 10. In Figure 10-a. the airfoil is completely in front of the flow and leads to the formation of the vortexes behind it. This increases the dissipations in the turbine. In Figures 10-b, since the airfoil is almost positioned in the direction of the flow, the low-pressure regions become weaker, and consequently the dissipation decreases.

IOP Conf. Series: Materials Science and Engineering 1196 (2021) 012035 doi:10.1088/1757-899X/1196/1/012035

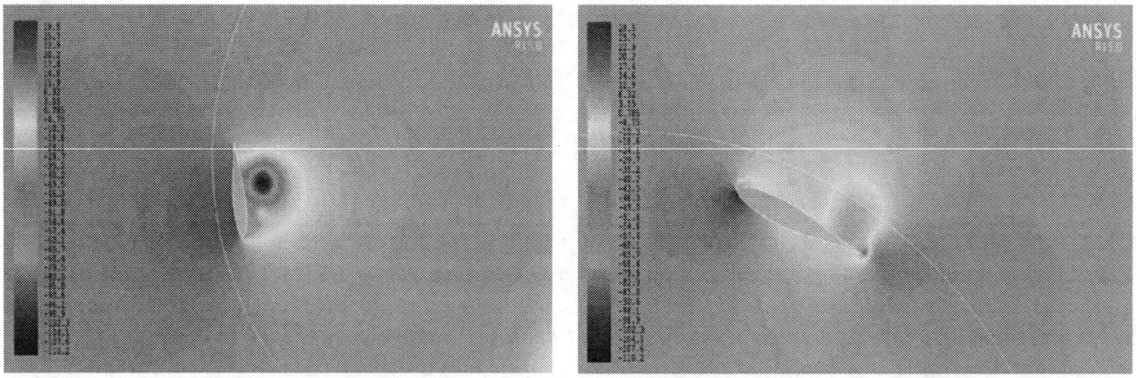

a. The pressure distribution around the blade that is perpendicular to the flow direction.

b. The pressure distribution around the blade which is moving against the direction of the flow.

Figure 10. The pressure distribution around the airfoil.

4.2. Results of the 3D simulation

Considering the 2D simulation results, among the examined airfoils, the S1046 airfoil presented the best performance, thus, for a more exact evaluation, the 3D simulation results and their comparison with 2D results have been shown in Figure 11. Overall, it can be said that the 2D and the 3D results are very close, and the 2D results can be a considerably good estimation for the designing process. As was observed in the 2D results, the 3D results show that increasing the TSR raises the performance, however, the inclination of this rise in the 3D simulation mode is lower than it was in the 2D simulation mode. The reason is that in the 3D simulation mode, besides the dissipations behind the airfoil, due to flow transportation from high-pressure to low-pressure regions, and forming wing tip vortex as illustrated in Figure 13, more energy dissipations happen. These dissipations increase as the TSR rises, and that is the reason for increasing the difference between 2D and 3D results in higher TSRs. By comparing the 2D and 3D results, overall, it can be concluded that the 2D results can properly show the behaviour patterns of the airfoil, but for calculating the maximum coefficient of performance, the effects of the 3D flow should be considered. Figure 12 shows the velocity magnitude contours in TSR of 1. Similar to the 2D results that were presented in Figure 8, the presence of the turbine leads to velocity reduction in the downstream of the flow.

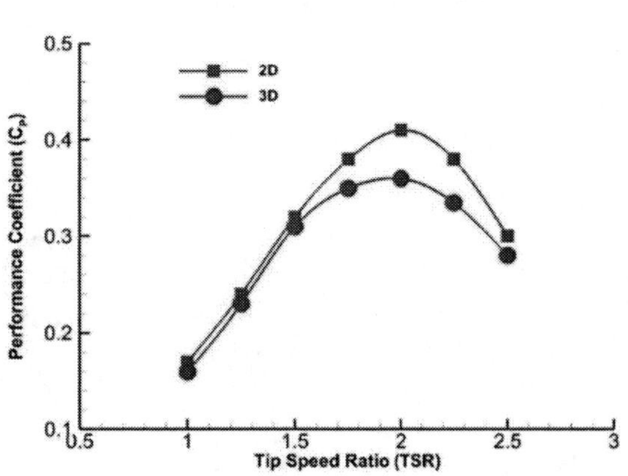

Figure 11. The comparison of the 2D and 3D results for the designed turbine with the S1046 airfoil.

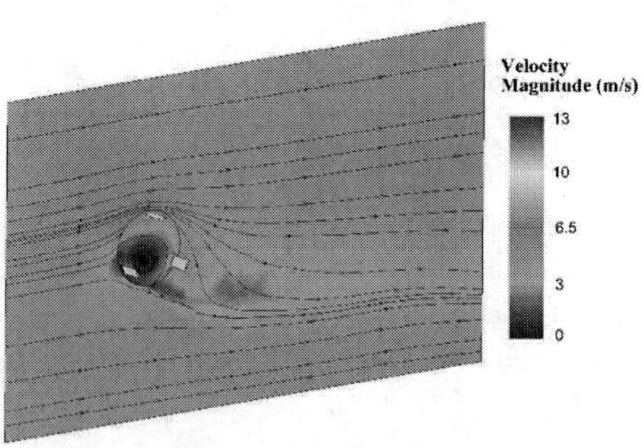

Figure 12. The velocity contour and surface streamline in the 3D simulation of the S1046 turbine.

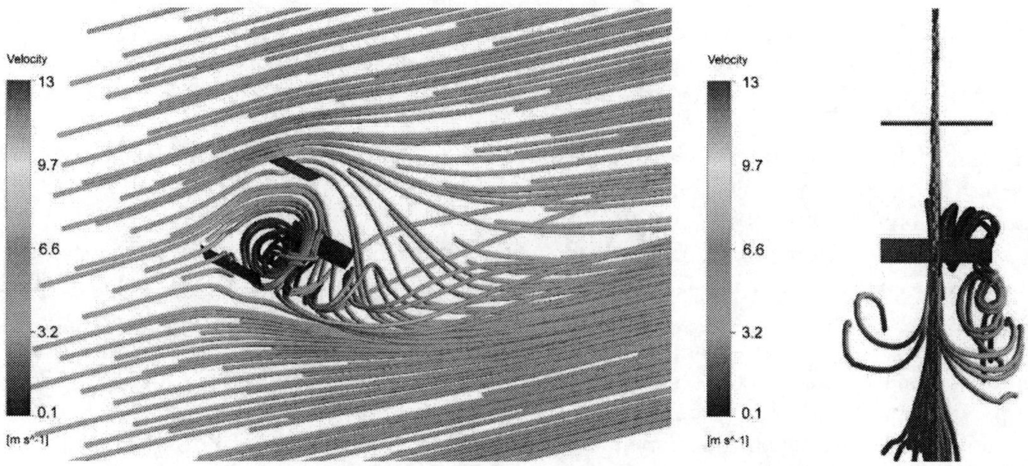

Figure 13. Forming wing tip vortex, Left: side view, Right: Front view.

5. Conclusion

In the current study, five different airfoils, including NACA 63415, NACA 0018, AH93W215, FXLV152, and S1046 airfoils were numerically examined in different external wind speeds. The simulations were performed in an unsteady state in 2D and 3D simulation models and using the k-ω SST turbulence model. The simulation results show that by increasing the TSR, the coefficient of performance increases for all of the airfoils until it reaches a maximum value. After that, a further increase in the TSR decreases the coefficient of performance. The maximum performance coefficient value for each of the airfoils was different and this maximum value was obtained in a proper TSR that was specific for each of the airfoils. The maximum coefficient of performance for the S1046 airfoil was almost 40%, which in that TSR, was almost 10% higher than the next airfoil in terms of the highest performance coefficient. The 3D simulation results for the S1046 airfoil show that the 2D and 3D results are in good compatibility, thus the results of the 2D simulations can be used in the estimation of the performance and designing process to optimize the wind turbines.

References:

[1] G.M. Farias, M.A.B. Galhardo, J.R.P. Vaz, J.T. Pinho, A steady-state based model applied to small wind turbines, J. Brazilian Soc. Mech. Sci. Eng. 41 (2019). https://doi.org/10.1007/s40430-019-1704-0.

[2] K.A.R. Ismail, T. Canale, F.A.M. Lino, Parametric analysis of Joukowski airfoil for 10-kW horizontal axis windmill, J. Brazilian Soc. Mech. Sci. Eng. 40 (2018). https://doi.org/10.1007/s40430-018-1119-3.

[3] IEA, Global Energy Review 2020, Iea. (2020). https://www.iea.org/reports/global-energy-review-2020.

[4] K.A. Sunny, N.M. Kumar, Vertical Axis Wind Turbine: Aerodynamic Modelling and its Testing in Wind Tunnel, Procedia Comput. Sci. 93 (2016) 1017–1023. https://doi.org/10.1016/j.procs.2016.07.305.

[5] S. Santhakumar, I. Palanivel, K. Venkatasubramanian, An experimental study on the rotational behaviour of a Savonius wind turbine for two-lane highway applications, J. Brazilian Soc. Mech. Sci. Eng. 40 (2018) 1–12. https://doi.org/10.1007/s40430-018-1158-9.

[6] A. Rezaeiha, H. Montazeri, B. Blocken, Characterization of aerodynamic performance of vertical axis wind turbines: Impact of operational parameters, Energy Convers. Manag. 169 (2018) 45–77. https://doi.org/10.1016/j.enconman.2018.05.042.

[7] F. Porté-Agel, M. Bastankhah, S. Shamsoddin, Wind-Turbine and Wind-Farm Flows: A Review, Springer Netherlands, 2020. https://doi.org/10.1007/s10546-019-00473-0.

[8] J. He, X. Jin, S. Xie, L. Cao, Y. Wang, Y. Lin, N. Wang, CFD modeling of varying complexity for aerodynamic analysis of H-vertical axis wind turbines, Renew. Energy. 145 (2020) 2658–

2670. https://doi.org/10.1016/j.renene.2019.07.132.
[9] C. Liang, H. Li, Effects of optimized airfoil on vertical axis wind turbine aerodynamic performance, J. Brazilian Soc. Mech. Sci. Eng. 40 (2018). https://doi.org/10.1007/s40430-017-0926-2.
[10] M.T. Bouzaher, M. Hadid, D. Semch-Eddine, Flow control for the vertical axis wind turbine by means of flapping flexible foils, J. Brazilian Soc. Mech. Sci. Eng. 39 (2017) 457–470. https://doi.org/10.1007/s40430-016-0618-3.
[11] M. Biglarian, M. MomeniLarimi, B. Ganji, A. Ranjbar, Prediction of erosive wear locations in centrifugal compressor using CFD simulation and comparison with experimental model, J. Brazilian Soc. Mech. Sci. Eng. 41 (2019) 1–10. https://doi.org/10.1007/s40430-019-1610-5.
[12] A. Mobin, H. Akimoto, Y. Hara, Comparative CFD analysis of Vertical Axis Wind Turbine in upright and tilted con fi guration, Renew. Energy. 85 (2016) 327–337. https://doi.org/10.1016/j.renene.2015.06.037.
[13] M. Ragheb, A.M. Ragheb, Fundamental and Advanced Topics in Wind Power, Fundam. Adv. Top. Wind Power. (2012). https://doi.org/10.5772/731.
[14] R. Howell, N. Qin, J. Edwards, N. Durrani, Wind tunnel and numerical study of a small vertical axis wind turbine, Renew. Energy. 35 (2010) 412–422. https://doi.org/10.1016/j.renene.2009.07.025.
[15] R. Lanzafame, S. Mauro, M. Messina, 2D CFD modeling of H-Darrieus Wind Turbines using a transition turbulence model, Energy Procedia. 45 (2014) 131–140. https://doi.org/10.1016/j.egypro.2014.01.015.
[16] W. Tian, B. Song, Z. Mao, Numerical investigation of wind turbines and turbine arrays on highways, Renew. Energy. 147 (2020) 384–398. https://doi.org/10.1016/j.renene.2019.08.123.
[17] J.F. Van Der Auweraert, Modeling of wind turbine wake with a sliding mesh, (2015).
[18] M. De Lellis, R. Reginatto, R. Saraiva, A. Trofino, The Betz limit applied to Airborne Wind Energy, Renew. Energy. 127 (2018) 32–40. https://doi.org/10.1016/j.renene.2018.04.034.
[19] M. a. Sayed, H. a. Kandil, A. Shaltot, Aerodynamic analysis of different wind-turbine-blade profiles using finite-volume method, Energy Convers. Manag. (2012). https://doi.org/10.1016/j.enconman.2012.05.030.
[20] M.H. Sadr, D. Badiei, S. Shams, Developments of a semiempirical dynamic stall model for unsteady airfoils, J. Brazilian Soc. Mech. Sci. Eng. 41 (2019) 1–15. https://doi.org/10.1007/s40430-019-1907-4.
[21] P. Marsh, D. Ranmuthugala, I. Penesis, G. Thomas, The influence of turbulence model and two and three-dimensional domain selection on the simulated performance characteristics of vertical axis tidal turbines, Renew. Energy. 105 (2017) 106–116. https://doi.org/10.1016/j.renene.2016.11.063.
[22] Y.T. Lee, H.C. Lim, Numerical study of the aerodynamic performance of a 500W Darrieus-type vertical-axis wind turbine, Renew. Energy. 83 (2015) 407–415. https://doi.org/10.1016/j.renene.2015.04.043.

AUTHOR INDEX